DECOLONIZING THE MAP

THE KENNETH NEBENZAHL, JR., LECTURES IN THE HISTORY OF CARTOGRAPHY

PUBLISHED IN ASSOCIATION WITH THE HERMON DUNLAP SMITH CENTER
FOR THE HISTORY OF CARTOGRAPHY, THE NEWBERRY LIBRARY

Series Editor, James R. Akerman

ALSO IN THE SERIES

*Maps: A Historical Survey of
Their Study and Collecting*
by R. A. Skelton

Five Centuries of Map Printing
by David Woodward

British Maps of Colonial America
by William Patterson Cumming

Mapping the American Revolutionary War
by J. B. Harley, Barbara Bartz Petchenik,
and Lawrence W. Towner

Art and Cartography: Six Historical Essays
edited by David Woodward

*Monarchs, Ministers, and Maps:
The Emergence of Cartography as a Tool of
Government in Early Modern Europe*
edited by David Buisseret

*Rural Images: Estate Maps in the
Old and New Worlds*
edited by David Buisseret

*Envisioning the City:
Six Studies in Urban Cartography*
edited by David Buisseret

*Cartographic Encounters: Perspectives on Native
American Mapmaking and Map Use*
edited by G. Malcolm Lewis

*The Commerce of Cartography:
Making and Marketing Maps in Eighteenth-
Century France and England*
by Mary Sponberg Pedley

Cartographies of Travel and Navigation
edited by James R. Akerman

*The Imperial Map: Cartography
and the Mastery of Empire*
edited by James R. Akerman

*Ancient Perspectives: Maps and Their Place in
Mesopotamia, Egypt, Greece, and Rome*
edited by Richard J. A. Talbert

CARTOGRAPHY FROM COLONY TO NATION

DECOLONIZING THE MAP

Edited by

James R. Akerman

THE UNIVERSITY OF CHICAGO PRESS

CHICAGO AND LONDON

The University of Chicago Press, Chicago 60637
The University of Chicago Press, Ltd., London
© 2017 by The University of Chicago
Published 2017.
Printed in the United States of America

26 25 24 23 22 21 20 19 18 17   1 2 3 4 5

ISBN-13: 978-0-226-42278-7 (cloth)
ISBN-13: 978-0-226-42281-7 (e-book)
DOI: 10.7208/chicago/9780226422817.001.0001

Library of Congress Cataloging-in-Publication
Data

Names: Akerman, James R., editor.
Title: Decolonizing the map : cartography from
   colony to nation / edited by James R. Akerman.
Other titles: Kenneth Nebenzahl, Jr., lectures in
   the history of cartography.
Description: Chicago : The University of Chicago
   Press, 2017. | Series: Kenneth Nebenzahl, Jr.,
   lectures in the history of cartography | Includes
   bibliographical references and index.
Identifiers: LCCN 2016030112 | ISBN
   9780226422787 (cloth : alk. paper) | ISBN
   9780226422817 (e-book)
Subjects: LCSH: Cartography—Political aspects. |
   Decolonization.
Classification: LCC GA108.7 .D44 2017 | DDC
   526—dc23 LC record available at https://lccn
   .loc.gov/2016030112

♾ This paper meets the requirements of ANSI/NISO
   Z39.48–1992 (Permanence of Paper).

CONTENTS

v

ACKNOWLEDGMENTS

The eight chapters in this volume were originally presented as the Seventeenth Kenneth Nebenzahl, Jr., Lectures in the History of Cartography at the Newberry Library in late 2010. Throughout the book's gestation, the authors have exhibited exemplary forbearance and have responded to repeated editorial requests both cheerfully and constructively. The editor wishes to acknowledge in particular the unusual amount of effort the authors expended in obtaining images and permissions for the book's complex illustration program. Several anonymous readers offered invaluable critiques and timely advice on both the substance and the structure of the volume, while giving us the confidence to move forward. The University of Chicago Press has been unwavering in their support for this project, and its staff both collegial and professional at every turn. The enduring material and moral support of the Newberry Library, the host of the Nebenzahl Lectures since 1966, is foundational. Without this extraordinary community of creative librarians, curators, teachers, and readers dedicated to free inquiry in the humanities, there would be no Nebenzahl Lectures. In the autumn of 2016 the Nebenzahl Lectures celebrated their fiftieth anniversary. This collection is dedicated to Ken and Jossy Nebenzahl in honor of their kind and enduring sponsorship of this series and the scholarship it has promoted.

INTRODUCTION

James R. Akerman

The idea for the Seventeenth Kenneth Nebenzahl, Jr., Lectures in the History of Cartography (2010), "Mapping the Transition from Colony to Nation," the basis for this book, emerged from the Fifteenth Nebenzahl Lectures, "The Imperial Map" (2004), published under the same name by the University of Chicago Press in 2009. *The Imperial Map*'s broad, if episodic, examination of how modern imperial powers used mapping to conquer and manage their colonies and to promote and affirm their imperial identities, as well as—somewhat antithetically—its contemplation of the limits of imperial mapping suggested an obvious counterpoint and successor. This volume publishes the result. At the simplest (and titular) level it considers the roles mapping has played in the passage from colony to nation—or, if you will, from dependent to independent state. The seven chapters all concern the engagement of mapping in the long and clearly unfinished process of decolonization and the parallel process of nation building from the late eighteenth century to the mid-twentieth century. A few scholars have addressed this issue. However, the subject has not been examined systematically or comprehensively. The first chapter, by Ray Craib, offers a wide-ranging introduction to the subject, identifying salient features of the relationship between mapping and decolonization, and placing them within the context of recent theoretical debates about the nature of decolonization itself. There follow seven case studies examining this relationship over more than two centuries and on three continents. Needless to say, this volume can only suggest the general contours of this vast subject and propose some of its historical problems and dilemmas.

When I first conceived of this series I imagined that the papers would focus on twentieth-century mapping and would be associated primarily with African and Asian countries emerging from European imperialism. It soon became clear that the questions I wanted to ask of the mid-twentieth-century decolonization could not, or should not, be separated from the revolutions and emergent nationalisms of the nineteenth century. Indeed, the broad theme of decolonizing the map could easily have embraced accounts of mapping and its relationship to nineteenth-century nationalist movements for independence in Eastern Europe, and their echoes in post-Soviet times. In the end, the papers focus on Latin America, Africa, and Asia. I made a calculated decision as well—perhaps not theoretically defensible—to exclude America north of Mexico, already the subject in English of a large literature. There are other notable gaps and biases in chronological and geographic coverage—the former French colonies in Southeast Asia and Africa, for example, are not treated here, while former British and Spanish colonies are the subject of three articles apiece. Even so, the book covers a broad chronological and geographic context, and while there are commonalities to be found in the mapping and decolonization from the late eighteenth through the twentieth centuries—such as the iconic and propagandistic use of maps—the several contributions show the relationship between mapping and decolonization, like the concept of decolonization itself, to be multifaceted and defiant of summary. Our goal in assembling this volume has not been to offer a comprehensive treatment, but rather to move toward an understanding of the processes at work in decolonizing the map and to suggest where there may be common issues across time and geography, while respecting the distinctiveness of each context. In much the same way that *The Imperial Map* asked whether there is and what constitutes imperial cartography, so this collection asks whether there is and what is a decolonizing cartography.

The authors were asked to consider, each within the context of their areas of specialty, how peoples and states emerging from colonization use maps to define, defend, and administer their polities and territories; to develop their national identities; and to establish their place in the community of nations. The papers, both as presented orally and in enlarged form here, offer many partial answers to these questions, but both the editor and the audience were struck by what cartographic history reveals about the limits of decolonization as well as the inadequacy of the term.

At the heart of Craib's sweeping review are the ongoing theoretical and geopolitical challenges to the very idea of decolonization. While acknowledging the widespread agreement among scholars and political observers that the simple equation of independence with decolonization is problematic, he shows nevertheless that the problematic nature of the concept is perhaps what makes it a meaningful line of inquiry into the character of mapping in the later modern world. If, as he acknowledges, "the contours of colonialism still configure the globe," so, too, it is undeniable that the uneven retreat from empire has recast the meaning of these contours in the conduct of human affairs.

More to the point here, his excursions into the broader scholarship of decolonization always come back to questions of geography and spatial practice, and therefore, fundamentally, maps and mapping. This is perhaps most explicit in his examination of the ways in which the boundary lines and place names inscribed on maps reflect and enforce both continuity and differentiation in decolonizing regions. Warning that though the power of imperial mapping and its influence on a decolonizing world are considerable, they were never as absolute as it is most often imagined. Incorporations, creolizations, resistance, and countercartographies, shaped the character of imperial cartography and set the decolonization of the map in motion well before independence. By the same token, Craib continues, decolonizing the map has been neither monolithic nor complete. Internal colonialism—the disproportionate distribution of power among elites, indigenes, regions, and ethnic groups within decolonizing states—has been expressed and reinforced by official and popular cartography, as have contests among political ideologies and political economies. Craib concludes by showing that, elusive as the concept is, the significance of decolonization as a lens through which we might come to understand cartographic history in the modern world is underscored by the extent to which the historians of cartography over the past thirty years have sought to decolonize the field itself.

Magali Carrera's chapter demonstrates that mapping is engaged in the formation of decolonizing national identities well before legal independence. The immediate context is late colonial Mexico, but the general observation is no less valid in preindependence Ghana, Egypt, and India-Pakistan, the subject of later chapters in this collection, as it is no doubt in many other African states and in French Indochina. What sets New Spain apart from these other contexts is that its late colonial identity emerged from settlement colonialism, where indigenous populations were largely or partially displaced, their survivors living alongside substantial populations of migrants from the imperial center and their descendants, who became—in Mexico, as in much of the Americas—the most important factor in the formation of an emergent national identity. Eighteenth-century Spanish colonial authorities saw in the mapping of New Spain opportunities for economic development and assertion of state control that were imperfectly developed in Spain itself. While published maps asserted the identity and authority of the viceroyalty in a manner consistent with European geopolitical idioms, they also fostered the emergence of a New Spanish (and ultimately Mexican) identity distinct both from Spain and the global Spanish Empire. The surveying and mapping of land tenure, for example, entailed a nuanced and complicated dialogue between European and indigenous mapping traditions well into the eighteenth century. Finally, Carrera argues that narrative mapping, in both textual and graphic form, plays in admixture with modern European mathematical cartography a critical role in an emergent criollo mapping in the latter eighteenth century, weaving New Spanish colonial history into Iberian sociogeographic traditions emphasizing local community. The tripartite entanglement of Spanish, native, and criollo cartographies, she concludes, was readily translated into a hybridized official and scholarly cartography of an independent Mexico.

Similarly, but at the moment of revolution, in chapter 3 Lina del Castillo shows how mapping reflected competing ethnic and political visions of nationality in an independent Colombia/Gran Colombia. At the most superficial level this is a story of the competition between maps of two "Colombias": a pan–South American federal Colombia (or "Colombia Prima") promoted most prominently by the Venezuelan Francisco Miranda, and a more pragmatically conceived and narrowly defined "República de Colombia," or Gran Colombia (composed of the territories of modern Colombia, Venezuela, Ecuador, Panama, and parts of Guiana and Brazil) supported in the later stages of the revolution by Simón Bolívar in a move to consolidate his power. Through detailed analysis of competing maps, their delineations of boundaries, emphasis of specific places, and omissions, Del Castillo shows how Bolívar's Liberator Party took control of the early cartographic image of Colombia and, just as important, the mapped narrative of the Colombian revolution to discredit the revolutionary pedigree of political and military competitors. Del Castillo shows how the mapping of early Colombia represented who would get credit for its revolution, which aspects of revolutionary history would be preserved by a national history, and which would be forgotten.

Extending this process in Latin America into the later nineteenth and twentieth centuries, Jordana Dym next explores how the often arbitrary nature of colonial boundaries can destabilize the territorial expression of national identities for a century and more after independence. She shows that delineation of Guatemala's national geo-body (the often iconic geographic outline of a state; see Thongchai 1994) remained unstable for two centuries due to prolonged territorial negotiation with adjacent states. Here, as in many other places in Latin America, there was no clear continuity between colonial political divisions and postcolonial states, no consistent principle to govern this transition. The geopolitical chaos that often resulted played out in many different types of maps. Dym highlights especially the contradictory mapping of the relationship between British Honduras/Belize and Guatemala as well as the instability of Guatemala's border with Mexico. Here again, factors of ethnicity and class come into play. Overarching specific territorial issues, Dym argues, was a shift in the social reach of the mapping involved, from a period of decolonization largely managed by and for criollos, elites, and the state to more democratized forms of mapping in the spheres of education, marketing, and tourism.

The next four chapters move across the Atlantic and shift our focus to states that achieved independence in the twentieth century. In different cartographic spheres, topographic mapping and national atlas publication, McGowan and Culcasi trace the enduring influence of cartographic practices brought to Africa by European colonizers lingering in postcolonial mapping, even as African administrations, surveyors, and cartographers took over.

In chapter 5, Jamie McGowan takes a long view of the role local intermediaries and native technocrats played in creating and mapping a British Gold Coast identity that became Ghana in 1957. McGowan begins with an outline of the career of George Ekem Ferguson, a mixed-race colonial operative whose combined skills both as mapmaker and political intermediary helped define the territory and political identity of the future Ghana. Ferguson's partial African descent gave him an access to indigenous political relationships no British colonial official could hope to achieve, producing diplomatic successes that extended and solidified British control of the greater Gold Coast. His maps were the direct expression both of his British training and his African diplomatic sensibilities.

Ferguson's influence as a model in the twentieth century was limited, however, as colonial authorities organized formal surveys of the Gold Coast based on European, and more specifically British, models. Though Africans played a considerable role of as surveyors, draftsmen, and surveyor's assistants, they were valued because they were a source of cheaper, semiprofessional labor, rather than for their intermediary status. Although their participation was increasingly unacknowledged and unrecorded, the training of African surveyors and their proportional role in the mapping of the colony nevertheless increased. In the aftermath of the Second World War, mapping agencies were largely Africanized, forming the core of both personnel and methodologies that would account for a smooth transition to Ghanaian control of its now national topographic and cadastral mapping. The methods and standards on the new national mapping agencies were indistinguishable from European ones, a fact underscored by the belief, among African surveyors whose careers spanned both regimes, that surveying had nothing to do with politics—even as they ascended to positions of management in postcolonial Ghana.

In like manner, Karen Culcasi traces the application of Western mapping practices, most especially in the production of national atlases, to the creation of an Egyptian national identity during that country's protracted emergence as a modern nation-state. She argues that the Egyptian case is complicated not only by its long history of quasi independence from the Ottoman Empire, France, and Britain, but also by its unstable relationship to secular nationalism, internal minorities, pan-Arabism, and Islamic movements of the late twentieth century. Culcasi shows that the first national atlases and topographic projects created after nominal independence in 1922 remained essentially British projects, created mostly by British cartographers and in accordance with British standards. Even a second edition of the first national atlas, the *Atlas of Egypt,* published in 1958, after the Officer's Revolution, was unchanged, with the exception of its rhetorical trappings and erasure of marks of British authorship. Thereafter, the cartographic assertion of a postcolonial Egyptian identity was strident but not univocal. It was complicated by the promotion of a pan-Arabic identity in Egyptian atlases and maps, as well as by the countermappings of minorities such as Coptic Christians and Nubians. Culcasi shares with Dym a concern for the public projection of national identity, but whereas Guatemalan cartographers grappled with an unstable geo-body, Culcasi shows that, despite the territorial stability of postcolonial Egypt, its cartography has struggled with a national identity buffeted by social, ethnic, and religious pressures from both within and beyond its boundaries.

The contradictions and conflicts generated by the twentieth-century rush to carve out and validate national territories from former colonies lie at the heart of Sumathi Ramaswamy's interrogation of the enduring crisis created by the partition of British India. She sets the stage by recalling the halfhearted and almost banal way in which the Radcliffe line separating the independent states of Pakistan and India was conceived and committed to paper, strangely without any significant involvement of the considerable cartographic and geographic institutions the colonial power had at its disposal. Predictably, the partition itself prompted immediate cartographic reactions, both Pakistani and Indian, but precise, official public representations of the border were scarce.

Ramaswamy then argues that the boundary line and the concept of partition have been both duplicated and reconceived by literary and artistic mappings, as well as propagandistic and other forms of popular cartography. While these mappings have to some extent naturalized the geo-bodies of India, Pakistan, and Bangladesh, seen as a whole, and sometimes explicitly, they also challenge the legitimacy of the partition, and with it one of the most intractable problems of the decolonizing world map.

To the same general point, chapter 8, by Thomas Bassett, argues that mapping for travelers and tourists has actively contributed to the survival of internal colonialisms within nominally independent, decolonized states. Bassett shows how during the apartheid era, and often beyond it, South African road and street maps masked or entirely erased the presence of indigenous African communities within their own homeland in the name of better serving their audience of predominantly white motorists. I am struck by the similarity to the way that cover art on road maps published in the United States in the 1920s–50s depicted African Americans and Indians as part of the landscape and whites, only, as motorists—that is, as the only American citizens truly liberated by the geographic mobility the automobile offered (Akerman 2006).

Bassett finds in the history of road mapping in apartheid and postapartheid South Africa a rich context for examining the persistence of racial prejudices and the racialized organization of space in a country that has experienced a protracted and ambiguous decolonization. South African road mapping offers a dramatic example of how the spatial practices of internal colonialism utterly destroy any simple historical equation between the nominal achievement of sovereignty and the spatial expression of decolonization, if we mean by that the achievement by formerly colonized peoples of some measure of liberty and power over territory. One might expect road mapping, with its navigational intent, to be a cartographic genre that is largely depoliticized. Bassett shows powerfully that racial and colonial politics are present even in this realm, and perhaps more perniciously so, because the eye and mind of the map user are likely to be quite unaware of them.

In his introduction to *The Imperial Map* Matthew Edney struggled mightily with the idea of an "imperial map." There is nothing intrinsically imperial, he argued, about any map: what makes a map imperialistic is not its specific content or form, but how and for whom they are deployed. Imperial maps do not constitute some complex of related mapping genres. Neither are they functionally distinct from other maps that exert power. Edney was particularly concerned with the similarities and dissimilarities of imperial mapping and mapping implicated in the formation and maintenance of nations and states. Clearly the two are intertwined, and any distinction, he argued, was one enforced not by differences in form or content, but by the sociopolitical and geographic separation of the peoples and polities involved. Imperial mapping is that of territories and polities by peoples and interests removed—emotionally, morally, and spatially—from the territories and peoples mapped, who have relatively little say in *how* and *why* they are mapped (Edney 2006).

It follows that decolonizing the map would entail processes and practices by which colonized peoples become more engaged or reengaged in mapping their own spaces and territories. Inasmuch as a major portion of mapping in the modern world by governmental or other agencies is implicated in the modern state system, mapping as an aspect of decolonization correlates with movements for, and the achievement of, the independence of colonies and colonized peoples. This commonsense narrative is reinforced by the almost universal attention newly independent states pay to the production of new maps and atlases that affirm their independence and identity. But this narrative is both complicated by the nuances of individual cases and, in some instances, undermined by them. The cartography of a nation-state emergent from a colonial regime only throws off with difficulty, if ever, the implied inequality of imperial mapping, and decolonization is often experienced unevenly and restrictively by peoples and social groups within a new state.

Collectively, the chapters that follow show that decolonized mapping is forefronted primarily by the operation of colonial and former colonial elites or other privileged groups, creoles (criollos), other intermediaries between indigenous communities, and authorities within the colony. For the most part this mapping adopts and adapts the cartographic idioms and practices used by European imperial powers, even where employed by countermapping generated by minorities within the new nation-states. Drawing new boundaries and creating new geo-bodies, naming and renaming places and territories, powerful as they may be in shaping public opinion and social spaces on the ground, either fail to achieve their desired effect or produce unintended consequences. The cartographic and geopolitical record of the past two centuries shows that, in mapping their new states, decolonizing communities distinguish themselves from their former colonizers and consolidate new identities only gradually and incompletely.

REFERENCES

Akerman, James. 2006. "Twentieth-Century American Road Maps and the Making of a National Motorized Space." In *Cartographies of Travel and Navigation*, edited by James R. Akerman, 157-206. Chicago: University of Chicago Press.

Edney, Matthew H. 2009. "The Irony of Imperial Mapping." In *The Imperial Map: Cartography and the Mastery of Empire*, edited by James R. Akerman, 11-45. Chicago: University of Chicago Press.

Thongchai Winichakul. 1994. *Siam Mapped: A History of the Geo-body of a Nation*. Honolulu: University of Hawaii Press.

CHAPTER ONE

CARTOGRAPHY AND DECOLONIZATION

Raymond B. Craib

I. INTRODUCTION

I will begin with a rather plain but instructive map: map 6 from James Francis Horrabin's *An Atlas of Empire*, published in 1937 (fig 1.1). The map, created and published on the eve of Italy's invasion of Abyssinia, is a blunt representation of European colonial possessions in Africa. The image reveals visually the following statistic: six European states covering a total area of 660,000 square miles "own[ed] close on 11 and ½ million square miles in Africa."[1] Such statistics led Horrabin, a working-class educator and innovative socialist geographer, to proffer a remarkably succinct definition of Europe: it was, he wrote, "a group of States holding colonial possessions in other continents."[2] Two decades later, Frantz Fanon would take this observation to its logical, and historicized, conclusion: "Europe is literally the creation of the Third World."[3]

At their imperial apex, colonial powers (including not only European states but also the United States and Japan) laid claim to more than three-quarters of the world's landmass. Yet no more than a half-century later this equation had been inverted as anticolonial struggles, domestic agitation in the metropoles,

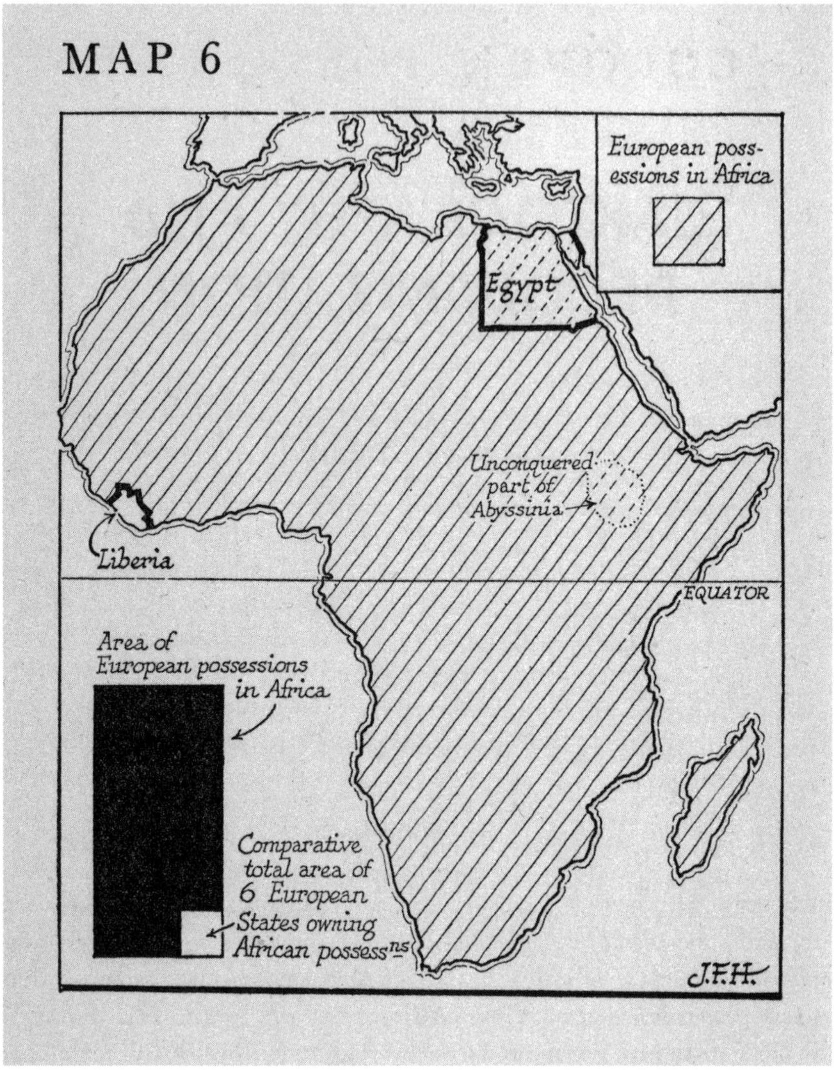

MAP 6

European poss-
essions in Africa

Egypt

Unconquered
part of
Abyssinia

Liberia

EQUATOR

Area of
European possessions
in Africa

Comparative
total area of
6 European
States owning
African possess^{ns}

J.F.H.

FIGURE 1.1. Africa possessed. European domination of Africa. Map 6 from J. F. Horrabin, *An Atlas of Empire* (London: Knopf, 1937).

an international war against fascism (colonialism's boomerang effect, in Aimé Césaire's powerful rendering), and international pressures of various kinds fostered the demise of most forms of colonial rule in most of Africa, Asia, Oceania, and the circum-Caribbean.[4] One would expect such dramatic political transformations to have serious repercussions in the realm of cartography. Certainly the "military red" of British domination no longer spread itself across the globe. If in 1898 Rand McNally could celebrate the British Empire in an atlas by noting that "the real magnitude of the vast empire of Great Britain can not be fully comprehended until it is studied in a series of GOOD MAPS, such as those contained in this volume," by the early 1960s even the most ardent imperialists had to recognize that the empire's sun was indeed setting.[5] By 1962 Britain's colonial empire, which had formerly spanned the globe, appeared as little more than a hodgepodge of bloody outposts on the maps of the annual British Colonial Office Lists (fig. 1.2).[6]

The magnitude of how much had changed in the wake of two world wars and shifts in global resource accumulation can be conveyed with a few statistics: the continent of Africa contained only four independent countries in 1948; by 1965 there were thirty-seven.[7] The United Nations counted fifty-one member states in 1945; three decades later it had three times that number.[8] The pace of change was so rapid that the cartographic division for the US National Geographic Society, on its September 1960 atlas plate of Africa, felt compelled to note the following: "Boundaries on this map reflect the political situation as of July 15, 1960, the day the map went to press."[9] Much had changed. But there were continuities, a reality nowhere better captured than in the fact that theories of neocolonialism developed in tandem with decolonization in the 1960s.

In this chapter I take a synoptic view of cartography and decolonization in the twentieth century and make occasional reference to nineteenth-century processes of decolonization in the Americas. The essay is intentionally broad and synthetic: it looks at colonization and decolonization across the globe and across nearly two centuries, although the primary focus is on the twentieth century. While cognizant of the fact that the particularities of decolonization varied according to time and space—something that is clear in the subsequent essays in this volume—I have taken this wide-angle perspective in order to put the historical experiences and processes of decolonization in Africa and Asia, in North America and Latin America, in Oceania and Southwest Asia into conversation, comparison, and relation to one another. Doing so highlights certain patterns and continuities but, of course, at the risk of eliding differences and disjunctures. The essay builds on a growing body of secondary literature on cartography, colonialism, and postcolonialism in order to highlight general themes and points of comparison and contrast, while at the same time plumbing particular primary sources (maps of various kinds) that illustrate such processes or trends. While it highlights certain themes and points, it invariably also suffers from elisions, exceptions, and generalizations. In this respect the essay resembles not only a map—which is inherently selective—but also "decolonization" itself which was, as historian Prasenjit Duara observes, "neither a coherent event [. . .] nor a well-defined phenomenon."[10]

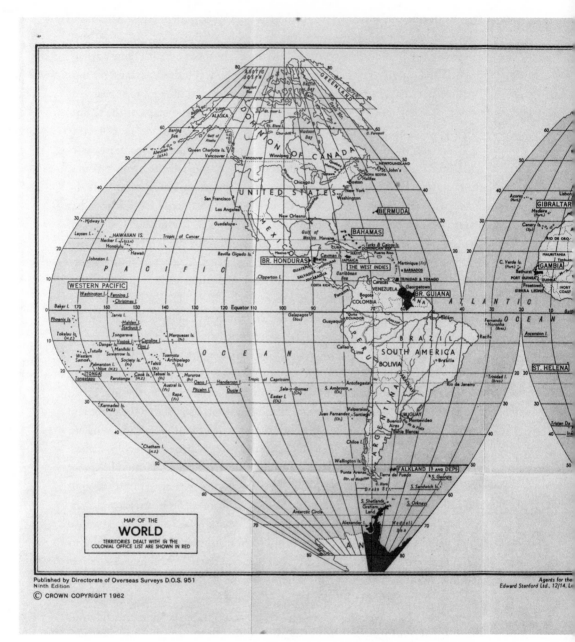

FIGURE 1.2. Bloody outposts. Directorate of Overseas Surveys, *Map of the World*. Insert in *The Colonial Office List, 1962* (London: Her Majesty's Stationery Office, 1962).

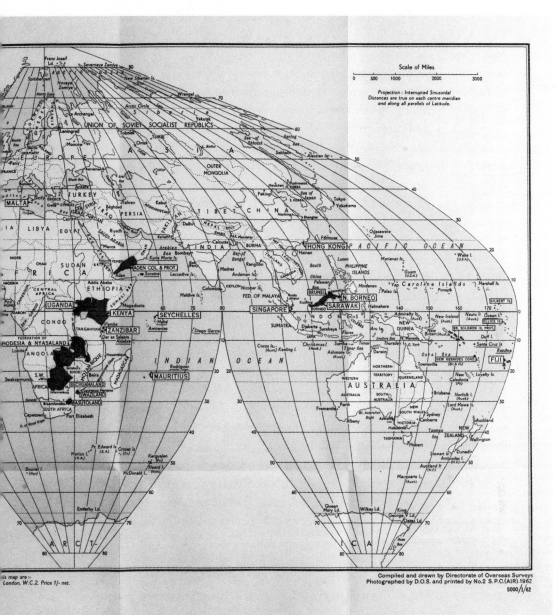

Compiled and drawn by Directorate of Overseas Surveys
Photographed by D.O.S. and printed by No.2 S.P.C.(AIR).1962
5000/1/62

FIGURE I.2. *Continued*

The conceptual, as well as geographic, approach to colonialism and decolonization I take here is also somewhat capacious. While my emphasis is on colonialism in the formal sense—that is, formal political control—I do, at certain points and particularly toward the end of the essay, also take up the question of decolonization and neocolonialism or informal empire. My intent is not to conflate what are, and should remain, useful distinctions—between, say, "imperialism" and "colonialism"—regarding forms of external control and/or domination. At the same time, and as will become apparent throughout the essay, such distinctions are rarely clear cut, and some attention to the range of relationships, often unequal and involving a variety of coercive mechanisms, that might be plotted on a "colonial" spectrum is warranted.[11] "Informal" does not seem adjectively adequate to capture the ruthlessness with which, say, the United States engaged the Caribbean and Central America in the twentieth century.[12] But more to the point, informal empire was never far removed from its more heavy-handed partner, which is why when anticolonial leaders, intellectuals, and masses living under the yoke of colonial rule spoke of independence or emancipation, they often did so with a vocabulary informed as much by socialism (and, as I will argue below, anarchism) as by nationalism or anticolonialism per se.[13]

I wrote this essay as part of a series of talks on mapping the transition from colony to nation, but many colonies did not make such transitions. In a recent grand work on empire, historians Jane Burbank and Fred Cooper observed that nation-states were not the only alternative to empire: federation, confederation, and a host of other possibilities existed.[14] Similarly, historian Jeremy Adelman notes for the specific case of Ibero-America, "no single vision of postcolonial sovereignty filled the vacuum left behind. . . . The nation was not prefigured by colonialism to herald its demise."[15] Moreover, what exactly independence means is a contentious question. What is the relationship between decolonization and independence? Independence from what? And for whom? And how do such issues manifest themselves in the realm of cartography? To what degree does looking at decolonization through the lens of the history of cartography affect or force us to rethink our understandings of processes of decolonization and vice versa?

II. LINES, MINDS, AND NAMES

The contours of colonialism still configure the globe.[16] Decolonization led to the dramatic, and still contentious, partition or territorial reconfiguration of some formerly colonized lands, most notably in what had been British India and in Southwest Asia. In other cases, colonial territorial configurations outlived their creators. While newly independent states would frequently reconfigure internal jurisdictional boundaries, the international political boundaries inscribed by colonial powers oftentimes remained intact, despite having been imposed arbitrarily, as a matter of administrative convenience, or based upon flawed and ethnocentric principles. Thus, even if they bore little relation to political, social, linguistic, or ecological relationships on the ground, such spatial constructs—and the abbreviated history they carried with them— persisted, both as a consequence of and an impetus for modern cartography.[17] As geographer Matthew Sparke has aptly noted, "Cartography is part of a reciprocal or, better, a *recursive* social process in which maps shape a world that in turn shapes its maps."[18] Thus it was that the membership of the Organization of African Unity, created in 1966, agreed to abide by the boundaries established by the colonial powers in the years following the Berlin Conference of 1885.[19] Precedent existed in such matters: while the boundaries of many of the fledgling republics of Latin America born of independence movements in the early nineteenth century would change over the course of that century, leaders of those movements initially applied the legal principle of *uti possidetis juris* (Latin for "as you possess under law") to ensure that existing colonial contours would serve as the international boundaries for their new states.[20] Yet those contours—and the criteria used to determine them—were themselves not easily determined or agreed upon, and boundary changes and conflicts were commonplace through much of the nineteenth century.[21] Moreover, in many instances the newborn states shared with their Spanish predecessor an inability to assert administrative control beyond the bounds of the centers of political life. To use the possessive when discussing Mexico's far north or Chile's or Argentina's Patagonia is to engage in a useful, but fictional, cartographic shorthand. These were regions where the postcolonial state had little purchase until the arrival of global capital and the development of new transportation and military technologies later in the century.

The fact that formerly colonized subjects did not escape the boundaries created by colonial powers does not, of course, mean that newly formed states were somehow inauthentic or colonies in postcolonial drag. Boundaries may be static (and even that is open to debate at some level) but their meanings are not. Regardless of how dramatically they obscured or overwrote other existing territorialities, boundaries created by colonial powers acquired a reality and a meaning over time to many living within their bounds. The intervening decades or centuries between the creation of a colonial territory and its demise were never mere interregnums, nor did the violence of colonialism and the imposition of new property regimes and ethnic identities necessarily prevent colonized subjects from developing affective attachments to new territorialities, new properties, and new identities. Whatever their initial artificiality, the passage of time gave them weight.

Thus, while one can readily agree that postcolonial national maps naturalize what are historical and social colonial constructions, it is important to recognize that they do not remain solely ideological and repressive fictions. The anticolonial imaginary was often geographic, and such spaces often constituted what were perceived to be necessary geographies of decolonization. He overstated the case, but writer Albert Memmi was not entirely amiss when he argued that "being oppressed as a group the colonized *must* necessarily adopt a *national* and *ethnic* form of liberation."[22] Not surprisingly, in most cases the national and ethnic form of liberation took shape in part through reference to colonially imposed boundaries. That such was the case highlights the fact that processes of decolonization, like colonization itself, rarely unfolded among equals. At least in the twentieth century, the colonized intelligentsia and the individuals who in many instances led anticolonial movements and/or composed the first generation of postcolonial leadership were often educated in the metropole or by metropolitan teachers, and so the political systems, models, and aspirations they held were often ones associated with the metropole.[23]

If many postcolonial states did not—and perhaps did not want to—escape the territorial boundaries established by their colonial predecessors, anticolonial leaders and the inheritors of independence could at least decolonize their spaces in other ways, including rewriting the landscape toponymically and recasting the past. Like novels, art, and music, maps became a means through which to re-present and develop a postcolonial identity liberated from colonial determinations. Maps and atlases thus helped perform the hard cultural work of decolonizing the land, the past, and, in the famous phrase of Kenyan intellectual Ngugi wa Thiong'o, the mind.[24]

Take, for example, the case of toponyms on maps. Decolonization meant that colonizers would no longer dictate the terms of representation: neither politically nor, in this instance, poetically. Naming constituted a basic strategy of imperial and colonial control.[25] Thus, with decolonization a veritable tide of toponymic change washed much of the globe, at least in the twentieth century. In some cases only the country name changed: British Honduras may now be Belize, but while there one can still travel through Roaring Creek, Bamboo Camp, and Double Head Cabbage. In other cases, new administrations oversaw name changes not only to their nations but also to cities, towns, districts, streets, and natural features. New names simultaneously overturned those imposed by colonizers and fostered a renewed interest in, and affirmed the validity of, local and regional dialects and national languages. Thus, in many instances spellings, rather than names per se, changed. Whether it be the resuscitation of the hamzah (') and the macron (ī) in Oceanic place names or the slight modification to Calcutta, a quintessentially colonial city, the idea was to shift language away from an imposed Europeanization. Toponyms and languages matter not solely for the imprint they leave and the impression they make but also for the knowledges they hold, the identities they evoke, the history they convey. Place names are often mnemonic devices that sustain genealogies, ground cultures, and write histories.[26] In all of these cases, language becomes a key site of anticolonial struggle. As the Irish patriarch in *Translations*, Brian Friel's remarkable play on the English mapping of Ireland in the 1840s, remarks: "English [. . .] couldn't really express us."[27]

The names applied to regions or nonstate spaces can challenge colonialist frames of reference and further efforts to decolonize not only the mind but the past. Take for example the subtle but powerful difference in perspective that might be evoked by using Oceania or Pacific Islands. Tongan intellectual Epeli Hau'ofa poignantly noted, "There is a world of difference between viewing the Pacific as 'islands in a far sea' and as 'a sea of islands.' The first emphasizes dry surfaces in a vast ocean far from the centers of power. . . . The second is a more holistic perspective in which things are seen in the totality of their relationships."[28] Hau'ofa's vision here is emphatically decolonial and cartographic: it is a rethinking of the relation of water, land, and sky and a conceptualization of the space of the ocean itself as connective tissue such that oceanic states no longer appear as small and isolated outposts too remote to achieve economic independence from wealthier nations and too dispersed to meaningfully cohere.[29] Although it is important to note that indigenous names are not themselves somehow "neutral" in their meaning and representation, Hau'ofa has history on his side. Many of Oceania's inhabitants have long traversed and settled swaths of ocean as if they were a liquid prairie. Outrigger canoes—varying in length from nine feet for vessels used in lagoons to thirty feet for double-hulled vehicles with sails intended for long-distance, interisland travel—turn the ocean into a conduit rather than a barrier, a vast plain webbed by routes to island configurations "arranged in blocks or groups rather than widely spaced, isolated specks of land."[30] Outrigger canoe navigators, orienting themselves through reference to stars and horizon points, sidereal compass technology, and a deeply learned ability to read the ocean and sky, traveled vast distances, in the process establishing fishing and hunting grounds,

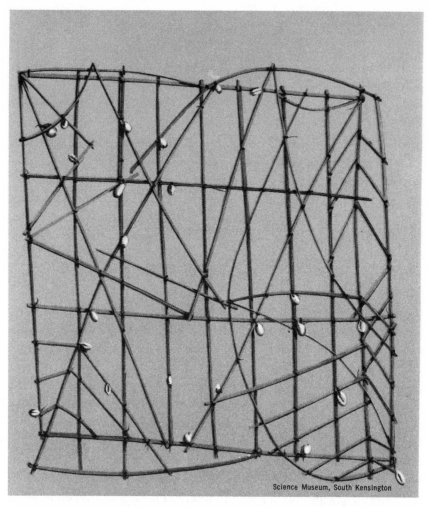

Science Museum, South Kensington

FIGURE 1.3. Water world. A Marshall Island stick chart for long-distance oceanic canoe voyaging. The chart depicts islands, ocean swells, and currents. Majuro, Marshall Islands, ca. 1920. Made of wood and shells, 67 × 108 cm. Courtesy of Library of Congress Geography and Map Division, Washington, DC.

settlements, and trade and communication networks.[31] Their maps—or stick charts, such as the one shown in figure 1.3—represented patterns of ocean currents and swells in relation to island formations. From this horizontal perspective, the ocean was less a limit or endpoint of a culture or society than a full-fledged part of it, a point captured by none other than navigator and explorer James Cook, who felt compelled to pose the following rhetorical question in his journal in 1778: "How shall we account for this nation spreading itself so far over the Vast ocean?"[32]

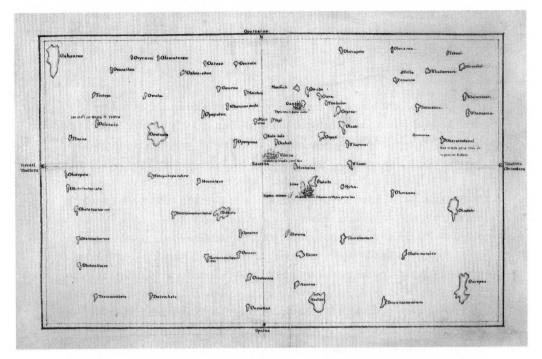

FIGURE 1.4. Tupaia's world? The Society Islands, James Cook, from an original chart by Tupaia. The chart is a drawing by Cook based on the information and an initial image made by Tupaia. The scale is impressive: according to David Turnbull, the space represented in the map here is approximately the size of the continental United States.

Cook had already made some effort to learn an answer when he took on board the *Endeavor* in 1769 a navigator and priest named Tupaia from the island of Raiatea (near Tahiti). Tupaia died while on board but at least a portion of his geographic knowledge was put down on parchment and copies made, including one in the work of Johann Rheingold Forster in his *Observations Made during a Voyage around the World* (1779) (fig. 1.4). So here is Tupaia's map . . . or is it? Anthropologist Margaret Jolly captures the quandary well:

> Though he is the author, this map is *not* his indigenous view. We will never know the details of that view, but his vision was likely rather differently "situated knowledge." I suspect it located the observer not soaring high above the islands, powerfully riding on the confident coordinates of longitude and latitude, plotting a changing global position relative to east and west, north and south, but rather lying low in a canoe, looking up at the heavens, scanning the horizon for signs of land, and navigating the powerful seas with the embodied visual, aural, olfactory, and kinesthetic knowledge passed down

through generations of Pacific navigators. His knowledge would have been communicated to other Tahitians through genealogical stories and chants, through the materials of the canoe and the sails, and through the embodied practice of navigation.[33]

The point here is not to argue over whether Tupaia did or did not create the map but rather to emphasize that Hau'ofa's "Oceania" asks us to acknowledge—historically and linguistically—such theoretical and practical knowledge, not as a means to articulate radical difference but in order to move beyond colonial legacies that have cast "a nation spread over a vast ocean" as little more than isolated and distant relations stranded on *tierra firme . . .* as little more than so many "Pacific Islands."

As well as place names, postcolonial administrations and populations have produced maps and atlases—national, cultural, and political—as part of a broader effort to connect a place to a past. They helped foster (whether as resurrection or invention) a nationalist historical narrative. Cartographers in places diverse as Mexico, Thailand, Turkey, and Iran inscribed—and reinscribed—a distant past onto modern maps of the nation in an effort to stress the nation's temporal longevity and cultural coherence, or to visualize and legitimize irredentist aims.[34] Similarly, Indian intellectuals in turn-of-the-century British India recast the map of the colony as a female adorned with archaic and culturally specific items in order to link history and geography and thereby "foster the sentiment of belonging and possession."[35] More recently, players for the Chilean soccer club Palestino—founded by Palestinian immigrants in 1920—sport not only jerseys with the colors of the Palestinian flag, but also numbers on their shirts in which, controversially, the "one" is in the shape of Palestine's geographic configuration prior to the creation of Israel.

By the mid-twentieth century such efforts were often pursued through the production of national atlases, what geographer Mark Monmonier has called the "symbol of national unity, scientific achievement, and political independence" par excellence.[36] Atlases served as a powerful means of cultural and political self-representation. Frequently designed for broad public consumption and reflecting concerted efforts to inculcate a particular image of a place and its past in order to forge collective unity, national atlases combined modern technological forms and formats with attention to cultural specificity and particularity. Thus atlases would often celebrate (or, at times, idealize) pasts that preceded the colonial era, narrating how the country took shape beyond the confines of colonialism.

These creations frequently carry with them the specter of invented tradition. As historian Jeremy Black has noted, they can suffer from a "habit of seeking to portray the long-term history of states whose territorial extent and ethnic composition were often the work of European conquerors and therefore relatively recent."[37] As important a point as this is, some caution is warranted here. For one, we should not too quickly jump to the conclusion that always everywhere Europeans made colonial boundaries and affiliations while their subjects merely watched. Subjects they may have been; passive they were not. Second, such an argument might be taken to mean that therefore these efforts at decolonial history and place making are somehow contrived. Well . . . yes, but the whiff of artifice lingers around all nation-states, regardless of their purported longevity. In fact, one might suggest a corollary: the more ancient a past claimed, the stronger the smell. Third, such a perspective tends to forget that the formation of modern European states often unfolded in relation to colonial expansion, to the degree that one might argue that the nation-states of Europe were in fact at least partially a product of the colonies. And finally, there *was* a past (a history, if not History) prior to the arrival of the colonials. In the aftermath of decades or generations of colonial rule, how does one write that past? More to the point, how does one write such pasts—violently occluded, suppressed, or ignored by colonizers—in deeply political and persistently neocolonial contexts? On what space do such pasts get to unfold? The very creation of such nationalist pasts—of Histories that locate recent states deep in time—was in part an effort to situate the colonized in historical time; an effort to respond to, and overcome, colonizers' models of the world that had cast the colonized as essentially timeless and premodern.[38] Rather than anachronistic and essentialist projections, such historical geographies might more productively be seen as simultaneous efforts to account for those neglected pasts and to engage in a dialectical process of cultural recreation in the wake of generations of colonial rule and cultural, social, and political oppression.[39]

III. OCCUPIED: INTERNAL COLONIALISM

The more troubling aspect of such efforts is not their supposed historical and spatial infidelities. It is, rather, their potential for social and political exclusion. Who wields such visions of the past and to what end? Decolonization did not necessarily entail political, social, or cultural equality for all those living within a given empire's or nation-state's bounds. Leaders of anticolonial movements could be more intent on protecting and perpetuating the interests of a particular class or sector of society than on implementing broad social change. In this they shared the experience of European elites who, during and after the French Revolution, used notions of national sovereignty largely as a means to hold on to their status.[40] Other examples abound, from Lat Dior and his Wolof kingdom in Senegal to the political powers in postcolonial Vanuatu to the marginalization of Berber speakers in northwest Africa.[41] Or take the Creole elite who garnered independence from Spain in early nineteenth century Latin America, men such as Agustín de Iturbide; some have argued that they can hardly have been considered colonial subjects and that the transition from Spanish to national rule heralded little change for the majority of the indigenous and poor mestizo inhabitants.[42] A significant number of the anti-imperialists in Cuba and Puerto Rico who sought liberation from Spanish rule still held dreams of a nation purged of blacks.[43] And lest we forget, the Age of Liberty was also the Age of Slavery. In the United States liberation from imperial rule perpetuated slavery rather than hastened its demise: remaining British colonies would abolish slavery some three decades prior to its end in the United States.[44] Such lessons were not lost on Frantz Fanon, who, a century later, would strike an angry and suspicious tone when envisioning Algerian and African independence: "The changeover will not take place at the level of the structures set up by the [national] bourgeoisie during its reign, since that caste has done nothing more than take over unchanged the legacy of the economy, the thought, and the institutions left by the colonialists."[45] At the expense of the majority of the populace, they "inherit the unfair advantages which are a legacy of the colonial period."[46]

Clearly there were changes in the shift from colonial to independent rule. In most of Latin America, formal caste distinctions disappeared; in the wake of the ravages of a decade of wars, new opportunities arose for peasants and Indians (often one in the same) to regain or expand landholding; slavery was abolished, as early as the 1820s in some areas, and by the 1850s in all countries except Brazil and Cuba. Even so, for the mass of working people, urban and rural, there was often limited change in the social hierarchy after independence. We might speak of independence for particular polities, but for many inhabitants of former colonies this did not necessarily translate into self-rule. Similarly, for long-marginalized ethnic, cultural, and linguistic minorities, the transfer from colonial to national rule could appear as no transfer at all, being little more than a shift from external to internal colonialism. The Creole elites of much of Spanish America may have celebrated their nations' native pasts, as Mexican cartographer Antonio García Cubas did here in an 1885 plate from his *Atlas* (fig. 1.5), but they seemed to be, at best, at a loss when it came to their

FIGURE 1.5. Distant Indians. Antonio García Cubas, "Carta histórica y arqueológica," from his *Atlas pintoresco e histórico de los Estados Unidos Mexicanos* (México City: Debray, 1885). The map is surrounded by illustrations of a number of prominent archaeological sites as well as objects from Mexico's National Museum. Image courtesy of David Rumsey Map Collection (www.davidrumsey.com).

indigenous contemporaries.[47] The Mapuche of Chile and the native peoples of Argentina's Patagonia and Chaco regions all confronted expansionist states in the late nineteenth century that, allied with international capital, were arguably just as rapacious and ruthless as their colonial predecessor, if not more. The same could be said regarding the struggle for sovereignty in what became the United States.

In the twentieth century, similar cases abound: take, for example, South Africa, where members of the minority white population created atlases that purported to represent the whole country even as the regime of apartheid found its fullest expression.[48] The Portuguese colony of East Timor declared independence in late November 1975; in December of that same year, Indonesian forces invaded. The Bougainville Rebellion of 1988 pitted an armed insurgency seeking independence against the military forces of Papua New Guinea, itself only recently liberated from Australian rule in 1975. Ethiopia's annexation of Eritrea in 1962 ushered in language and educational policies that smacked of old colonial habits to many Eritreans. The decades-long war that followed eventually ended with recognition of Eritrean independence in 1993. In Sudan the three southernmost provinces were home to a missionary-educated elite cultivated by British colonial machinations that sought to "retribalize" Sudan. This elite found itself with little access to the state and rebelled against the Khartoum-based government at the very moment of Sudanese independence.[49] The simultaneous persistence and reworking of colonial technologies, epistemologies, and categories of difference is captured vividly in a 1969 map, a sheet of which is shown in figures 1.6a and 1.6b. The Sudanese authorities took a 1946 British colonial map of tribal and provincial boundaries and reinscribed it thickly with racial categories and boundaries. The Sudanese Survey Department labeled those groups of the southern provinces, still resisting Khartoum, as "Negroid," and those in the remaining provinces, with two exceptions, as "Arabic." One of those exceptions was Darfur, the populace of which the Survey Department categorized as "Negroid" but situated within an administrative space deemed "Arabic."

The intricacies of such anticolonial movements around the globe during the 1950s and 1960s inspired marginalized populations in countries that had long since been formally decolonized, including the United States. This should hardly come as a surprise: were the differences between the racial hierarchies that held sway in many colonies and the virulent Jim Crow laws in the United States really that marked?[50] Thus, by the 1950s and 1960s, in the midst of a "cold war," black militants in the United States questioned the narrative teleology of "freedom" proffered by a nominally liberal state that still tolerated segregation and racial violence; instead, they found inspiration in the racial tolerance purportedly advocated by Soviet Russia. The equation of social justice with racial justice—of the intimate link between racial and class prejudice—explains Stokely Carmichael's appreciation of Fidel Castro as "the blackest man in the Caribbean."[51] Meanwhile, in the late 1960s and 1970s, Chicano nationalists spoke of "internal colonialism" and identified Aztlán as the ancestral homeland of the Aztecs and, therefore, as the ancestral home of people of Mexican descent living in, as Rudolfo Acuña provocatively phrased it, an "occupied America" (figs. 1.7 and 1.8).[52] They needed no further explanation

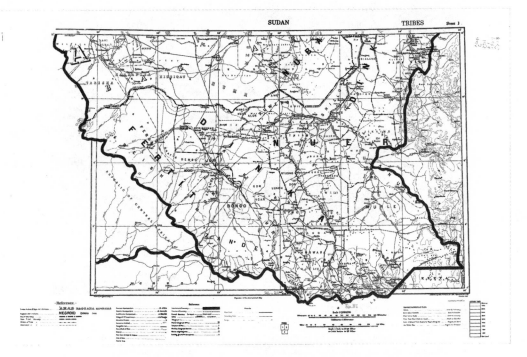

FIGURE 1.6A. A new Sudan? This sheet shows the southernmost areas of Sudan with the racialized boundary marker. Sudan Survey Department, *Sudan: Tribes, Sheets 1–3*. 1:2,000,000. Khartoum: Sudan Survey Department, October 1946, corrected 1969. Courtesy of Map Collection, Olin Library, Cornell University.

— Reference —

'Araba, Nubian & Beja and divisions ...... 'ĀRAB ᴮᴬᴳᴳᴬᴿᴬ ᴴᴬᵂᴬᶻᴹᴬ

Negroid and divisions ...... NEGROID   DINKA TWIJ

Racial Boundary ....

Main Tribal Boundary ....

Division of Tribe   "   "   ....

Subdivision   "   "   ....

Reference

Province Headquarters .... ⊕ JUBA

District Headquarters .... ⊗ Amadi

Sub District Headquarters .... ⊙ Maridi

Village of 1st Importance .... ○ Li Rangu

Christian Mission .... ⓑ

Contours (Metric) ....

Navigable river ....

River, Wadi or Kher ....

Swamp ....

Major Areas developed for Irrigation ....

Areas of Basins ....

Reservoir Areas ....

Reference

International Boundary ....

Provincial Boundary ....

Council Boundary Surveyed ——— Unsurveyed

Railway ....

Telegraph line ....

Post & Telegraph Office .... ℗

Telephone Office .... T

Wireless Telegraph Station .... ✢

Marine sighting area ....

Landing ground for aeroplanes ....

FIGURE 1.6B (detail). Sudanese legends. The legend shows the categories used for boundaries on the postcolonial map.

when philosopher Jean-Paul Sartre wrote that Americans (by which he meant the United States) "have their colonies 'at home' in their own country," or when educational theorist Paulo Freire, after teaching and living in the United States, explained that "third world" was a political rather than geographic category.[53]

FIGURE 1.7. Occupied ground. Emilio Aguayo, *Somos Aztlán*. Mural at the Seattle Ethnic Cultural Center. The contemporary boundaries of the states of California, Arizona, Colorado, New Mexico, and Texas are highlighted and then overwritten with the words "somos Aztlán" ("We are Aztlán"), while descendents of Aztlán confront "The Society," draped in a white cloak reminiscent of the Ku Klux Klan and wielding a scythe of racism and oppression. Courtesy of Oscar Rosales Castañeda. Photograph by Oscar Rosales Castañeda. Reproduced with permission.

IV. AN EMPIRE OF NATION-STATES

All of this raises a question: To what degree is the nation-state itself a perpetuation, rather than supercession, of colonialism? A "poisoned gift," as Michael Hardt and Antonio Negri put it?[54] As numerous authors have observed, the nation-state form was not necessarily the only—or primary—option for anticolonial movements. Yet too frequently, as Fred Cooper notes, "because we know that the politics of the 1940s and 1950s did indeed end up producing nation-states, we tend to weave all forms of opposition to what colonialism did into a narrative of growing nationalist sentiment and nationalist organization. That the motivations and even the effects of political action at several junctures could have been something else can easily be lost."[55] Scholars have made similar arguments for insurgent politics in nineteenth-century Latin America.[56] Lost are the alternative political possibilities that circulated that did not take the liberal state as their starting and ending point. Indeed, simultaneous with late nineteenth-century anticolonial struggles a remarkable, cosmopolitan, and transnational movement opposed to the imperialism of both colonial

FIGURE 1.8. Return. Gilbert (Magu) Sánchez Luján, *Trailing los Antepasados* (2000). The artist positions the globe with South America at the top, a reversal mirrored in the reverse migration of Mexican migrants to the mythical homeland of Aztlán. In Virginia M. Fields and Victor Zamudio-Taylor, eds., *The Road to Aztlan: Art from a Mythic Homeland* (Los Angeles: Los Angeles County Museum of Art, 2001), 18. Reproduced with the kind permission of Otoño Luján.

states and nation-states reached its apex: anarchism. The most important leftist movement of the era, anarchism was the primary oppositional force challenging capitalism, imperialism, oligarchy, and autocracy, and it linked anticolonial leaders across hemispheres, from Manila to Havana, from Cairo to Bengal, from Edo to Moscow.[57] Anarchism would confront opposition from both the left and the right as the twentieth century progressed, particularly by the 1930s, and not surprisingly so given the devotion of both liberals and communists to the nation-state.

Thus, nationalism—at times, but certainly not always, a virulent ethnonationalism—often emerged as the counter to colonial rule. New forms—domestic forms, for lack of a better word—of colonialism arose, ones that sought to force "an isomorphism between national populations and territorial domains."[58] New minoritized populations were created; others persisted through the transfer of power. This meant for many already marginalized, or newly marginalized, populations that the nation-state appeared on the horizon as little more than another form of labor capture, cultural repression, resource exploitation, and territorial expropriation. James Scott argues:

> the encounter between expansionary states and self-governing peoples [that occurred in Southeast Asia] . . . is echoed in the cultural and administrative process of "internal colonialism" that characterizes the formation of most modern Western nation-states; in the imperial projects of the Romans, the Hapsburgs, the Ottomans, the Han, and the British; in the subjugation of indigenous peoples in "white-settler" colonies such as the United States, Canada, South Africa, Australia, and Algeria; in the dialectic between sedentary, town-dwelling Arabs and nomadic pastoralists that have characterized much of Middle Eastern history.[59]

In such cases, the nation-state appears as a form of colonial persistence, a perpetuation of, rather than emancipation from, colonial ontologies.[60] Indeed, in certain cases one would be hard pressed to find the difference between colonial and national cartographic techniques and the manner in which they were applied, particularly in relation to long-marginalized populations whose relationships to any state form—whether colonial or national—were invariably conflictual and unequal.[61] Historian Jeremy Black, one of the few writers to have attempted to wrestle broadly with decolonization and cartography, writes that "the cartographic processes and devices employed elsewhere in the world are very much those of the West. Western science and techniques remain central, and there has been no real attempt to revive or devise different cartographic methods."[62]

I am sympathetic to Black's point—indeed, it goes to the heart of both the loose abstraction and brute materiality of colonialism, and I will pursue its implications in a moment—but it does require at least some qualification. It is worth recalling, for one, that the technologies used by colonizing powers did not necessarily develop prior to, or in isolation from, acts of colonization and the activities of the colonized themselves. Claims of technological diffusion—that located the origins and development of "science" in the metropole—were (and remain) integral components of self-serving colonial ideologies that emphasized "difference" in order to legitimate subjugation, despite the fact that such "differences" were produced through, rather than prior to, colonial relations. (Indeed, such claims were essential components to the very constitution—the very fabrication—of something called "the West.") These were retrospective claims that suppressed the realm of interchange, overlap, and reciprocal interaction—the entire world of dialogic, transcultural engagement—that characterized how colonial science often functioned and developed on the ground.[63] Colonial cartographic routines drew upon local practices, knowledges, and mapping operations; meanwhile, colonized peoples participated in mapping enterprises and populated cartographic bureaucracies, as agronomists, as surveyors and draftsmen, as informants and guides, and in a multitude of other roles.[64]

Indeed, this helps to explain to some degree the remarkable continuities in colonial and postcolonial cartographic technologies, practices, personnel, and modes of representation. Postcolonial governments around the globe often took the cadastral, geodetic, and topographic surveys of previous colonial administrations and put them to use for new political and social programs. Continuities in personnel were common also, whether it be in the composition of survey crews in postrevolutionary Mexico or in postcolonial Ghana.[65] In Nigeria, Northern Rhodesia/Zambia, and Swaziland, the British Directorate of Overseas Surveys continued working on large-scale topographic surveys after independence.[66] This should not be surprising. The same technologies and skills that went into generating colonial boundaries or creating settler properties were used to affirm postcolonial boundaries, to expropriate or confirm properties, or to undertake vast programs of land reform.[67] While such reforms were not always necessarily associated with formal decolonization, they were often part of a larger set of initiatives undertaken by many young (and some not so young) governments to assert power and ownership over national assets and to fulfill the promise of self-determination and sovereignty. This was as true for, say, postrevolutionary Mexico or Bolivia—long independent from formal colonial rule—as it was for only recently liberated Nigeria or Tanzania.

Colonial cartographic technologies, regardless of purported origin, could be used for myriad purposes. Technologies were rife with subversive possibility. Kingdoms, states, and populaces could, and did, appropriate various mapping technologies associated with colonizing powers to their own ends—in some cases to stave off colonial encroachments, in other cases to expand their own territorial bounds, and yet in other instances to defend highly localized forms of land tenure or sovereignty. In other cases such technologies were used to visually represent an alternative nationalist imagination to that plotted out by both colonial and postcolonial regimes, as Sumathi Ramaswamy has demonstrated for the case of Lemuria, or to purport historical or legal territorial rights, as seen in the Lakota logo-map shown in figure 1.9.[68] Although such technological and epistemological appropriations (forced appropriations, perhaps) could be cast as part of a narrative of loss, a shift in perspective might suggest we see them as, in Ramaswamy's felicitous phrasing, "hijackings," a term

FIGURE 1.9. Hijackings. Lakota logo-map, by the Black Hills Alliance of South Dakota. This illustration comes from a lapel button and is meant to visually emphasize the size and boundaries of territory promised the Sioux in the 1868 Ft. Laramie Treaty. Courtesy of Zoltán Grossman.

that draws our attention to the dialogic and politically empowering manner in which such technologies and epistemologies can be used.[69] Rather than being passively infiltrated, colonized subjects here become active agents and historical subjects.[70] And, if one is to take the recent arguments of Kapil Raj to their logical conclusion, we might similarly argue that the British in certain cases hijacked methods and theories from their soon-to-be colonial subjects.[71] There is, then, a dialectic—epistemologically, technologically, and artistically—at work that makes it difficult to narrate the history of cartography as a story solely of domination and resistance.

Having said that, colonialism was never a relationship of equals, and the power imbalances cannot be ignored. Strategies of cartographic appropriation, as geographers Joel Wainwright and Joe Bryan have argued, "do not reverse colonial social relations so much as rework them."[72] Colonized populations found themselves in situations in which they had little choice but to engage to some degree with the idiom of the state and its emphasis on property and contract, thus becoming physically and epistemologically enclosed.[73] Cartographic technologies may have been useful in carrying out radical land reform programs, for example, but the basic premises of modern land surveying left little room for forms of engaging with the land that did not conform to the specifics of nominally Western property norms. Different ways of understanding, occupying, and working space were made to disappear—largely through the law—and different ways of representing space dismissed in courts as unscientific, unreliable, or simply illegible. For example, connections to and tenure in land among some aboriginal peoples of Australia is expressed and determined through "knowledge of dreamings," in which "stories, songs, dances and sacred objects relating to the dreamings are the very title deeds. [. . .] In contrast with the commodity view of rights to land, an Aboriginal person's rights to land are not capable of being bought and sold, because the self cannot be traded."[74] Or take the case brought to court in 1987 by the Gitksan and Wet'suwet'en peoples in the Canadian province of British Columbia, regarding issues of self-government and territorial jurisdiction, in which they drew on very different conceptualizations and practices of space and cartography from that proffered by the Canadian state.[75] They used oral traditions and songs as a means to evoke a historical geography. Both in form and content these challenged the standard and imposed criteria for geographic and historical evidence followed by the court.[76] Their efforts were initially not successful, revealing the difficulties faced by populations attempting to challenge colonialism's legacies outside of its own epistemological formations (fig. 1.10).

Thus, as Curtis Berkey notes, "the challenge for Native peoples and their lawyers is to develop maps that incorporate traditional ways of knowing, while at the same time adhering to cartographic procedures that are acceptable in an alien legal system. In this way, courts may be persuaded that the truth contained in Native maps is worthy of the same respect as the official maps of the government."[77] Participatory and community mapping projects, at times controversial, have highlighted both the promise and problem of trying to navigate through one of the most persistent realities of colonialism: that colonizers unequally set the terms of ontological and epistemological engagement.[78] Such projects can simultaneously confront and challenge colonial logics and generate a dialogue between tradition and modernity, among the past, present, and future—factors that are an integral part of any collectivity. "By definition," writes Joe Bryan with respect to participatory community mapping, "maps present indigenous traditions in new ways, to new audiences, creating new understandings of community and territory."[79] The point is essential if we are to move beyond stale arguments that celebrate stasis as the sine qua non of authenticity. Holding "culture" or "custom" in a state of stasis is no less a colonial view. The question of "custom" is never outside the political context in which it is invoked. This is not to suggest it is deployed purely instrumentally but that "custom," much like "modernity," is a relational and contextual concept.[80] More bluntly, determining what is native and what is foreign is always a fraught—that is to say, political—exercise. Thus, instead of equating the authentic with the pure or unsullied, a strategy of conservative enclosure if there ever was one, or, conversely, suggesting no such thing as authenticity is possible, we might instead think about authenticity in the terms eloquently laid out by anthropologist Sherry Ortner:

> Authenticity is another highly problematized term, insofar as it seems to presume a naïve belief in cultural purity, in untouched cultures whose histories

The Law vs. Ayook
Written vs. Oral History

FIGURE 1.10. Song spaces. "The Law vs. Ayook: Written vs. Oral History." Don Monet shows competing epistemologies in a case pitting First Nations peoples against the Canadian government. The elders are singing or chanting their community's territoriality while tomes of texts and topographic maps sit between them and the judge. From Don Monet and Skanu'u, *Colonialism on Trial: Indigenous Land Rights and the Gitksan and Wet'suwet'en Sovereignty Case* (Philadelphia: New Society Publishers, 1991), 41. Copyright Don Monet. Reproduced with permission.

are uncontaminated by those of their neighbors or of the west. I make no such presumptions; nonetheless, there must be a way to talk about what the Comaroffs call "the endogenous historicity of local worlds" [. . .] in which the pieces of reality, however much borrowed from or imposed by others, are woven together through the logic of a group's own locally or historically evolved bricolage.[81]

Such a perspective is useful in avoiding the pitfalls of a static and romanticized ethnocentrism (what Michael Watts has trenchantly critiqued as the romance of community) that obsesses over spatial fixity rather than mobility and separation rather than connection.[82] Rather than lamenting change with an implicit narrative of decline, such an understanding of authenticity works to wrestle with the constant struggle between continuity and change, how it unfolds, and on whose terms. It situates the colonized as subjects and agents of their own history, as active participants in historical time. Jon Parmenter captures the stakes:

> Persistent conceptions of early American space as a surface upon which Europeans acted and Native peoples reacted have yielded narratives that obscure the contemporaneous temporalities and heterogeneities of space for non-European actors. Such accounts fail to acknowledge the coeval yet vastly different experience of the Iroquois (and other indigenous) peoples. Subordinated to the self-producing narrative of the emergence of the United States, the Iroquois have been denied their own historical trajectories and effectively (though artificially) held still while others have done the moving.[83]

As well as being held historically still, so too have non-European actors been held geographically still. For example, despite the fact that Oceania inhabitants were (and are) quite mobile, in the historiography it is the Europeans who do the moving and the Oceania natives who stay in place.[84] Movement on their part tends to be read as an indicator of declension, not expansion. It is recognizable state actors who occupy center stage, who move historically and geographically, who act: "the" British, "the" Spanish, "the" French, and so forth. The inexorable logic of manifest destiny leaves little room for non-state spaces or different imperial spaces. Yet postcontact "indigenous imperialism" is hardly an oxymoron. The Sioux were a dominant power in the northern plains for much of the nineteenth-century, asserting control over spaces where others also lived (the Pawnee, Ponca, Otoe, and Omaha, among others).[85] Thus, one might look at the Black Hills Alliance logo-map shown above as simultaneously—and not paradoxically—an imperial and anti-imperial representation. Or take Pekka Hamalainen's recent, trenchant study, which shows in compelling detail how it was in fact the Comanche who were the dominant empire of the south-central plains of North America in the eighteenth and nineteenth centuries, not the European settlers. In a form of what he calls "reverse colonialism," the Comanche from the period 1750 to 1850 were able to "expand, dictate and prosper, and European colonists [forced to] resist, retreat and struggle to survive."[86] Hence the need for a map of a place of which few have heard: "la Comanchería" (fig. 1.11). Whether because they were fixated on the birth of nations, mired in exceptionalist extravaganzas, or obsessed with the fetishizing of cultural difference, historians have ignored the possibilities that indigenous polities could be imperial, powerful, political—in a word, human—and shape in proactive ways the modern history of North America. In effect, the past has been colonized by an empire of nation-states.[87]

V. FIVE-SIXTHS OF THE WORLD

Has the present, too, been in some sense colonized by the nation-state? To what degree has the understanding of the nation-state as the antithesis of the colonial state masked a reworking of, rather than an end to, imperial relations? Let me return again to the work of J. F. Horrabin, who, writing in 1937, captured the dilemma succinctly:

> Modern imperialism . . . does not always proceed by the method of armed conquest and annexation of territory. It has discovered that economic "penetration," leaving the political independence of the penetrated country nominally intact, is occasionally sufficient for its purposes. In a very real sense, therefore . . . such "penetrated" (and dependent) countries may be accounted colonial possessions of other states. But to have included all these in our catalogue would have been to map *five-sixths of the world*.[88]

I share Horrabin's frustrations. Exactly where to draw the line in examining colonialism and its purported demise is not easy. To expand the definition too far is to make the very concept of "decolonization" analytically anemic. It is to also risk conflating different forms of imperialism and subsequent forms of decolonization. At the same time, to limit the discussion to *political* independence—and the transition from formal colonial status to formal

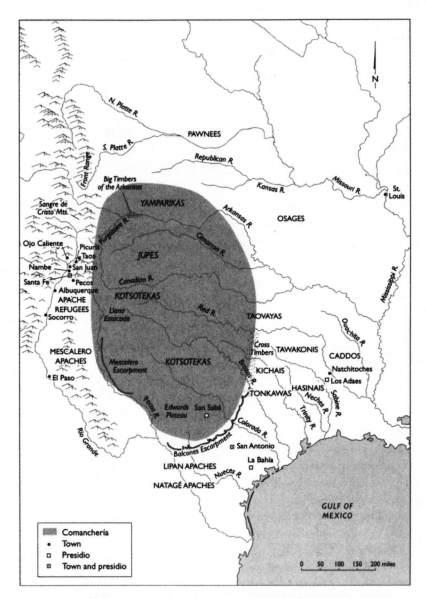

FIGURE I.II. Indigenous empire. La Comanchería, from Hamalainen, *The Comanche Empire* (New Haven: Yale University Press, 2008). Reproduced with permission of Yale University Press.

nation-state status—would be intellectually naive and conceptually dubious. It would be, in the pithy words of the authors of a classic statement on the problem, "like judging the size and character of icebergs solely from the parts above the water-line."[89] Independence and decolonization are not necessarily the same thing, and all too often decolonization could take shape as largely a rearticulation of forms of dependence: that is to say, of forms of colonial power.[90]

Even the claim of political independence raises questions. After all, colonial holdings were never homogeneously administered or defined. There were "repertoires of imperial power": dominions, commonwealths, and protectorates were all imperial designations that, to varying degrees, sought to blend subordination and autonomy.[91] These may have avoided the most oppressive of colonial relations, but they remained forms of imperial oversight nonetheless and should not be excluded from discussions of decolonization, as historian Anthony Hopkins has persuasively argued.[92] The creation of the Commonwealth was in large part meant to sustain beneficial relations of economic assistance and favorable terms of trade in the midst of postwar decolonization.[93] But it also provided perhaps a bit of psychological relief for the dislocation generated by empire's demise. Evoking his childhood in Essex and along the Thames, historian Simon Schama recalled how he would watch "the ships move purposefully out to sea toward all those places colored pink on our wall map at school, where bales of kapok or sisal or cocoa beans waited on some tropical dock so that the Commonwealth (as we had been told to call it) might pretend to live up to its name."[94] Certainly there was much more pink to satisfy the eye on a map of the Commonwealth than there would be on a map of the colonies, as the Colonial Office seems to have concluded. Figure 1.12 shows a map from the 1964 Colonial Office List. Rather than the lonely colonies meagerly scattered on its map of two years earlier (see fig. 1.2), here the image of the global place and reach of England is reasserted by emphasizing the Commonwealth member states. The map, not surprisingly, bears a striking resemblance to those produced around 1910, at the apex of Britain's imperial reach. A comforting image, perhaps, especially for those who still found in empire much about which to crow.[95] The chairman of council of the United Empire Society in 1958 had the following to say when explaining the society's name change, from the United Empire Society to the Royal Commonwealth Society: "This is an exciting and challenging moment in our history. Our proposed name change is *not a retreat from Empire*. It is a proud assertion of what British Imperialism has created, namely a great commonwealth of free nations."[96]

But of course, not all former colonies were reborn as free nations. Canada's trajectory from colony to full independence spanned decades, culminating finally in 1982, followed shortly thereafter by the publication, in 1987, of the remarkable *Historical Atlas of Canada*.[97] The point is myriad forms of political life thrive both between and outside the poles of nation-state and colony.

Stewart Firth's accounting of the status of many island communities in Oceania is revealing:

> Guam is officially an "organized unincorporated territory" seeking to become a "commonwealth" of the United States; American Samoa is a "US unorganized unincorporated territory"; the Northern Mariana Islands are a US commonwealth that, by many people's reckoning, is still part of the US Trust Territory of the Pacific Islands; Tonga is an independent kingdom; Tokelau is New Zealand territory; the Cook Islands are a self-governing state in free association with New Zealand; Wallis and Futuna are an overseas territory of France; Kiribati is an independent republic; and so on.[98]

It would be tempting to see in this a kind of truncated decolonization: that is, to view such myriad political forms cynically, as yet more instantiations of colonial rule reconfigured in such a way so as not to offend enlightened sensibilities. Certainly this is indeed partially the case. The United Nations Special Committee on Decolonization as recently as 2009 still considered Guam to be a colony.[99] Its history had been wedded to that of other former Spanish possessions which came under the control of the United States in the aftermath of the Spanish-Cuban-American War of 1898 and occupied revealingly peripheral locations on one map in Rand McNally's postwar *Imperial Atlas* (fig. 1.13).

What to do with such possessions, and their populations, proved a thorny issue for the self-described "empire of liberty." In the case of Puerto Rico, the Supreme Court, in a moment of inspired verbal gymnastics, determined that the populace of newly acquired Puerto Rico was "foreign [. . .] in a domestic sense."[100] To this day the status of many former colonial possessions in the Pacific remains complicated. Many have been and continue to be defined by the strategic military objectives of their former masters: Japan, France, the United States, and, to a lesser degree, England. Cold War concerns of the United States ensured that archipelagoes such as Micronesia would not garner full independence but rather a kind of loose autonomy—a Compact of Free Association—overseen by the strategic needs of the United States, even if the US secretary of war insisted that the islands "are not colonies; they are outposts" (fig. 1.14).[101]

But such a position should not be held without some measure of recognition for the voices of the formerly colonized themselves, who thought carefully and critically about the positives and negatives of full political independence. Burbank and Cooper have observed that "as recently as the 1950s influential

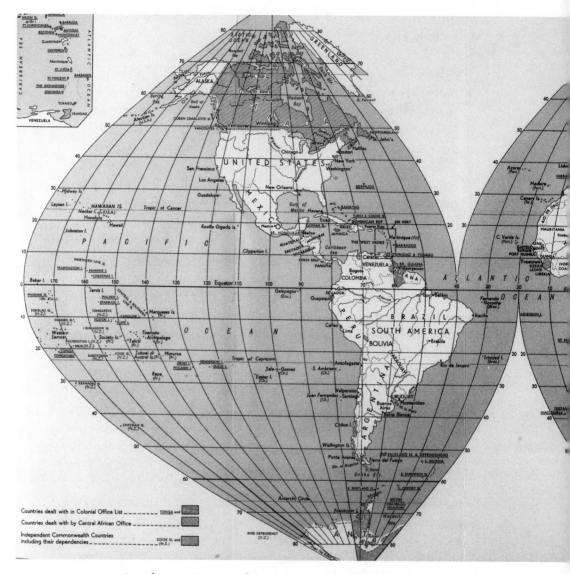

FIGURE 1.12. A comfort map. Directorate of Overseas Surveys, *Map of the World*. Insert in *The Colonial Office List, 1964*. (London: Her Majesty's Stationery Office, 1964).

leaders in French West Africa argued that confederation in which France and its former colonies would be equal participants was preferable to the breakup of empire into independent nation-states."[102] Debates over the status of Puerto Rico frequently revolve around whether the island would in fact become more dependent on the United States if it were to garner political independence. Similar debates characterize parts of Oceania that were formerly under

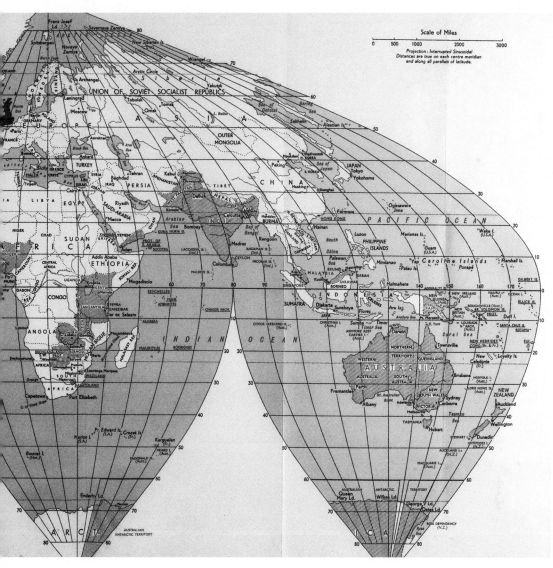

FIGURE 1.12. *Continued*

French rule, where any discussion of independence has to consider the impact it would have on subsidies from the metropole.[103] In these cases, those who have the most at stake in the debate recognize that colonialism is a relation of dependence built on more than political foundations. The economic power of multinational corporations, international monetary institutions, and industrial metropoles can mean economic dependence persists despite political indepen-

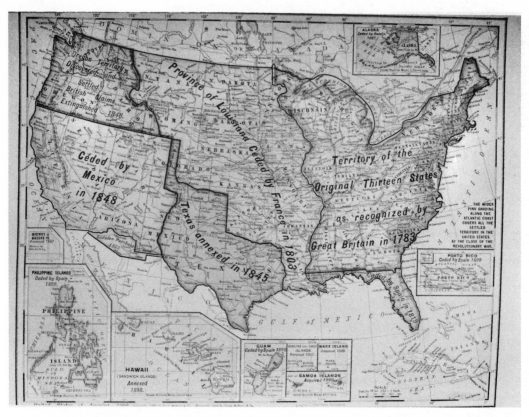

FIGURE 1.13. Manifest colonization. Rand McNally's historical map of US expansion. Rand McNally, *New Imperial Atlas of the World* (Chicago: Rand McNally, 1905).

dence and should force us to question the value of terms such as "decolonization" or "postcolonial."[104] In other words, and despite the knotty analytical and definitional questions it provokes, it is worth considering colonialism as something more than a *political* relation of asymmetry.

Indeed, at the very height of decolonization, in the 1950s and 1960s, intellectuals in parts of Africa and Latin America began to write of *neo*colonialism, questioning the determinations of what constituted the end of colonial rule through the lens of international political economy. Was it purely an issue of political self-determination and formal structures of political power? Had colonial rule collapsed with the flight of the colonial powers? To what degree had national autonomy severed the unequal relationship of dependency between metropole and periphery, or north and south, or colonizer and colonized? How could, for example, the promise of self-determination be fulfilled if confronted with the strategic threat of capital flight?[105] Critics of capitalism

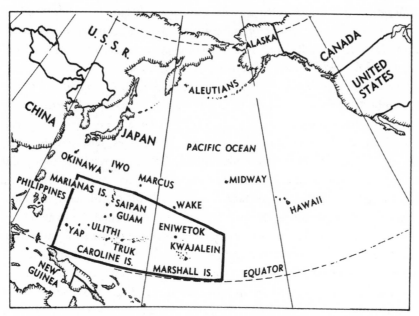

The heavy line circles former Jap islands asked by the U.S.

FIGURE 1.14. Outposts, not colonies. US claims over "formerly Jap islands" in 1947. A short article next to the map noted the following: "The United States presented to the United Nations a plan calling for UN trusteeship over 623 former Japanese-mandated islands, with the U.S. as administrator. [. . .] The United States made it clear that regardless of the United Nations' decision, it will retain control over the islands to ensure American security." Detail from News Map of the Week, Inc., "World News of the Week," Vol. 9, No. 26 (Monday, February 24, 1947). Courtesy of Olin Library Map Collection.

have long stressed its deeply implicated relationship with colonialism. Thus, even as colonial empires crumbled, a variety of voices argued that the demise of colonialism might be called into question by the persistence of capitalism. Within Latin America intellectuals such as Raúl Prebisch (1946), Andre Gunder Frank (1967), Eduardo Galeano (1971), and Fernando Henrique Cardoso and Enzo Faletto (1969) generated historical analyses and interpretations that offered trenchant explanations for a world still deeply colonized by tracing historical networks and spatial grids of capitalist dependency, in the process highlighting the structural relationship between "development" and "underdevelopment."[106]

That such perspectives came out of Latin America is no surprise. There, despite a century of formal political independence, countries had struggled to free themselves of the yoke of European and, increasingly, US economic and political power.[107] So much so, in fact, that Frantz Fanon invoked Latin

America's historical experience in his writings on Algerian decolonization: "The national bourgeoisie of certain underdeveloped countries has learned nothing from books. If they had looked closer at the Latin American countries they doubtless would have recognized the dangers which threaten them."[108] To be clear, there was no "scramble for Latin America" comparable to the "scramble for Africa," in which European powers annexed and carved up the continent.[109] Even so, politically freed from colonial rule, Latin American statesmen and elites had to contend (or ally themselves) with new imperial formations, namely, British and subsequently US commercial hegemony and its attendant political machinations and occasional interventions. Despite the US-issued Monroe Doctrine of 1823, it was initially the British who were most powerful in "postcolonial" Spanish America. They owned the large homes in many of Spanish America's primary cities; maintained men-of-war off the coasts of various nation-states to protect their commerce; negotiated the best treaties of friendship, commerce, and navigation with newly emergent nation-states in return for political recognition; and mapped the lands and resources to which they would informally lay claim. Indeed, the musings of a British foreign secretary in 1824 were both prescient and a cogent summation of the state of things: "Spanish America is free, and if we do not mismanage our affairs sadly, she is English."[110] And to some degree "she" was, at least until the latter half of the nineteenth century when US hegemony grew substantially in the hemisphere.

US hemispheric hegemony brought with it a more recognizably colonial set of relations at times, most dramatically with the Spanish-Cuban-American War of 1898, the subsequent occupation of Cuba and the Spanish Pacific, and the annexation of Puerto Rico. US readers followed the paths of empire in an array of cartographic and ethnographic publications—with revealing titles such as the *American Colonial Handbook* and *Our Islands and Their People*—assembled to help the populace assimilate these new possessions.[111] Roosevelt's 1904 Corollary to the Monroe Doctrine, which legitimized US intervention in Latin American countries to protect business interests or collect debts, drew on the language of paternalism and barbarism in affirming the neocolonial relationship. So too did Roosevelt's secretary of state, Elihu Root. The two hemispheres complemented each other, he noted: the people in the south needed North American manufactures, and North Americans needed the south's raw materials. "Where we accumulate," remarked Root, "they spend. While we have less of the cheerful philosophy which finds happiness in the existing conditions of life, they have less of the inventive faculty which strives continually

FIGURE 1.15. Debt crisis. Charles L. "Bart" Bartholomew's cartoon appeared in the *Minneapolis Journal* ca. 1903. It portrays the purported lack of fiscal responsibility of various Latin American countries and serves as a valuable reminder that debt crises are not particularly new. Its appearance so close to Roosevelt's corollary to the Monroe Doctrine is telling. Reproduced from John Johnson, *Latin America in Caricature* (University of Texas Press), 41.

to increase the productive power of men." Cartographic cartoons translated the point for a broader public (fig 1.15).

US imperial aspirations affected other countries in the hemisphere in more subtle, if just as important, ways. Mexico, for example, could never shrug off its proximity to the northern colossus, and patterns of US development and investment exercised substantial influence in the country. Mexico's revolution (1910–1920) was one that in some ways sought to surpass the limits imposed by the geopolitical and economic realities of early globalization and the age of empire.[112] Revolutionaries sought not only political change but also the elim-

FIGURE 1.16. Revolutionary surveyors. Diego Rivera, sketch from *Ilustración para la Primera Convención de la Liga de Comunidades Agrarias y Sindicatos Campesinos del Estado de Tamaulipas, 1926.* N.p., 1926.

ination of the latifundia (or large landholding) that had characterized much of the countryside both prior to and after independence. But they also sought to curtail the influence of foreign capital, investment, and ownership in the country. Foreign capital's presence was manifest in the north, where owners and managers settled, intermarried, and hobnobbed with regional elites. In the Yucatán Peninsula, such presence was more latent but felt nonetheless. Much of Mexico's populace carried with it histories and memories of repeated foreign interventions, by the US military, Napoleon III's troops, and gringo Rangers, among others. "Neocolonial" is too strong a term for the relationship, but the perception of foreign control and influence existed nonetheless.[113] The revolutionary governments, in an assertion of sovereignty, assumed control over all mineral and subsoil rights and maintained the right to expropriate foreign companies. They also engaged in a far-reaching agrarian reform. In all of this, maps were crucial, and the revolutionary state hastened to train a new generation of land surveyors and agronomists loyal to the state and its revolutionary project. Along with rural schoolteachers, surveyors and agronomists became the pivotal figures in a new era characterized by its cultural and economic nationalism, as can be discerned from their presence in much post-revolutionary nationalist art (fig. 1.16).[114]

Other revolutions followed, revolutions that might be situated within the language of anticolonialism or national liberation.[115] Guatemala's democratic aperture and October Revolution of the 1940s, and the agrarian reform law 900 passed by Jacobo Árbenz in 1952, sought to put an end to the egregious and arbitrary power of an entire class of agrarian lords, led by the US-based United Fruit Company. While there was little in terms of foreign enterprises to expropriate, the reforms to land and mining in Bolivia in the early 1950s were justified in part by President Victor Paz Estenssoro as the beginning of the end to four hundred years of oppression.[116] During the 1959 overthrow of Fulgencio Batista's US-coddled regime in Cuba, revolutionaries self-consciously evoked the anticolonial insurgencies of a half century prior. Such revolutions occurred in tandem with, and drew inspiration from, anticolonial insurrections elsewhere, building on an earlier transcontinental precedent set in 1927 in Brussels with the first meeting of the League Against Imperialism (attendees included future leaders of independence movements from around the globe as well as a number of Latin American intellectuals, such as Víctor Rául Haya de la Torre, from Peru, and the Mexican José Vasconcelos, who spoke on behalf of Puerto Rico.)[117]

Such efforts after World War II ran parallel to, and often up against, new-fangled attempts to create structures that would ensure a system of free trade and capitalist longevity—structures, such as the United Nations, the World Bank, and the International Monetary Fund, that effectively proved Karl Polanyi's point that the invisible hand of the free market depends on the long arm of state intervention.[118] They also confronted a bipolar world in which the Soviet Union and the United States sought to shape the political and economic relations of countries in order to ensure control over and access to a range of primary resources, without being encumbered by formal colonial rule.[119] Thus, Mohammad Mossadeq's attempted nationalization of oil in Iran and Jacobo Árbenz's agrarian reform efforts in Guatemala both met with an increasingly common form of Cold War neocolonial response: CIA-sponsored coups d'état (1953 and 1954, respectively) and the imposition of authoritarian regimes committed to protecting foreign direct investment and the interests of foreign

FIGURE 1.17A. Guerrillas in our midst. In this Central America wall map published by the Civic Education Service in 1966 for classroom use, the author writes, under the heading "Sleepy Republics Begin to Stir," that "Communists, organized from nearby Cuba, are active in several of the countries. They are a particularly big threat in Guatemala. Two different guerrilla bands are spreading terror in that land. One, led by a half-Chinese, owes its allegiance to Peking." *Central America, Headline-focus wall map 13*, vol. 3, number 13, 1966. Copyright Scholastic. Reproduced with permission. Courtesy of Olin Library Map Collection, Cornell University.

multinationals and domestic elite partners. The social and political conflict that ensued in the wake of these imperial intrusions was, in the maps pinned to the walls of US schools, cast as a product of Soviet ambition (figs. 1.17a and 1.17b).

Cartography was not immune to or removed from such questions. New structures, combined with the quickening collapse of old imperial formations, saw the emergence of new spatial conceptualizations and mental cartographies. Former geographic frameworks premised upon colonial empires and continental divisions gave way to ones shaped by Cold War geopolitics. Area studies paradigms, which compartmentalized the world according to world regions of strategic importance with a certain amount of supposed cultural coherence, and the three-worlds paradigm, first used in 1952 and which organized

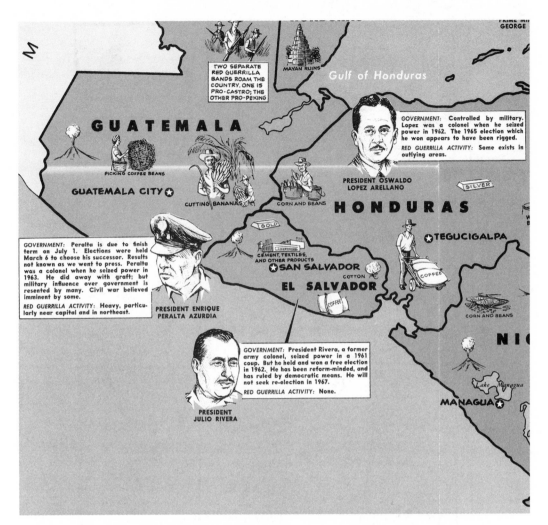

FIGURE 1.17B (detail). Leaders of each Central American country are highlighted, followed by a comment on "Red Guerrilla Activity" in the country.

the globe according to political ideologies and presumed stages of development, both achieved institutional and political validity and shaped subsequent mappings of the world.[120] These paradigms did not go unchallengedl, as intellectuals from both the global north and the global south offered alternative, and implicitly decolonial, images of their own. As well as the world systems maps already mentioned, one could include here Joaquín Torres-García's 1936 inverted sketch of the map of South America, which derived from his broader project to forge a universal artistic aesthetic that neither rejected nor privileged European models (fig. 1.18); or Arno Peters's alternative global projections,

FIGURE 1.18. An area study. Joaquín Torres-García's south-at-the-top map of 1936.

which critiqued traditional cartographic projections on the grounds of Euro-centrism; or Buckminster Fuller's nonoriented (and nonaligned) Dymaxion map of 1946: all reveal that dominant cartographic and geographic frameworks were never hegemonic and that maps were used repeatedly as a means to challenge particular ways of "thinking" the world.[121]

Decolonization (and the cold war) was lived daily, not only in the colonies but also "at home." French philosopher Henri Lefebvre, after World War II, spent much of his life writing about and analyzing what he called "the colonization of everyday life." Decolonization of the periphery was, for Lefebvre, unfolding simultaneously with a kind of recolonization of the metropole, especially in the realm of everyday life, with the expansion of the market, new patterns of consumption, the rise of cybernetics and information systems, and bureaucratized systems of planning associated with what he termed the state mode of production. The state form and monopoly capitalism had invaded every space—public or private, work or leisure—and effectively colonized everyday life.[122] Lefebvre, inspired by anarchist forbears, hinted at how a more authentic form of decolonization might be found in the ideas and practices of *autogestion* (self-management or workers' control).[123] The Situationists in the 1950s and 1960s, inspired and mentored by Lefebvre even as they fell out with him over time, sought to find ways to contest and subvert what they saw as the alienating forces of capitalism and bureaucracy, in part through their spatial practices. This included efforts to subvert conventional cartographic forms of representation, such as the well-known map by Guy Debord entitled *The Naked City* (fig. 1.19). Drawing on the Situationists' emphasis on psycho-geography and *détournement* (on wandering, reroutings, digressions), the map seeks to unsettle standard, purportedly transparent forms of representation and instead juxtapose images and words removed from their original context, in a kind of "hijacking" of original material.[124] The map critiques not only the claims to realism of standard cartographic representation at the time but also attempts to disassemble the bird's-eye perspective, the overview, so standard in most maps. Here the view is an overview but one "piecing together an experience of space that is actually terrestrial, fragmented, subjective, temporal and cultural."[125] The Situationist map—in some sense a countermap—exposes and subverts the close tie between Cartesian perspective and capitalism.[126]

VI. A COLONIAL PRESENT?

So where does this all leave us? My point here is not to suggest an absolute continuity between colonial and postcolonial states but rather to reiterate a basic point: the unequal relationships that have structured the globe exist on a continuum. An overemphasis on decolonization or postcolonialism can obscure how such forms of domination have been adjusted rather than abolished.[127] It is hardly coincidental that the high point of decolonization coin-

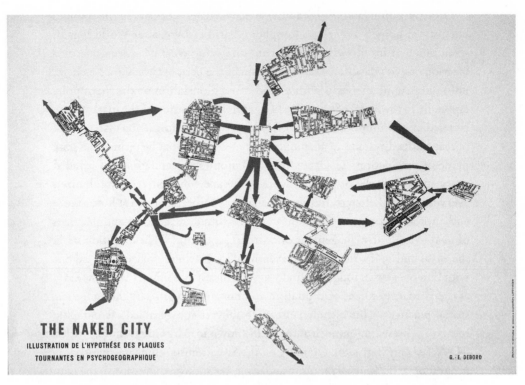

FIGURE 1.19. Disassemblage. Guy Debord's cartographic effort to capture the experience of Parisian urban space and to challenge the grid overlays of postwar planners.

cided politically with a "cold war" and economically with the rapid expansion of multinational—and increasingly transnational—corporations and, by the 1970s, post-Fordist production methods.[128] And for the peoples of much of Oceania the cold irony, as the website of the University of Hawai'i's Center for Pacific Islands Studies notes, is that "political power was restored to colonized peoples [of Oceania] just when the significance of the sovereign nation-state was declining in the face of unprecedented levels of global interdependence."[129]

Such trends continue into our colonial present. The purported end of the Cold War gave rise to suggestions that, as political scientist Francis Fukuyama put it, the end of history itself was at hand: that liberalism had emerged as the motor of historical progress, assigning alternatives—communism and socialism, among others—to the dustbin of history.[130] Fukuyama's geopolitical vision is staunchly cartographic and uncannily colonial: former colonial powers by and large are where history has come to its end; meanwhile, the rest of the world remains woefully "historical."[131] The conflation of geography and temporality so crucial to the colonial enterprise reappears yet again.

Still, could it be the case that generally the last of colonialism's remnants have been dismantled? Historian Charles Maier has suggested that the collapse of the Soviet Union meant the end of the last of the classic land-based empires. Others have argued that residents in postsocialist Eastern and Southern Europe have engaged in a process of "institutional, disciplinary and psychological decolonization . . . from state socialism," one that involves various remappings, particularly of property.[132] Perhaps . . . although others have drawn on the language of colonialism to make sense of rapid changes taking place in the former Soviet Union and its ex-satellites. Anthropologist Katherine Verdery shows in detail, for the case of Aurel Vlaicu, Transylvania (modern-day Romania), the unpredictable and messy process of transition from a state-run to a free-market economy—and the concomitant emphasis on creating or restoring private property under the auspices of the IMF and the World Bank—and how it has enabled "new forms of western colonization through transferring expertise, employing Eastern Europe's cheap skilled labor force, and flooding markets hitherto closed to western products."[133] Across the breadth of former Soviet states, local and regional elites have used privatization schemes to gather and consolidate expanses of land, even as the state immerses itself further in, rather than withdraws from, economic life. Similar processes have been underway since the 1980s in parts of Latin America, most notably in Mexico, where, with the implementation of the North American Free Trade Agreement in 1994, the government declared land redistribution officially at an end. Lands distributed to petitioners after the revolution could now be sold, rented, and circulated on the market.[134] Despite the claims that such changes constitute the withdrawal of the state from the economy, the fact is such initiatives have spurred the creation of new bureaucracies tasked with the mapping and titling of such properties.

Such moves are part of a longer trend. The Keynesian paradigm that developed in the wake of the Depression and World War II—alongside processes of decolonization—has yielded now to an aggressive free-market absolutism, referred to in short-hand as "neoliberalism." State industries are being dismantled, food subsidies abolished, tariffs and protectionist measures removed, currencies devalued, and the public sector abandoned. Institutions created under the Bretton Woods agreement have transformed themselves into international creditors with substantial power over the political and economic destinies of states around the globe, regardless of their purported "independence." (The first time I looked at Bart's cartoon [fig. 1.15], I thought not of Roosevelt's Corollary or the Monroe Doctrine, but of the IMF and the long history of debt crises brought on by strategic lending and interest rate manipulations.) If colonialism seems to cast a long shadow, it may be because it still stands there in the full light of day. Tax-free zones created in parts of Africa have been justified, according to the BBC, by economists with a particularly colonial cartographic logic: they "argue that free trade zones are particularly suited to African countries which were created under colonial occupation when land was divided up, often with little regard for the economic sustainability of the newly created plot."[135] Little wonder that the gross national product of many African countries is no larger than the annual sales of some multinationals (fig. 1.20). Multinationals—in finance, technology, and an array of diversified holdings—and some individuals have acquired a political heft and financial girth comparable to that of the high imperial states of yore. Have we returned to the era of the company-state? Perhaps we never left it. Now we just call them multinationals. Regardless, endeavors long considered the purview of states only due to the exorbitant expense—such as space exploration—have now been taken up by individuals whose personal wealth often exceeds the GDP of many smaller nations.[136] What map will capture that?

To be sure, claims regarding the nation-state's demise have been greatly exaggerated—after all, it is taxpayer money and US state intervention that opened up Iraq to an array of multinational corporations, with the US First Cavalry Division leading the charge in "Operation Adam Smith."[137] Operation Adam Smith helped implement L. Paul Bremer's and the Coalition Provisional Authority's "100 orders," which transformed Iraqi law, including dramatic changes in copyright and patent law and the setting of a legal framework to facilitate the transfer of public commons to private hands.[138] Geographic information system (GIS) technologies and renewed attention to the power of "geography" in foreign policy circles have accompanied these shifts, creating new controversies over mapping, countermapping, and forms of purported geopiracy.[139] Just how central a role new cartographic technologies play in all of this is indicated by a particularly blunt appraisal of one US military official: "GIS is how civilians spell IPB [Intelligence Preparation of the Battlefield]."

At the same time, such technologies, like their predecessors, can be used to resist, deflect, or otherwise undermine dominant paradigms. That is, they have counterhegemonic potential. Cartographers have produced radical atlases and have crafted maps that draw public attention to everything from new processes of militarization to patterns of migration, flows of capital, and movements of production.[140] Others have developed mapping practices that attempt to avoid the hierarchies of knowledge production and intervene meaningfully and politically. For example, the countercartographies collective based at the

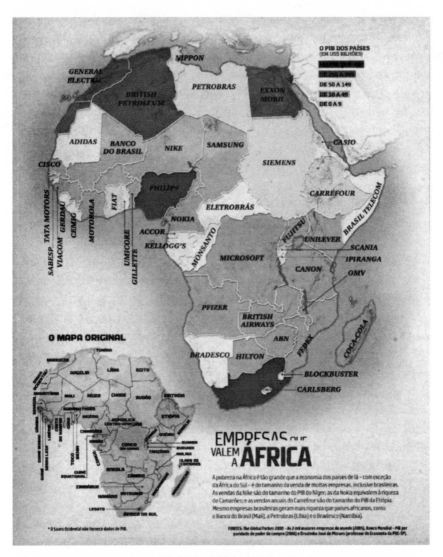

FIGURE 1.20. The company-state. One of a series of maps produced by the Brazilian magazine *Super-interesante*. This one maps equivalences between multinational companies' annual sales and the gross national product of a country. For example, "sales for Nike are equivalent to the gross national product of Niger." "Empresas que valem a Africa," reproduced in Frank Jacobs, *Strange Maps: An Atlas of Cartographic Curiosities* (New York: Penguin, 2009), 190, quoted text from 190.

University of North Carolina in Chapel Hill has sought to generate maps that capture "the changing landscape of labor when production is geographically diffused."[141] Open source tools and other mechanisms have, to some degree, democratized cartography or at least, in the words of Jeremy Crampton and John Krygier, "undisciplined it."[142] Whether or not this constitutes a form of "decolonizing the map"—akin to the efforts of indigenous communities who countermap, or the Situationists and their anti-Cartesian spaces—is an open question but certainly one worth asking, inasmuch as it raises again the perenniel issue of the originary relationship between modern cartography and colonial expansion.

I may, by this point, have stretched the term "colonial" (and therefore "decolonial") beyond any useful measure. Perhaps that is a worthwhile endeavor in that it suggests to us that the very terms "colonial" and "decolonial" may simply not be sufficient—analytically, historically, politically—for making sense of the long processes under discussion here; that such terms are more usefully understood as aspects of a larger problematic—capitalism, perhaps, or modernity.[143] But even if one were not to go that far, even if we were to recognize (as we should) that there were serious and meaningful repercussions to legal independence, that it *meant* something, the cases just discussed at the minimum serve as a reminder of the persistence of colonialism as something more than a system of overextended political rule. A final series of maps may help emphasize the point: these are maps that show the "proportion of worldwide Gross Domestic Product measured in US$ equalised for purchasing power parity," for years 1500, 1900, and 1960 (figs. 1.21–23).[144]

Confronted with such images, one can understand the skepticism expressed by the *Midnight Notes Collective* in 1990: "'Colonial emancipation' is a phrase that, if anyone has the bad taste to bring it up, can only cause derision."[145]

ACKNOWLEDGMENTS

My thanks to my fellow participants in the Nebenzahl Lecture Series, where this was first delivered: Tom Bassett, Magali Carrera, Lina del Castillo, Karen Culcasi, Jordana Dym, Jamie McGowan, and Sumathi Ramaswamy. I am indebted to Jim Akerman, David Bernstein, Ryan Edwards, Durba Ghosh, Dan Magaziner, Jon Parmenter, Jim Scott, Suman Seth, and Robert Travers for reading and commenting on this essay at various stages. I owe a special debt of gratitude to Ben Brower, David Hanlon, Dane Kennedy, Alan Knight, and Ted Melillo, for very detailed critiques of the penultimate draft. My thanks also to

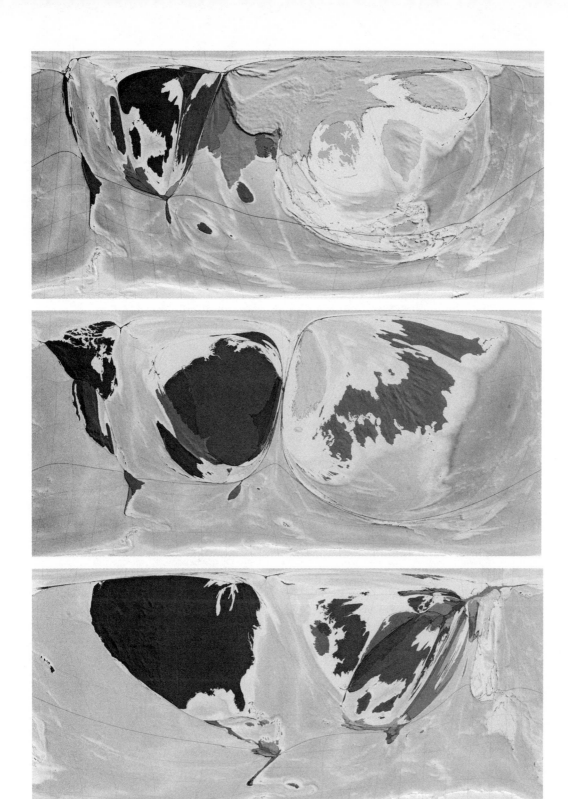

FIGURES 1.21–23. Parity and the longue durée. Proportion of worldwide gross domestic product, in US dollars, equalized for purchasing power parity, for the years 1500, 1900, and 1960. Copyright world-mapper. http://www.worldmapper.org/. Reproduced with permission.

the two anonymous referees for the Press, who pushed me in generous and productive ways. I have not taken all of their advice, but I hope they see the results of their efforts here. Portions of this essay appeared in an abbreviated version prepared for volume 6 of *The History of Cartography* (Chicago: University of Chicago Press, 2015) at the kind invitation of Mark Monmonier.

NOTES

1. J. F. Horrabin, *An Atlas of Empire* (New York: Knopf, 1937), 13.

2. Ibid., 3. For a succinct biography of Horrabin, his work, and the description of him as a working-class educator, see Leslie W. Hepple, "Socialist Geography in England: J. F. Horrabin and a Workers' Economic and Political Geography," *Antipode* 31 (January 1999): 80–109.

3. Frantz Fanon, *The Wretched of the Earth* (New York: Grove Press, 1968), 102; for a broad expansion on Fanon's assertion, see Edward Said, *Culture and Imperialism* (New York: Vintage, 1993).

4. Aimé Césaire, *Discourse on Colonialism* (New York: Monthly Review Press, [1955] 2000). See also the discussion in Robert Young, *White Mythologies: Writing History and the West* (London: Routledge, 1991), 39. I am grateful to Ben Brower for pointing out the specificity of Césaire's phrase "boomerang effect."

5. *Rand McNally & Co.'s New Standard Atlas of the World* (Chicago: Rand, McNally & Co., 1898). Capitals in the original.

6. *The Colonial Office List, 1962* (London: Her Majesty's Stationery Office, 1962), map insert.

7. "Africa 1965," news map by Edwin Sundberg (News Syndicate Co., 1965); E. J. Hobsbawm, *The Age of Extremes: A History of the World, 1914–1991* (New York: Vintage, 1996), 344.

8. Mark Monmonier, "The Rise of the National Atlas," *Cartographica* 31, no. 1 (1994): 1–15.

9. National Geographic Magazine, "Africa: Atlas Plate 54, September 1960."

10. Prasenjit Duara, "The Decolonization of Asia and Africa in the Twentieth Century," in *Decolonization: Perspectives from Now and Then*, ed. Prasenjit Duara (London: Routledge, 2003), 1; and more broadly Patrick Wolfe, "History and Imperialism: A Century of Theory, from Marx to Postcolonialism," *American Historical Review* 102, no. 2 (April 1997): 388–420. Hence the value of proceeding as much by "comparison" as by "analysis," a point made cogently by Paul Feyerabend in his *Against Method: An Outline of an Anarchistic Theory of Knowledge* (London: Verso, [1972] 2010).

11. I am indebted to Dane Kennedy, Alan Knight, and one of the press's anonymous readers for pushing me on the distinctions.

12. Fernando Coronil, "After Empire: Reflections on Imperialism from the Americas," in *Imperial Formations*, ed. Ann Laura Stoler, Carole McGranahan, and Peter Perdue (Santa Fe, NM: SAR Press, 2007); Greg Grandin, *The Last Colonial Massacre: Latin America and the Cold War* (Chicago: University of Chicago Press, 2004); and Walter LaFeber, *Inevitable Revolutions: The United States in Central America*, 2nd ed. (New York: W. W. Norton, 1993).

13. See Duara, "The Decolonization of Asia and Africa in the Twentieth Century," 8; Aijaz Ahmad, *In Theory: Classes, Nations, Literatures* (London: Verso, 1992), passim; and Vijay Prashad,

The Darker Nations: A People's History of the Third World (New York: New Press, 2007). On anarchism and anticolonialism, see especially Benedict Anderson, *Under Three Flags: Anarchism and the Anticolonial Imagination* (London: Verso, 2005); as well as Michael Schmidt and Lucien van der Walt, *Black Flame: The Revolutionary Class Politics of Anarchism and Syndicalism* (Oakland: AK Press, 2009); and Lucien van der Walt and Steven Hirsch, eds., *Anarchism and Syndicalism in the Colonial and Postcolonial World, 1870–1940: The Praxis of National Liberation, Internationalism, and Social Revolution* (Leiden: Brill, 2010).

14. Jane Burbank and Fred Cooper, *Empires in World History: Power and the Politics of Difference* (Princeton: Princeton University Press, 2010), 10–11.

15. Jeremy Adelman, *Sovereignty and Revolution in the Iberian Atlantic* (Princeton: Princeton University Press, 2006), 396.

16. Hobsbawm, *Age of Extremes*, 208; Charles Maier, "Consigning the Twentieth Century to History: Alternative Narratives for the Modern Era," *American Historical Review* 105, no. 3 (June 2000): 807–31.

17. For varied perspectives on colonial and postcolonial boundary making see D. G. Burnett, *Masters of All They Surveyed: Exploration, Geography and a British El Dorado* (Chicago: University of Chicago Press, 2001); Thongchai Winichakul, *Siam Mapped: A History of the Geo-body of a Nation* (Honolulu: University of Hawaii Press, 1994); Lucy Chester, *Borders and Conflict in South Asia: The Radcliffe Boundary Commission and the Partition of Punjab* (Manchester: Manchester University Press, 2009); Paula Rebert, *La Gran Linea: Mapping the United States–Mexico Boundary, 1849–1857* (Austin: University of Texas Press, 2001); Eric Worby, "Maps, Names and Ethnic Games: The Epistemology and Iconography of Colonial Power in Northwestern Zimbabwe," *Journal of Southern African Studies* 20, no. 3 (1994): 371–92; and Donald S. Moore, *Suffering for Territory: Race, Place and Power in Zimbabwe* (Durham: Duke University Press, 2005). For a superb study attentive to the social history of borders, see Bernardo Michael, *Statemaking and Territory in South Asia: Lessons from the Anglo-Gorkha War (1814–1816)* (London: Anthem Press, 2012).

18. Matthew Sparke, *In the Space of Theory: Postfoundational Geographies of the Nation-State* (Minneapolis: University of Minnesota Press, 2005), 12.

19. Raymond Betts, *Decolonization*, 2nd ed. (New York: Routledge 2004), 2, 55; Jeffrey Stone, *A Short History of Cartography in Africa* (Lewiston, ME.: E. Mellen Press, 1995), 77.

20. Jorge Domínguez et al., *Boundary Disputes in Latin America* (Washington, DC: United States Institute of Peace, 2003).

21. See Tamar Herzog, "Historical Right to Land: Politics, History and Law in Latin America" (paper presented at conference "Rethinking Space in Latin America," Yale University, 2014). For close study of one example of postcolonial boundary conflicts see the analysis of the boundary between Ecuador and Peru in Eric J. Lyman, "War of the Maps," in *Mercator's World*, at http://www.ericjlyman.com/mercators2.html (last accessed August 23, 2010); and, with particular emphasis on the relationship between science and nationalism in nineteenth-century Ecuador, Ana María Sevilla Pérez, "El Ecuador en sus mapas: Estado y nación desde una perspectiva espacial" (doctoral thesis, FLACSO [Ecuador], 2011). On the Argentine-Chilean boundary and Patagonia, see Richard O. Perry, "Argentina and Chile: The Struggle for Patagonia, 1843–1881," *Americas* 36, no. 3 (1980): 347–63; and, for an astute social history, Alberto Harambour-Ross, "Borderland Sovereignties: Postcolonial Colonialism and State Making in Patagonia (Argentina and Chile, 1840s–1922)" (Ph.D. dissertation, Stony Brook University, 2012). For how the Argentine state used a prison as a means to set the bounds of state power in Patagonia, see Ryan

Edwards, "From the Depths of Patagonia: The Ushuaia Penal Colony and the Nature of 'the End of the World,'" *Hispanic American Historical Review* 94, no. 2 (May 2014): 271–302.

22. Albert Memmi, *The Colonizer and the Colonized*, expanded edition (Boston: Beacon Press, [1957] 1965), 39; see also Benedict Anderson, *Imagined Communities: Reflections on the Origin and Spread of Nationalism*, rev. ed. (London: Verso Press, 1991); Frantz Fanon, *The Wretched of the Earth*, esp. "On National Culture"; Edward Said, *Culture and Imperialism*, esp. chap. 3; and Seamus Deane, "Introduction," in *Nationalism, Colonialism and Literature*, by Terry Eagleton, Frederic Jameson, and Edward W. Said (Minneapolis: University of Minnesota Press, 1990); and Terry Eagleton, "Nationalism: Irony and Commitment," in ibid.

23. Hobsbawm, *Age of Extremes*, 202; John D. Kelly and Martha Kaplan, "'My Ambition Is Much Higher Than Independence': US Power, the UN World, the Nation-State, and Their Critics," in Duara, *Decolonization*, 131–51; David Chappell, "A 'Headless' Native Talks Back: Nidoish Naisseline and the Kanak Awakening in 1970s New Caledonia," *Contemporary Pacific* 22, no. 1 (Spring 2010): 37–70; and more generally Dipesh Chakrabarty, *Provincializing Europe: Postcolonial Thought and Historical Difference* (Princeton: Princeton University Press, 2000); and Ranajit Guha and Gayatri Chakravorty Spivak, eds., *Selected Subaltern Studies* (London: Oxford University Press, 1998).

24. Ngugi wa Thiong'o, *Decolonising the Mind: The Politics of Language in African Literature* (London: James Currey, 1986). See also Ashis Nandy, *The Intimate Enemy: Loss and Recovery of Self under Colonialism*, 2nd ed. (London: Oxford University Press, [1983] 2010). On decolonizing knowledge, for Latin America see Aníbal Quijano, "Coloniality of Power, Eurocentrism and Latin America," *Nepantla: Views from the South* 1, no. 3 (2000): 533–80; more broadly, Graham Huggan, "Decolonizing the Map: Post-colonialism, Post-structuralism and the Cartographic Connection," in *Interdisciplinary Measures: Literature and the Future of Postcolonial Studies* (Liverpool: Liverpool University Press, 2008), chap. 1; the essays collected as part of the special issue entitled "Back to the Future: Decolonizing Pacific Studies," *Contemporary Pacific* 15, no. 1 (Spring 2003): 1–148; and Linda Smith, *Decolonizing Methodologies: Research and Indigenous Peoples*, 2nd ed. (London: Zed Books, 2013). For an excellent essay on the importance of new flags and anthems, among other things, to cultural decolonization in self-governed dominions of the British Empire, see Anthony Hopkins, "Rethinking Decolonization," *Past and Present* 200 (August 2008): 211–47.

25. A subject much commented upon but nowhere explored with more nuance than in Paul Carter, *The Road to Botany Bay: An Exploration of Landscape and History* (Chicago: University of Chicago Press, 1987).

26. See, as a sampling, the perspectives offered in Renee Pualani Louis, "Indigenous Hawaiian Cartographer: In Search of Common Ground," *Cartographic Perspectives* 48 (Spring 2004): 7–23; and the related discussion in Jay T. Johnson, Renee Pualani Louis, and Albertus Pradi Halmono, "Facing the Future: Encouraging Critical Cartographic Literacies in Indigenous Communities," *ACME: An International E-Journal for Critical Geographies* 4, no. 1 (2005): 80–98; Keith Basso, *Wisdom Sits in Places: Landscape and Language among the Western Apache* (Albuquerque: University of New Mexico Press, 1995); and Epeli Hauʻofa, "Pasts to Remember," in *We Are the Ocean: Selected Works* (Honolulu: University of Hawaii Press, 2008), 60–79. See also the discussion of alternative modes of writing and recording history in David Hanlon, "Beyond 'the English Method of Tattooing': Decentering the Practice of History in Oceania," *Contemporary Pacific* 15, no. 1 (Spring 2003): 19–40.

27. Brian Friel, *Translations* (London: Faber & Faber, 1981), 25. I am aware of the critiques leveled at Friel's play, especially by J. H. Andrews, whose own work overturns any mistaken impressions left in the reader regarding the mapping of Ireland. See J. H. Andrews, *A Paper Landscape: The Ordnance Survey of Nineteenth-Century Ireland* (Oxford: Oxford University Press, 1975); and Andrews, "Notes for a Future Edition of Brian Friel's *Translations*," *Irish Review* 13 (Winter 1992/93): 93–106. Having said that, a fair defense of Friel's work can be made by noting that too literal a reading of the play makes for a poor translation.

28. Epeli Hau'ofa, "Our Sea of Islands," *Contemporary Pacific* 6, no. 1 (1994): 148–61, quoted passage on 152–53.

29. Ibid., 150. For discussions of Hau'ofa's work see Margaret Jolly, "Imagining Oceania: Indigenous and Foreign Representations of a Sea of Islands," *Contemporary Pacific* 19, no. 2 (2007): 508–45; Arif Dirlik, "There Is More in the Rim Than Meets the Eye: Thoughts on the "Pacific Idea,"" in *The Postcolonial Aura: Third World Criticism in the Age of Global Capitalism* (Boulder. CO: Westview Press, 1997), 129–45; Paul D'Arcy, *The People of the Sea: Environment, Identity and History in Oceania* (Honolulu: University of Hawai'i Press, 2006), esp. the conclusion; David Hanlon, "The Sea of Little Lands: Examining Micronesia's Place in 'Our Sea of Islands,'" *Contemporary Pacific* 21, no. 1 (2009): 91–110; and the special section dedicated to his work in *Contemporary Pacific* 22, no. 1 (Spring 2010): 101–23. More broadly, see Philip E. Steinberg, *The Social Construction of the Ocean* (Cambridge: Cambridge University Press, 2001); and Matt Matsuda, *Pacific Worlds: A History of Seas, Peoples, and Cultures* (Cambridge: Cambridge University Press, 2012).

30. Bruce G. Karolle, *Atlas of Micronesia* (Mangilao, Guam: Guam Publications, 1987), 30. Here he is discussing the perceptions of Satawalese and Pulutawese navigators, from Micronesia.

31. "The canoe," one student of the subject has written, "is conceived as stationary beneath the equally fixed position of the stars and sun. The sea flows past, and the islands move astern." Ibid., 30–31. The classic study is David H. Lewis, *We, the Navigators: The Ancient Art of Landfinding in the Pacific* (Honolulu: University of Hawai'i Press, [1972] 1974). See also the film *Sacred Vessels: Navigating Tradition and Identity in Micronesia*, directed by Vicente Diaz (Pacific Islanders in Communication, 1997).

32. Cited from David Turnbull, "Cook and Tupaia, a Tale of Cartographic Méconnaissance?," in *Science and Exploration in the Pacific: European Voyages to the Southern Oceans in the Eighteenth Century*, ed. Margarette Lincoln (Woodbridge, Suffolk, UK: Boydell Press, 1998), 120.

33. Jolly, "Imagining Oceania," 509; emphasis in the original. For slightly different emphases regarding the map and the question of authorship, see Gordon R. Lewthwaite, "The Puzzle of Tupaia's Map," *New Zealand Geographer* 26 (April 1970): 1–19; and Turnbull, "Cook and Tupaia."

34. Thongchai, *Siam Mapped*; Raymond B. Craib, "A Nationalist Metaphysics: State Fixations, National Maps and the Geo-historical Imagination in Nineteenth-Century Mexico," *Hispanic American Historical Review* 82, no. 1 (February 2002): 33–68; Bulent Batuman, "The Shape of the Nation: Visual Production of Nationalism through Maps in Turkey," *Political Geography* 29, no. 4 (May 2010): 220–34; and Firoozeh Kashani-Sabet, *Frontier Frictions: Shaping the Iranian Nation, 1804–1946* (Princeton: Princeton University Press, 1999).

35. Sumathi Ramaswamy, "Maps and Mother Goddesses in Modern India," *Imago Mundi* 53 (2001): 97–114, cited passage on 100. See also Sumathi Ramaswamy, *The Goddess and the Nation: Mapping Mother India* (Durham: Duke University Press, 2010).

36. Monmonier, "Rise of the National Atlas" (n. 8 above), 1.

37. Jeremy Black, *Maps and History: Constructing Images of the Past* (New Haven: Yale University Press, 1997), 186.

38. For a trenchant discussion see Fouad Makki, "Imperial Fantasies, Colonial Realities: Contesting Power and Culture in Italian Eritrea," *South Atlantic Quarterly* 107 (Fall 2008): 735–54; as well as Ranajit Guha, *History at the Limit of World-History* (New York: Columbia University Press, 2003). For complementary perspectives, see Johannes Fabian, *Time and the Other: How Anthropology Makes Its Object* (New York: Columbia University Press, 1983); Eric R. Wolf, *Europe and the People without History* (Berkeley: University of California Press, 1982); Fanon, "On National Culture," in *The Wretched of the Earth* (n. 3 above); William Roseberry and Jay O'Brien, eds., *Golden Ages, Dark Ages: Imagining the Past in Anthropology and History* (Berkeley: University of California Press, 1991); and James C. Scott, *The Art of Not Being Governed: An Anarchist History of Upland Southeast Asia* (New Haven: Yale University Press, 2010). For a discussion that repeatedly hints at the simultaneous processes of homogenization and fragmentation that structured the universe of possibilities within which decolonizing states could be born, see the essays of Henri Lefebvre collected in *State, Space, World: Selected Essays*, ed. Neil Brenner and Stuart Elden (Minneapolis: University of Minnesota Press, 2009).

39. Nicholas Thomas, *In Oceania: Visions, Artifacts, Histories* (Durham: Duke University Press, 1997), 9; Margaret Jolly, "Specters of Inauthenticity," *Contemporary Pacific* 4, no. 1 (1992): 49–72.

40. Adelman, *Sovereignty and Revolution* (n. 15 above), 346. More broadly on this question I have found especially useful the critiques of postcolonial theory and third world literatures by Ahmad, *In Theory* (n. 13 above); and Neil Larsen, *Determinations* (London: Verso, 2001); and Larsen, *Reading North by South* (Minneapolis: University of Minnesota Press, 1995).

41. Fred Cooper, *Colonialism in Question* (Princeton: Princeton University Press, 2003), 27; Jolly, "Specters of Inauthenticity," 54.

42. J. Jorge Klor de Alva, "The Postcolonization of the (Latin) American Experience: A Reconsideration of 'Colonialism,' 'Postcolonialism,' and 'Mestizaje,'" in *After Colonialism: Imperial Histories and Postcolonial Displacements*, ed. Gyan Prakash (Princeton: Princeton University Press, 1994).

43. Burbank and Cooper, *Empires in World History* (n. 14 above), 293.

44. Ibid. For a compelling treatment of the paradox of liberty and slavery, see Greg Grandin, *The Empire of Necessity: Slavery, Freedom and Deception in the New World* (New York: Metropolitan Books, 2014).

45. Fanon, *The Wretched of the Earth*, 176.

46. Ibid., 152.

47. Rebecca Earle, *The Return of the Native: Indians and Myth-Making in Spanish America, 1810–1930* (Durham: Duke University Press, 2007); and Luis Villoro, *Los grandes momentos del indigenismo en México* (Mexico City: Secretaría de Educación Pública, 1987). On archeology see Cristina Bueno, "Forjando patrimonio: The Making of Archaeological Patrimony in Porfirian Mexico," *Hispanic American Historical Review* 90, no. 2 (May 2010): 215–46; and Stefanie Ganger, "Conquering the Past: Post-war Archaeology and Nationalism in the Borderlands of Chile and Peru, c. 1880–1920," *Comparative Studies in Society and History* 51, no. 4 (2009): 691–714. On the work of Antonio García Cubas, see Raymond Craib, *Cartographic Mexico: A History of State Fixations and Fugitive Landscapes* (Durham: Duke University Press, 2004), chap. 1; Magali Carrera, *Traveling*

from New Spain to Mexico: Mapping Practices of Nineteenth Century Mexico (Durham: Duke University Press, 2011); and Hugo Pichardo Hernández, "Hacia la conformación de una geografía nacional: Antonio García Cubas y el territorio mexicano, 1853–1912" (unpublished MA thesis, Universidad Nacional Autónoma de México, 2004).

48. Black, *Maps and History*, 166–67. See also Tom Bassett's essay in this volume.

49. Mahmood Mamdani, *Saviors and Survivors: Darfur, Politics and the War on Terror* (New York: Pantheon, 2009), 177. As I was in the process of revising this essay, South Sudan became an independent nation-state (July 9, 2011).

50. For a thoughtful intervention here, see Charles Maier, "Leviathan 2.0," in *A World Connecting, 1870–1945*, ed. Emily S. Rosenberg (Cambridge, MA: Belknap Press of Harvard University Press, 2012), 189–91.

51. Robert J. C. Young, *Postcolonialism: An Historical Introduction* (Oxford: Wiley-Blackwell, 2001), 223 (citing Winston James, *Holding Aloft the Banner of Ethiopia: Caribbean Radicalism in Early Twentieth Century America* [London: Verso, 1999], 246).

52. See Rudolfo Acuña, *Occupied America: The Chicanos Struggle toward Liberation* (San Francisco: Canfield, 1972); and Mario Barrera, *Race and Class in the Southwest: A Theory of Racial Inequality* (Notre Dame: University of Notre Dame Press, 1979); and, more recently, Armando Navarro, *Mexicano Political Experience in Occupied Aztlán: Struggles and Change* (Walnut Creek, CA: AltaMira Press, 2005). With some forty years' distance, Acuña's claim might appear a bit dated; a quick peek at Samuel Huntington's *The Clash of Civilizations and the Remaking of World Order* (New York: Simon & Schuster, 1998) will quickly dispel any such thought.

53. Jean-Paul Sartre, "The Third World Begins in the Suburbs," in *We Have Only This Life to Live: Selected Essays, 1939–1975*, ed. Ronald Aronson and Adrian van den Howen (New York: NYRB, 2013), 487; Paulo Freire, *The Politics of Education: Culture, Power and Liberation* (Westport, CT: Greenwood Press, 1985), 139–40.

54. Michael Hardt and Antonio Negri, *Empire* (Cambridge, MA: Harvard University Press, 2000), 132–34.

55. Cooper, *Colonialism in Question*, 18. See also Akhil Gupta, "The Song of the Nonaligned World: Transnational Identities and the Reinscription of Space in Late Capitalism," in *Culture, Power, Place: Explorations in Critical Anthropology*, ed. Akhil Gupta and James Ferguson (Durham: Duke University Press, 1997), 179–99; and the introduction and essays in Christopher J. Lee, ed., *Making a World after Empire: The Bandung Moment and Its Political Afterlives* (Athens: Ohio University Press, 2010).

56. See Florencia Mallon's discussion of Ada Ferrer's work on Cuba in "Pathways to Postcolonial Nationhood: The Democratization of Difference in Contemporary Latin America," in *Postcolonial Studies and Beyond*, ed. Ania Loomba et al. (Durham: Duke University Press, 2005), 273–92; Ada Ferrer, *Insurgent Cuba: Race, Nation and Revolution, 1868–1898* (Chapel Hill: University of North Carolina Press, 1999).

57. Anderson, *Under Three Flags* (n. 13 above), 54 and passim; Ilham Khuri-Makdisi, *The Eastern Mediterranean and the Making of Global Radicalism, 1860–1914* (Berkeley: University of California Press, 2010); Steven Hirsch and Lucien van der Walt, "Rethinking Anarchism and Syndicalism: The Colonial and Postcolonial Experience, 1870–1940," in van der Walt and Hirsch, *Anarchism and Syndicalism* (n. 13 above), xxxi–lxxiii; Maia Ramnath, *Decolonizing Anarchism: An Anti-authoritarian History of India's Liberation Struggle* (Oakland: AK Press, 2011); Sho Kinishi,

Anarchist Modernity: Cooperatism and Japanese-Russian Intellectual Relations in Modern Japan (Cambridge, MA: Harvard University Press, 2013); and Barry Maxwell and Raymond Craib, eds., *No Gods No Masters No Peripheries: Global Anarchisms* (Oakland, CA: PM Press, 2015).

58. Rosalind Morris, "Remembering Asian Anticolonialism, Again," *Journal of Asian Studies* 69, no. 2 (May 2010): 347–69, quote on 351.

59. Scott, *The Art of Not Being Governed*, 3.

60. Partha Chatterjee, *Nationalist Thought and the Colonial World: A Derivative Discourse* (Minneapolis: University of Minnesota Press, [1986] 1993).

61. A point made by Matthew Edney, "The Irony of Imperial Mapping," in *The Imperial Map: Cartography and the Mastery of Empire*, ed. James Akerman (Chicago: University of Chicago Press, 2009).

62. Black, *Maps and History* (n. 37 above), 186.

63. Acknowledging this is essential if we are to avoid inadvertently suggesting that the colonized are largely derivative (and the colonizer, conversely, largely autonomous). Chatterjee's *Nationalist Thought and the Colonial World* raises precisely such issues but does seem to suggest that there is no escaping some essential, autonomous West. For a cogent discussion, see Sumathi Ramaswamy's review of the book in *Journal of Asian Studies* 53, no. 3 (August 1994): 960–61. Similar critiques were leveled at Edward Said's *Orientalism*. More broadly, on seeing colonialism and the modern world as produced dialectically and through processes of transculturation, I have found especially useful Said, *Culture and Imperialism* (n. 22 above); Jean Comaroff and John Comaroff, *Of Revelation and Revolution*, vol. 1, *Christianity, Colonialism and Consciousness in South Africa* (Chicago: University of Chicago Press, 1991); John Comaroff and Jean Comaroff, *Of Revelation and Revolution*, vol. 2, *The Dialectics of Modernity on a South African Frontier* (Chicago: University of Chicago Press, 1997); and Mary Louise Pratt, *Imperial Eyes: Travel Writing and Transculturation* (London: Routledge, 1992).

64. For a critique of the argument that modern cartography—and science more generally—was a Western and rational construct imposed on "other" cultures, see Kapil Raj, *Relocating Modern Science: Circulation and the Construction of Knowledge in South Asia and Europe, 1650–1900* (New York: Palgrave MacMillan, 2007), esp. chaps. 2 and 6; but it should be read in conjunction with Matthew Edney, *Mapping an Empire: The Geographical Construction of British India, 1765–1843* (Chicago: University of Chicago Press, 1999), which offers a devastating critique of the very notion of rationality in the mapping of British India. I attempt to situate Raj's work explicitly in relation to the history of cartography in Craib, "Relocating Cartography," *Postcolonial Studies* 12, no. 4 (December 2009): 481–90. An excellent overview is provided by Suman Seth, "Putting Knowledge in Its Place: Science, Colonialism, and the Postcolonial," *Postcolonial Studies* 12, no. 4 (2009): 373–88.

65. On Mexico see Craib, *Cartographic Mexico*, chap. 7; on Ghana, see Jamie McGowan's essay in this volume.

66. Stone, *A Short History* (n. 19 above), 101, 132–33.

67. Hobsbawm notes that in the late 1940s nearly half the world's population lived in a society undergoing some kind of land reform. See Hobsbawm, *Age of Extremes* (n. 7 above), 354–55.

68. See, for examples, Thongchai, *Siam Mapped* (n. 17 above); and Sumathi Ramaswamy, "History at Land's End: Lemuria in Tamil Spatial Fables," *Journal of Asian Studies* 59, no. 3 (2000): esp. 586–89.

69. Ramaswamy, *The Goddess and the Nation* (n. 35 above), 9. For a very similar perspec-

tive, see Nancy Peluso's discussion of countermapping in Peluso, "Whose Woods Are These? Counter-mapping Forest Territories in Kalimantan, Indonesia," *Antipode* 27, no. 4 (1995): 383–406. For varied efforts to look at such processes historically, see Barbara Mundy, *The Mapping of New Spain: Indigenous Cartography and the Maps of the Relaciones Geográficas* (Chicago: University of Chicago Press, 1997); and Craib, *Cartographic Mexico*, chaps. 2 and 3.

70. On agency and subjectivity, see Michel-Rolph Trouillot, *Silencing the Past: Power and the Production of History* (Boston: Beacon Press, 1995), esp. chap. 1; on agency and passivity, see David A. Chappell, "Active Agents versus Passive Victims: Decolonized Historiography or Problematic Paradigm?," in *Voyaging through the Contemporary Pacific*, ed. David Hanlon and Geoffrey M. White (Lanham, MD: Rowman & Littlefield, 2000), 205–28.

71. Raj, *Relocating Modern Science*.

72. Joel Wainwright and Joe Bryan, "Cartography, Territory, Property: Postcolonial Reflections on Indigenous Counter-mapping in Nicaragua and Belize," *Cultural Geographies* 16, no. 2 (April 2009), 153.

73. The literature is substantial, but good starting points that wrestle with the complexities of countermapping include Peluso, "Whose Woods Are These?"; Don Monet and Skanu'u, *Colonialism on Trial: Indigenous Land Rights and the Gitksan and Wet'suwet'en Sovereignty Case* (Philadelphia: New Society Publishers, 1991); Johnson, Louis, and Halmono, "Facing the Future"; Bjorn Sletto, "We Drew What We Imagined: Participatory Mapping, Performance, and the Arts of Landscape Making," *Current Anthropology* 50 (2009): 443–76; Wainwright and Bryan, "Cartography, Territory, Property"; and Dorothy L. Hodgson and Richard A. Schroeder, "Dilemmas of Counter-mapping Community Resources in Tanzania," *Development and Change* 33, no. 1 (2002): 79–100.

74. Peter Gray, "Do the Walls Have Ears? Indigenous Title and Courts in Australia," *Australian Indigenous Law Reporter* 5, no. 1 (2000), http://www.austlii.edu.au/au/journals/AILR/2000/1.html (last accessed October 11, 2010). See also Jane Jacobs, *Edge of Empire: Postcolonialism and the City* (London: Routledge, 1996), esp. 113–15, who importantly notes that nonindigenous perspectives on aboriginality shaped the contours of the possible for various aboriginal groups at the time. I have also found valuable Paul Carter, *The Road to Botany Bay* (n. 25 above), esp. "A Wandering State"; and David Turnbull, *Maps Are Territories: Science Is an Atlas* (Chicago: University of Chicago Press, 1994).

75. Sparke, *In the Space of Theory* (n. 18 above), chap. 1; Monet and Skanu'u, *Colonialism on Trial*. For an excellent overview of mapping and history attentive to use perspectives, see Hugh Brody, *Maps and Dreams: Indians and the British Colombia Frontier* (Prospect Heights, IL: Waveland Press, [1981] 1998); as well as Paul Nadasdy, "Imposing Territoriality: First Nation Land Claims and the Transformation of Human-Environment Relations in the Yukon" (paper presented at the Cornell University Institute for the Social Sciences, September 13, 2013).

76. Sparke, *In the Space of Theory*, 16–29; Monet and Skanu'u, *Colonialism on Trial*.

77. Curtis Berkey, "Maps in Court," in *Mapping Our Places: Voices from the Indigenous Communities Mapping Initiative* (Berkeley: Indigenous Communities Mapping Initiative, 2005), 217.

78. Such projects can be riddled with tension. See, for example, the recent controversies surrounding the Bowman expeditions in Mexico, in particular the exchanges in the guest editorial section of *Political Geography* 29, no. 8 (November 2010); as well as the accusations of geopiracy presented at http://elenemigocomun.net/es/cat/mexico-indigena/; and Joe Bryan and Joel Wainwright, "Letter to the Association of American Geographers" (March 18, 2009), at http://

academic.evergreen.edu/g/grossmaz/HerlihyLetterSign.pdf. For an effort to see the expeditions as part of a broader tradition in geography, see Joel Wainwright, *Geo-piracy: Oaxaca, Militant Empiricism, and Geographical Thought* (New York: Palgrave Pivot, 2012); and the roundtable on Wainwright's book in *Dialogues in Human Geography* 4, no. 1 (March 2014).

79. Joe Bryan, "A World Where Many Fit," in *Mapping Our Places*, 218.

80. For an instructive set of exchanges, see Roger M. Keesing, "Creating the Past: Custom and Identity in the Contemporary Pacific," in Hanlon and White, *Voyaging through the Contemporary Pacific*, 231–54; Haunani-Kay Trask, "Natives and Anthropologists: The Colonial Struggle," in ibid., 255–63; Keesing, "Reply to Trask," in ibid., 264–67; Jocelyn Linnekin, "Text Bites and the R-Word: The Politics of Representing Scholarship," in ibid., 268–73; Jolly, "Specters of Inauthenticity," in ibid., 274–97; and Ben Finney, "The Sin at Awarua," in ibid., 298–330.

81. Sherry Ortner, "Resistance and the Problem of Ethnographic Refusal," *Comparative Studies in Society and History* 37, no. 1 (1995): 173–93, cited passage on 176.

82. See Michael Watts, "The Sinister Political Life of Community: Economies of Violence and Governable Spaces in the Niger Delta, Nigeria," in *The Romance of Community*, ed. Gerald Creed (Santa Fe, NM: SAR Press, 2007), 101–42; and Michael Watts, "Collective Wish Images: Geographical Imaginaries and the Crisis of Development," in *Human Geography Today*, ed. John Allen and Doreen Massey (Cambridge: Polity Press, 1999), 85–107. See also Noel Castree, "Differential Geographies: Place, Indigenous Rights and 'Local' Resources," *Political Geography* 23 (2004): 133–67.

83. Jon Parmenter, *The Edge of the Woods: Iroquoia, 1534–1701* (East Lansing: Michigan State University Press, 2010), 276. See also Carter, *The Road to Botany Bay*, esp. the chapter entitled "A Wandering State"; Liisa H. Malkki, "National Geographic: The Rooting of Peoples and the Territorialization of National Identity among Scholars and Refugees," in Gupta and Ferguson, *Culture, Power, Place* (n. 55 above), 52–74; Akhil Gupta and James Ferguson, "Beyond 'Culture': Space, Identity and the Politics of Difference," in ibid., 33–51; and Doreen Massey, *For Space* (Thousand Oaks, CA: Sage Publications, 2005), esp. 1–8; on authenticity, structure, and change I have found especially useful the brief but excellent discussion in Trouillot, *Silencing the Past*, chap. 5; as well as Marshall Sahlins, *Islands of History* (Chicago: University of Chicago Press, 1985); Greg Dening, *Islands and Beaches: Discourse on a Silent Land: Marquesas, 1771–1880* (Chicago: Dorsey Press, 1980); and Doreen Massey, "Places and Their Pasts," *History Workshop Journal* 39 (1995): 182–92.

84. Excellent studies that raise this point include Margaret Jolly, "On the Edge? Deserts, Oceans, Islands," *Contemporary Pacific* 13, no. 2 (Fall 2001): 417–66; Hau'ofa, "Pasts to Remember" (n. 26 above), 60–79; David Chappell, *Double Ghosts: Oceanian Voyagers on Euroamerican Ships* (Armonk, NY: M. E. Sharpe, 1997); and Gregory Rosenthal, "Life and Labor in a Seabird Colony: Hawaiian Guano Workers, 1857–1870," *Environmental History* 17 (October 2012): 744–82.

85. See especially Richard White, "The Winning of the West: The Expansion of the Western Sioux in the Eighteenth and Nineteenth Centuries," *Journal of American History* 65, no. 2 (September 1978): 319–43; and David Bernstein, "How the West Was Drawn: Indians, Maps, and the Construction of the Trans-Mississippi West" (Ph.D. dissertation, University of Wisconsin–Madison, 2011). I am indebted to David for his comments here.

86. Pekka Hamalainen, *The Comanche Empire* (New Haven: Yale University Press, 2008), 1. See also the recent studies by Parmenter, *The Edge of the Woods*; and Brian Delay, *The War of a Thousand Deserts: Indian Raids and the U.S.-Mexican War* (New Haven: Yale University Press, 2008).

87. For a powerful argument to this effect, see Scott, *The Art of Not Being Governed* (n. 38 above), passim but esp. chap. 6½ (yes, six and one-half!).

88. Horrabin, *An Atlas of Empire*, vi; my emphasis.

89. John Gallagher and Ronald Robinson, "The Imperialism of Free Trade," *Economic History Review*, 2nd ser., 6, no. 1 (1953): 1–15, quoted text on 1.

90. Mallon, "Pathways to Postcolonial Nationhood" (n. 56 above), esp. 278; and Aníbal Quijano, "Coloniality of Power" (n. 24 above).

91. Burbank and Cooper, *Empires in World History* (n. 14 above), 16–17, 300–301.

92. Hopkins, "Rethinking Decolonization" (n. 24 above), 211–47.

93. Dalzeil, *The Penguin Historical Atlas of the British Empire* (London: Penguin, 2006), 132–33.

94. Simon Schama, *Landscape and Memory* (London: HarperCollins, 1995), 5.

95. Hopkins observes that recent historiography has found that "Britain's aim in the years immediately following the Second World War was to reinvigorate the empire, not to abandon it." The historiography to which he refers includes William Roger Lewis, *Imperialism at Bay*; and Lewis, *The British Empire in the Middle East*. See Hopkins, "Rethinking Decolonization," 216.

96. "A Message from the New Chairman of Council, the Rt. Hon. the Earl De La Warr, G.B.E.," *United Empire* 49, no. 1 (January–February 1958): 1; my emphasis.

97. Hopkins notes that it was in 1982 that Canada acquired "full control over her affairs" when the "British parliament approved a new constitution for the country." Hopkins, "Rethinking Decolonization," 220. On the *Atlas*, see Sparke, *In the Space of Theory* (n. 18 above), chap. 1.

98. Stewart Firth, "Sovereignty and Independence in the Contemporary Pacific," *Contemporary Pacific* 1, no. 1–2 (1989): 75–96.

99. Alison Mountz, "The Enforcement Archipelago: Detention, Haunting, and Asylum on Islands," *Political Geography* 30 (2011): 118–28.

100. See the collected essays in Christina Duffy Burnett and Burke Marshall, eds., *Foreign in a Domestic Sense: Puerto Rico, American Expansion, and the Constitution* (Durham: Duke University Press, 2001); on the *Rand McNally Imperial Atlas* (1905), see Raymond Craib and D. Graham Burnett, "Insular Visions: Cartographic Imagery and the Spanish-American War," *Historian* 61, no. 1 (Fall 1998): 100–118.

101. See David Hanlon, *Remaking Micronesia: Discourses over Development in a Pacific Territory, 1944–1982* (Honolulu: University of Hawai'i Press, 1998). The quote by the US secretary of war is cited in Robert C. Kiste, "United States," in *Tides of History: The Pacific Islands in the Twentieth Century*, ed. K. R. Howe, Robert C. Kiste, and Brij V. Lal (Honolulu: University of Hawai'i Press, 1994), 229. For a remarkable wall map that charts US territory, see William Rankin, "The Territory of the United States: A Patchwork of Jurisdictions and Rights," available at http://www.radicalcartography.net/ (last accessed September 20, 2011).

102. Burbank and Cooper, *Empires in World History*, 10–11.

103. See, for example, the case of French Polynesia, discussed in Firth, "Sovereignty and Independence," 76.

104. For the case of Oceania, see Jolly, "Specters of Inauthenticity" (n. 39 above). More broadly see Dirlik, *The Postcolonial Aura* (n. 29 above).

105. Fanon observed, "The spectacular flight of capital is one of the most constant phenomena of decolonization." *The Wretched of the Earth* (n. 3 above), 103.

106. For examples of *dependista* arguments with varying degrees of sophistication and passion, see inter alia Eduardo Galeano, *The Open Veins of Latin America: Five Centuries of the Pillage*

of a Continent (New York: Monthly Review Press, [1975] 1997); Andre Gunder Frank, *Capitalism and Underdevelopment in Latin America: Historical Sketeches of Chile and Brazil* (London: Penguin, 1971); and Fernando Henrique Cardoso and Enzo Faletto, *Dependency and Development in Latin America* (Berkeley: University of California Press, 1979). An offshoot of dependency was Immanuel Wallerstein's much-debated world systems theoretic. See Wallerstein, *The Modern World-System: Capitalist Agriculture and the Origins of the European World Economy in the Sixteenth Century* (New York: Academic Press, 1974). For critiques see Robert Brenner, "The Origins of Capitalist Development: A Critique of Neo-Smithian Marxism," *New Left Review* 104 (1977): 25–92; and Steve J. Stern, "Feudalism, Capitalism, and the World-System in the Perspective of Latin American and the Caribbean," *American Historical Review* 93, no. 4 (October 1988): 829–72, which includes a response by Wallerstein. For a recent discussion in the context of postcolonial studies, see Timothy Brennan, "The Image-Function of the Periphery," in Loomba et al., *Postcolonial Studies and Beyond* (n. 56 above), 101–22. For an excellent overview, see Patrick Wolfe, "History and Imperialism" (n. 10 above).

107. For a good analysis, see Greg Grandin *Empire's Workshop: Latin America, the United States and the Rise of the New Imperialism* (New York: Metropolitan Books, 2006); and Gilbert Joseph, Catherine LeGrand, and Ricardo Salvatore, eds., *Close Encounters of Empire: Writing the Cultural History of U.S.–Latin American Relations* (Durham: Duke University Press, 1998), esp. Joseph's introduction.

108. Fanon, *The Wretched of the Earth*, 174.

109. Alan Knight, "Britain and Latin America," in *Oxford History of the British Empire*, vol. 3, *The Nineteenth Century*, ed. William Roger Lewis and Andrew Porter (New York: Oxford University Press, 1999), 141–42.

110. For an excellent overview, and for the quoted passage, see Knight, "Britain and Latin America." See also the classic formulation in Gallagher and Robinson, "The Imperialism of Free Trade."

111. Thomas Campbell-Copeland, *American Colonial Handbook: A Ready Reference Book of Facts and Figures, Historical, Geographical, and Commercial, about Cuba, Puerto Rico, the Philippines, Hawaii, and Guam*, 2nd ed. (New York, 1899); José de Olivares and William Smith Bryan, *Our Islands and Their People, as Seen with Camera and Pencil* (St. Louis: N. D. Thompson, 1899).

112. On seeing the era as one of early globalization, see Anderson, *Under Three Flags* (n. 13 above), passim; Maier, "Leviathan 2.0" (n. 50 above), 211–29, who discusses the case of Mexico comparatively; and Hobsbawm, *Age of Empire: 1875–1914* (New York: Vintage, 1989).

113. My suggestion here is not that the revolution was first and foremost an anti-imperial uprising. The debate around the causes of the revolution is sharp. For a perspective that stresses anti-imperialist aspects, see John Hart, *Revolutionary Mexico: The Coming and Process of the Mexican Revolution* (Berkeley: University of California Press, 1987); for an analysis that stresses largely the internal and agrarian roots of the revolution, and one which I find persuasive, see Alan Knight, *The Mexican Revolution*, 2 vols. (Lincoln: University of Nebraska Press, 1986). For an excellent collection that works between these poles, in many cases by taking up regional examples, see Daniel Nugent, ed., *Rural Revolt in Mexico: U.S. Intervention and the Domain of Subaltern Politics* (Durham: Duke University Press, 1998); as well as Gilbert M. Joseph, *Revolution from Without: Yucatán, Mexico and the United States, 1880–1924* (Cambridge: Cambridge University Press, 1982).

114. See Craib, *Cartographic Mexico* (n. 47 above), chap. 7; Craib, "The Archive in the Field: Document, Discourse and Space in Mexico's Agrarian Reform," *Journal of Historical Geography*

36 (October 2010): 411–20; Michael Ervin, "Statistics, Maps and Legibility: Negotiating Nationalism in Post-revolutionary Mexico," *Americas* 66, no. 2 (2009): 155–79; Ervin, "The Art of the Possible: Agronomists, Agrarian Reform, and the Middle Politics of the Mexican Revolution, 1908–1934" (Ph.D. dissertation, University of Pittsburgh, 2002); and Catherine Nolan-Ferrell, "Agrarian Reform and Revolutionary Justice in Soconusco, Chiapas: Campesinos and the Mexican State, 1934–1940," *Journal of Latin American Studies* 42, no. 3 (2010): 551–85.

115. On economic dependency and national liberation in Latin America in the mid-twentieth century, see Mallon, "Pathways to Postcolonial Nationhood" (n. 56 above), 274–75.

116. Grandin, *The Last Colonial Massacre* (n. 12 above); "Palabras de la Revolución Nacional, Dr. Victor Paz Estenssoro" (August 2, 1953), in *El libro blanco de la Reforma Agraria* (La Paz: Subsecretaria de Prensa, Informaciones, y Cultura, 1953), 123–24.

117. Prashad, *The Darker Nations* (n. 13 above), 19–21; Young, *Postcolonialism* (n. 51 above), 176–77.

118. Kelly and Kaplan, "My Ambition Is Much Higher Than Independence" (n. 23 above), 139; Sankaran Krishna, *Globalization & Postcolonialism: Hegemony and Resistance in the Twenty-First Century* (Lanham, MD.: Rowman & Littlefield, 2009), chap. 2; Hardt and Negri, *Empire* (n. 54 above); Karl Polanyi, *The Great Transformation: The Political and Economic Origins of Our Time*, 2nd ed. (Boston: Beacon Press, [1944] 2001); cf. Friedrich Hayek, *The Road to Serfdom*, 2nd ed. (London: Routledge, [1944] 2001). On the history of development see Gilbert Rist, *The History of Development: From Western Origins to Global Faith* (London: Zed, 1997); and James Ferguson, *The Anti-politics Machine: "Development," Depoliticization, and Bureaucratic Power in Lesotho* (Minneapolis: University of Minnesota Press, 1994).

119. On the United States, see Grandin, *Empire's Workshop*, 40.

120. Martin Lewis and Kären Wigen, *The Myth of Continents: A Critique of Metageography* (Berkeley: University of California Press, 1998); Hobsbawm, *Age of Extremes*, 357; Carl Pletsch, "The Three Worlds, or the Division of Social Scientific Labor, circa 1950–1975," *Comparative Studies in Society and History* 23, no. 4 (1981): 565–90; Susan Schulten, *The Geographic Imagination in America, 1880–1950* (Chicago: University of Chicago Press, 2001); and Prashad, *The Darker Nations*, 6–7.

121. On Torres-García's map, see Jennifer Jolly, "Reordering Our World," in *Mapping Latin America: A Cartographic Reader*, ed. Jordana Dym and Karl Offen (Chicago: University of Chicago Press, 2011).

122. See especially Lefebvre, *State, Space, World* (n. 38 above); Lefebvre, *Critique of Everyday Life*, trans. Gregory Elliott, 3 vols. (London: Verso, [1947, 1961, 1981] 2008); Guy Debord, *Society of the Spectacle*, trans. Donald Nicholson-Smith (New York: Zone Books, [1967] 1995); Andy Merrifield, *Henri Lefebvre: A Critical Introduction* (New York: Routledge, 2006), 9–10; and Kristin Ross, *Fast Cars, Clean Bodies: Decolonization and the Reordering of French Culture* (Cambridge, MA: MIT Press, 1995).

123. Lefebvre, "Theoretical Problems of *Autogestion*," in Lefebvre, *State, Space, World*, chap. 5.

124. David Pinder, "Subverting Cartography: Situationists and Maps of the City," *Environment and Planning A* 28 (1996): 405–27, quoted passage on 419. Also useful here, for comparison and to extend the argument, is Tim Ingold, *The Perception of the Environment*, rev. ed. (London: Routledge, 2011), esp. pt. II.

125. Simon Sadler, *The Situationist City* (Cambridge, MA: MIT Press, 1999), 82. More generally see Michel de Certeau, *The Practice of Everyday Life*, trans. Donald Nicholson-Smith

(Berkeley: University of California Press, 1994); and Abdelhafid Khatib, "Attempt at a Psycho-geographical Description of Les Halles," trans. Paul Hammond, originally published in *Internationale Situationniste*, no. 2 (1958), available at http://www.cddc.vt.edu/sionline/si/lehalles.html (last accessed on June 1, 2011).

126. The relationship between the grid and capitalism is examined with real brilliance, if not always in the exact way I have cast it here, by Denis Cosgrove, "Prospect, Perspective and the Evolution of the Landscape Idea," *Transactions of the Institute of British Geographers* 10, no. 1 (1985): 45–62; Brian Rotman, *Signifying Nothing: The Semiotics of Zero* (Palo Alto: Stanford University Press, 1993); and Ken Hillis, "The Power of Disembodied Imagination: Perspective's Role in Cartography," *Cartographica* 31, no. 3 (1994): 1–17.

127. See especially Dirlik, *The Postcolonial Aura* (n. 29 above), 54. See also Brennan, "The Image-Function of the Periphery" (n. 106 above); and Masao Miyoshi, "A Borderless World? From Colonialism to Transnationalism and the Decline of the Nation-State," *Critical Inquiry* 19, no. 4 (1993): 726–51.

128. I say "purported" because the Cold War was anything but in much of the global south: the experiences of Vietnam, Guatemala, Iran, Chile, and El Salvador, among others, are evidence enough of the murderousness of the Cold War in some locales.

129. http://www.hawaii.edu/cpis/oceania_1.html (accessed January 25, 2011).

130. Francis Fukuyama, *The End of History and the Last Man* (New York: Harper Perennial, 1993). For a powerful counter to liberal triumphalism, see Domenico Losurdo, *Liberalism: A Counter-history*, trans. Gregory Elliott (London: Verso, 2011).

131. See the useful discussion in Richard Warren Perry, "Rebooting the World Picture: Flying Windows of Globalization in the End of Times," in *Globalization under Construction: Governmentality, Law and Identity*, ed. Richard Warren Perry and Bill Maurer (Minneapolis: University of Minnesota Press, 2003), 324; and Debbie Lisle, *The Global Politics of Contemporary Travel Writing* (Cambridge: Cambridge University Press, 2006), 234–35.

132. John Pickles, "New Cartographies and the Decolonization of European Geographies," *Area* 37, no. 4 (2005): 355–64, cited passage on 359–60. The degree to which one can call the former USSR a "colonial power" is questionable. See Sharad Chari and Katherine Verdery, "Thinking between the Posts: Postcolonialism, Postsocialism and Ethnography After the Cold War," *Comparative Studies in Society and History* 51, no. 1 (January 2009): 6–34.

133. Katherine Verdery, *The Vanishing Hectare: Property and Value in Postsocialist Transylvania* (Ithaca, NY: Cornell University Press, 2003), 3–4; see also Jessica Allina-Pisano, *The Post-Soviet Potemkin Village: Politics and Property Rights in the Black Earth* (Cambridge: Cambridge University Press, 2008); and Chari and Verdery, "Thinking between the Posts."

134. Discrete cartographies—based on national or ethnic origin—are now deployed rhetorically to celebrate "multiculturalism," which seem in part a useful means for masking the dramatic transfers of wealth and power into the hands of "borderless" transnational corporations. Most recently see the insightful analysis of Nicholas Shaxson, *Treasure Islands: Tax Havens and the Men Who Stole the World* (London: Bodley Head, 2010).

135. http://news.bbc.co.uk/2/hi/business/8208254.stm. More broadly worth consideration here is the current epoch of land grabbing. See, for a useful introduction, Fred Pearce, *The Land Grabbers: The New Fight over Who Owns the Earth* (Boston: Beacon, 2012).

136. See the devastaing and succinct analysis in Mike Davis and Daniel Bertrand Monk,

"Introduction," in *Evil Paradises: Dreamworlds of Neoliberalism*, ed. Davis and Monk (New York: New Press, 2007); and Timothy Mitchell, "Dreamland," in ibid.

137. See Hardt and Negri, *Empire*; Joseph E. Stiglitz, *Globalization and Its Discontents* (New York: W. W. Norton, 2002). For "Operation Adam Smith," see Grandin, *Empire's Workshop* (n. 107 above), 160.

138. Lewis Hyde, *Common as Air: Revolution, Art and Ownership* (New York: Farrar, Straus & Giroux, 2010), 17; Edward D. Melillo, "Spectral Frequencies: Neoliberal Enclosures of the Electromagnetic Commons," *Radical History Review* 112 (Winter 2012): 147–61.

139. Robert Kaplan, *The Revenge of Geography*; Wainwright, *Geo-piracy* (no. 78 above); Craib, "The Properties of Counterinsurgency: On Joel Wainwright's *Geo-piracy*," *Dialogues in Human Geography* 4, no. 1 (March 2014): 86–91; and Jeremy Crampton, *Mapping: A Critical Introduction to Cartography and GIS* (Oxford: Wiley Blackwell, 2010).

140. For an excellent collection, Liz Mogel and Alexis Bhagat, eds., *An Atlas of Radical Cartography* (Los Angeles: Journal of Aesthetics & Protest Press, 2008). On hackitectura, see Sebastián Cobarrubias and John Pickles, "Spacing Movements: The Turn to Cartography and Mapping Practices in Contemporary Social Movements," in *The Spatial Turn: Interdisciplinary Perspectives*, ed. Barney Warf and Santa Arias (New York: Routledge, 2009), 36–58; and www.hackitectura .net (last accessed June 29, 2011).

141. Craig Dalton and Liz Mason-Deese, "Counter (Mapping) Actions: Mapping as Militant Research," *ACME: An International E-Journal for Critical Geographies* 11, no. 3 (2012): 439–66, quoted text on 440.

142. Jeremy Crampton and John Krygier, "An Introduction to Critical Cartography," *ACME: An International E-Journal for Critical Geographies* 14, no. 1 (2006): 11–33.

143. I am indebted to Sumathi Ramaswamy for raising this point with me.

144. http://www.sasi.group.shef.ac.uk/worldmapper/display.php?selected=162. These are maps that put into visual form the historical evidence marshaled by critics of modernization theory and developmentalist theorists, critics such as Raul Prebisch, who noted explicitly the *longue durée* of historical change in income inequality and distribution in formulating his import substitution industrialization policies. See Prashad, *The Darker Nations* (n. 13 above), 66–67. For an insightful discussion of how economists and historians have debated the history of wealth extraction under imperial rule and, therefore, the validity of dependency theory, see Brennan, "The Image-Function of the Periphery," 108–11.

145. "The New Enclosures," *Midnight Notes Collective* 10 (1990): 3, http://www.midnightnotes .org/pdfnewenc1.pdf (last accessed March 5, 2015).

CHAPTER TWO

ENTANGLED SPACES

MAPPING MULTIPLE IDENTITIES IN EIGHTEENTH-CENTURY NEW SPAIN

Magali Carrera

INTRODUCTION

By the end of the eighteenth century, approximately 5.5 million people, 60% of whom were indigenous, lived in the viceregal kingdom of New Spain, that is, colonial Mexico. Spread across a diverse landscape, this population was settled in 4,300 Indian pueblos as well as forty towns and twenty cities inhabited by Spaniards, criollos, Spaniards born in New Spain, mixed-blooded individuals, and a small number of indigenous people. Throughout the century, Spanish authorities attempted to address this sprawling spatial diversity and its perceived attendant administrative decentralization and economic inefficiencies through a series of bureaucratic reorganizations.

A manifestation of these plans was the intensified mapping of New Spain's territory for administrative purposes, supported by Spanish engineers and surveyors sent to New Spain. An example of such imperial planning was the division of New Spain's territory into administrative units known as intendancies.[1] In figure 2.1, the 1774 map of the Intendancy of Guadalajara depicts a bounded space anchored on the page overlaid with grid lines and dislocated from any

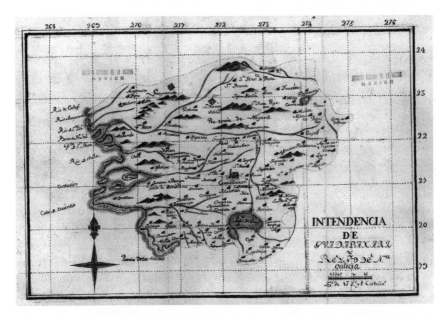

FIGURE 2.1. *Intendencia de Guadaxara y Reyno de Nueva Galicia*, 1774. Archivo General de la Nación, Instituciones Coloniales, Collecciones: Mapas, planos e ilustraciones. Source: *Correspondencia de Virreyes*, 1a. serie, vol. 50, exp. 6, fol. 360.

surrounding geography, without reference to the cultural and physical presence of inhabitants.

These efforts by Spanish authorities to intensify domination of New Spanish spaces introduced cartographic tools, as well as social and economic perspectives, that also supported resistance by viceregal subjects as active agents and historical subjects and, ultimately, the transition to nation. Consequently, along with imperial cartographic projects, there emerged other mappings by inhabitants, which asserted divergent viewpoints on the meaning, content, and function of the spaces of New Spain.[2] Such *localist mappings* were visual as well as narrative—meaning in written format—and articulated a body of observed and practical knowledge about New Spain's residents, environment, community, and history. For example, a 1767 sketch of a ranch and surrounding pueblos depicts a complex built and natural environment within which dwellers work the land (fig. 2.14).

In eighteenth-century New Spain, then, Spanish administrative cartography consistently excluded the cultural presence of local populations, while coeval, localist mapping persistently elaborated upon the inhabitants and their spaces in multiple cultural dimensions. Each of these mapping practices embedded differing views of spatial domination within their understandings; at the same time, each actively resisted the other's views of space and associated notions of domination.

Scholars tend to define domination and resistance as operating independently and in direct opposition. I frame domination *and* resistance as constructs that overlap on a continuum between two poles characterized as resistance in domination at one end and domination in resistance at the other. This perspective acknowledges that "dominating power is constantly fractured by the struggles of the subordinate"; while "moments of resistance are also constantly conditioned by structures of the dominant . . . neither domination nor resistance is autonomous."[3] Imperial power becomes "entangled spatially, since state power emerges in part from its territoriality, but this territory is never a homogeneous space."[4] Consequently, imperial and localist mappings in late viceregal Mexico were entangled on a continuum of domination and resistance.[5]

Throughout nineteenth-century Mexico, this continuum reverberated across the discourses of independence.[6] Antonio García Cubas (1832–1911), a highly respected Mexican geographer and cartographer, worked from the 1850s to the end of his life to give homogeneous cartographic form and geographic substance to his beloved *patria*.[7] Through his studies, he contributed to the nineteenth-century effort to reshape the kingdom of New Spain into the nation-state of Mexico using eighteen-century Spanish maps and documents as well as works by individuals such as Alexander von Humboldt, a Prussian geographer and naturalist, and José Antonio de Alzate y Ramírez, a Mexican scientist.

In his thought-provoking chapter, Raymond Craib comments on how the *X Carta histórica y arqueológica* page of García Cubas's *Atlas pintoresco e histórico* (1885; figure 1.5) displays the tension of decolonization and colonization in the formation of the nation-state. Indeed, this tension is present across the thirteen plates of *Atlas pintoresco* and linked to eighteenth-century entangled spatial perspectives. On the one hand, by presenting a thematic map surrounded by vignettes on each page, García Cubas exhibits the nation-state as now dominating former colonial spaces physically, economically, and culturally. Here, we see hundreds of images that confirm the efficacy of the nation—the bustling metropolis of Mexico City, active citizens, leaders of the independence movement, monumental landscapes, and abundant agricultural and mineral resources.

On the other hand, resistance to the imprint of the viceregal period is also present. Archeological sites and objects surround the map of *X Carta histórica y arqueológica*, which outlines the boundaries of the various states that con-

FIGURE 2.2. Antonio García Cubas, *XI Reyno de la Nueva Espana a principios del siglo XIX* (1885), from *Atlas pintoresco é historico de los Estados Unidos Mexicanos*, 63 × 80 cm (1885). Geography and Map Reading Room, Library of Congress.

stituted nineteenth-century Mexico, not precontact indigenous lands. These nationalist spaces are overlaid with the precontact migration routes of indigenous groups from the northern part of the country to the Valley of Mexico. The inset in the lower left corner shows the termination of the migration, that is, the founding of Tenochititlan, the Aztec-Mexica capital, the antecedent to Mexico City, the capital of the Republic of Mexico. And the map of the *XI Reyno de la Nueva Espana a principios del siglo XIX* page outlines New Spain, as it was divided into intendancies, Spanish administrative units, and territories in the eighteenth century (fig. 2.2). The graph in the lower left corner shows the area of each intendancy and its population. The coat of arms of New Spain floats above the map, while below it, those of the Hapsburg and Bourbon dynasties are placed on either side of an image of the central plaza of late eighteenth-century Mexico City. Surrounding the map are ninety-five portrait images of various individuals, from Columbus, Queen Isabel, King Ferdinand, Montezuma, and Cortés to Spanish viceroys through the end of the viceregal period.

The content of the *X Carta histórica* atlas page does not commemorate indigenous history. Rather, it recalibrates these cultures by certifying a lineage to the deep history that is prerequisite for a nation-state. Nor does García Cubas directly repudiate Spanish domination on the *XI Reyno de la Nueva Espana* page. Instead, while only in the sixty-first year of Mexican independence, he abridges 360 years of New Spain's viceregal administration into a single atlas page—assimilated as just a moment in the history of the nation.

Embedded in nationalist tenets and fervor, within the *Atlas pintoresco é historico* García Cubas deploys thematic maps and associated imagery to expound on the socioeconomic domination of the nation as well as realign and recalibrate the meaning of indigenous and viceregal history, spaces, and places. In this way, he reshapes viceregal themes by remapping them to the supposed—or imagined—linear continuity of a nationalist identity, agency, and subjectivity. These cartographic perspectives within García Cubas's *Atlas pintoresco é historico* provide an opportunity to understand nineteenth-century decolonization efforts as part of longitudinal processes and aligned to both Spanish efforts to strengthen their domination over American territory as well as entangled with the eighteenth-century New Spanish resolve to resist domination and form an identity that was distinct from Spain.

SPANISH MAPPINGS

Agustín de Ahumada y Villalón (ca. 1715–60), the Marqués de las Amarillas, had a distinguished military career. Gaining renown in the wars in Italy, he rose to the rank of lieutenant colonel in the Spanish Royal Guards. While Ahumada was serving as governor of the city of Barcelona, Fernando VI appointed the marqués to be the viceroy and capitan general of the viceroyalty of New Spain, serving from 1755 until his death in 1760.[8] The engraving titled *Mapa y Tabla Geografica de Leguas comunes, que ai de unos à otros Lugares, y Ciudades principales de la America septentrional*, dated 1755 and dedicated to the marqués, was probably prepared for his arrival in New Spain (fig. 2.3). The *Mapa y Tabla*, discussed in detail below, serves as a stepping-off point to examine the rapidly changing Spain that Ahumada left when he was appointed to his administrative post; these shifts continued during and after his tenure as viceroy and reverberated

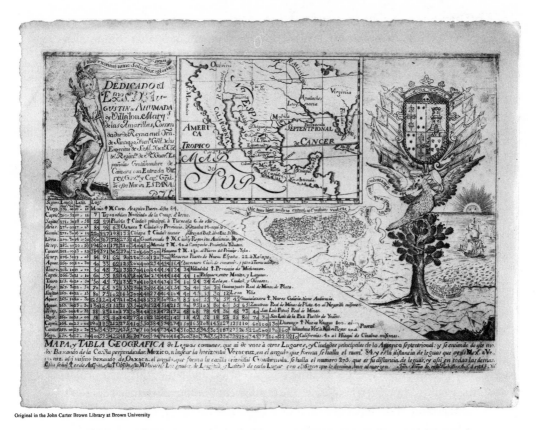

FIGURE 2.3. *Mapa y Tabla Geografica de Leguas comunes, que ai de unos à otros Lugares, y Ciudades principales de la America septentrional*, 1755. Courtesy of the John Carter Brown Library at Brown University.

into nineteenth-century Mexico. These important shifts reverberated across the Atlantic and are critical to understanding the transatlantic context of the *Mapa y Tabla* and other mappings in eighteenth-century New Spain.

By the end of the seventeenth century, Spain's transatlantic empire of earlier centuries had ebbed globally. The eighteenth century was marked by a series of wars that resulted in the loss of Spanish territory in Europe (War of Succession, 1701–14) as well as the Americas (Seven Years' War), the depletion of Spanish resources, and lost revenues. Further, Spain's diverse kingdoms—referred to by one scholar as the "Spains"—although united under the imperial crown, maintained their own cultural and economic distinctions.[9]

The eighteenth-century transfer of the Spanish throne from the Hapsburg to the Bourbon dynasty brought renewed reformist perspectives on the administration of Spain, which sought to better manage this political diversity through a more centralized government that required fluid communication of administrative instructions and directives to strengthen the management of its Iberian kingdoms. To facilitate the development of Spain's communications infrastructure, the state established educational institutions. In 1711, the Real Cuerpo de Ingenieros Militares (Royal Corp of Military Engineers) was established. That same year, standardization of map scales was implemented. By 1720, the Academia de Mathemáticas in Barcelona had become the foremost center for technical education. This institution offered full curricula in mathematics, topography, surveying, architecture, engineering, and the use of cartography by 1739.[10]

Within this shifting international situation and expanding educational agenda, a simple but striking local fact emerges: At midcentury, Spain was a nation of nine million people living in a country without functional roads. Spanish and European travelers complained about the appalling state of the existing dirt roads of the peninsula, which became impassible in winter. Bureaucrats saw this lack of easy communication between Madrid and the provinces as one cause of Spain's economic stagnation, and improved transport infrastructure became an essential element of plans for economic regeneration. To achieve linkage of all parts of Spain with Madrid, the condition of the roads would be addressed by midcentury with plans for a system of *caminos reales*, royal highways.[11] Grandiose in conception and scale, this road project was a failure, however, because of the insufficient number of trained engineers. Furthermore, although the road system was conceived as a centralizing project, its implementation was to occur at the regional level, which was impeded by feudal privileges and superseding local interests.[12]

The underlying objective of administrative centralization that directed the *caminos reales* endeavor was promoted in other reforms and projects in Bourbon Spain. In 1738, the Academia Real de Historia proposed an encyclopedic project, the *Diccionario histórico-crítico universal de España*. This initiative was conceived to include general history, ancient geography, modern geography, natural history, and chronology. Information was gathered from archives, libraries and provincial records. Additionally, officials tried to gather local reports, with varying levels of compliance. The *Diccionario* also included "mapas con la major exactitude," that is, detailed maps of each province, produced by Tomás López (1730–1802), a cartographer for the Spanish court, who had been trained in cartography and map engraving by Parisian mapmakers.[13]

In conception, then, the *Diccionario* was a comprehensive geographic and cartographic undertaking that sought more coherent understanding of the exact contours of Spain's territories, the interior content in terms of towns and cities, and information about local populations, economies, and resources.

Over the next six decades, regional reports providing this information as well as maps were produced and archived in the Academia. The conclusion of this project languished, however, in part due to the inattention and changing personnel of the Academia. The diverse and numerous studies produced for the *Diccionario* would finally be brought together at the end of the century, with the first volume published in 1802. Lopez's *Atlas geográfico de España*, the first atlas of Spain produced by a Spaniard, would not be published until 1810.

The *caminos reales*, *Diccionario*, and other projects bring into high relief the fact that in the eighteenth century, Spain was gathering extensive information about *itself* as part of an attempt to improve its social and economic conditions. In fact, then, the Bourbon administration was colonizing "the Spains," attempting to assert control over a territory, much of which still functioned under the more autonomous spatial, social, and political structures of previous centuries. These Bourbon plans to dominate Iberian spaces were entangled with various levels of provincial resistance.

This recolonization would be transatlantic, including efforts to bring the Spanish Americas into Spain's comprehensive economic and political web. Beginning in the sixteenth century, extensive questionnaires, which sought information about people, resources, landscape, and the progress of religious conversion, had been sent to New Spain for detailed response. The responses, known as *relaciones geográficas*, included written descriptions as well as some maps. Under the Bourbon regime, however, these information-gathering practices about New Spain intensified. Thus, despite more than two hundred years of Spanish domination, authorities did not have the comprehensive data—or a map using cartographic methodology—needed to administer New Spain centrally.[14] In 1741, Phillip V lamented that the Consejo de Indias still lacked critical information about the viceroyalties and mandated that the viceroys produce "noticias más individuales, y distintas del verdadero estado de aquellas provincias" (very detailed and specific reports about the true conditions of those provinces).[15] Such comprehensive reports were used to ascertain the economic and human resources of New Spain as well as identify resources for military advantage against possible aggression by other nations. Also, the Enlightenment interest in natural history and scientific classification, while not an originating impetus, was reflected in these inquires.[16]

Throughout the eighteenth century, then, Bourbon authorities sought to transform both their domestic and their colonial policies and practices in order to be more competitive with other European nations. Pedro Rodríguez de Campomanes (1723–1802), an economic adviser to Charles III, published two works, *Reflexiones sobre el comercio español a Indias* (1762) and *Discurso sobre la edu-*

cación popular de los artesanos y su fomento (1775), that described Spain—especially its economy—as in a state of decline. To improve this situation, Campomanes emphasized the need to refocus Spain's transatlantic economic interests from precious metals to raw materials, promoting the viceroyalties both as sources of raw materials and as markets for Spain's goods. He and other political writers recommended increased extraction of natural resources, cessation of industry that might compete with Spain, and levy-free trade with Spain.[17] As a result, the Bourbon administration tried to accomplish in the Spanish Americas what it could not fully realize in Spain: improved communications and transport infrastructure, increased economic expansion, and curbed local authority.

Cartography would support this infrastructural expansion as well as improve knowledge of vital coastal and border regions of the Spanish Americas.[18] This mapping effort, however, remained inconsistent despite the expanded training of engineers and the imposition of cartographic standards, such as the 1711 requirement of uniform map scales. This is evident in the *Atlas de América*, a 1791 compilation of maps from various sources prepared for the members of the Academia de Historia (likely by Tomás López) that were disparate in format, style, or scale because the *Atlas* was an assemblage of both original and derivative works.[19] It brought together maps of North America, with emphasis on coastal regions, including the peninsula of California, New England, Florida, Louisiana (with an inset map of New Orleans), and the Gulf of Mexico. The Caribbean was represented by maps of the islands of Cuba and its port of Havana, Santo Domingo, Jamaica, the Antilles, Barbados, and Antigua; South America by Juan de la Cruz Cano's *America Meridional* and the maps of the provinces of Caracas, Chile, Río Plata, the city of Cuzco, and various Atlantic ports. Curiously, but not surprisingly, the *Atlas* includes a print of the 1755 *Mapa y Tabla Geografica*. Like the *Diccionario* maps, this *Atlas* brought together disparate and outdated cartographic information about the Americas, the majority of which was focused on the coastlines and ports, with little attention to the interior lands. Clearly, even in the latter part of the eighteenth century, Spain could not collate coherent and standardized cartographic information about New Spain.

Redefining its territorial management, through scientific expeditions, cartography, the statistical studies of populations and resources, and the engineering of borders and forts, required the application of new technologies, and making maps became particularly important.[20] Examples of these intensified geopolitical mappings of New Spain include maps of the intendancies as well as numerous coasts and their harbors (fig. 2.1). This is to say that coastal maps of administrative units received greater attention than detailed mapping of the interior or the whole of New Spain. The reason for this is obvious: knowledge of coasts and ports was critically important for the transatlantic commercial exchange of raw materials and manufactured goods. Further, the ports and coasts needed to be protected from foreign intervention and pirates.[21]

Attentive to this broader transatlantic context, I return to the *Mapa y Tabla Geografica*, attributed to a Jesuit mapmaker, Ignacio Rafael Coromina. It is eight by twelve inches and divided into five sections.[22] At the far left, metaphorically representing New Spain, a female dressed in a long gown with a feathered headpiece holds an Indian weapon in her right hand and, in the left, a scroll penned with the dedication to Viceroy Ahumada that cites his titles and is initialed by P.J.E. or P.T.E. Above the scroll, a banner floats with the words "Luce tibi exoritur nunc Soli tunc gloria nobis" (With light, the sun's glory now rises up for you, then for us). Her right foot rests on a pot from which spill coins, probably refering to the mineral wealth of New Spain. Below this figure is a table; its left column provides the latitude and longitude of twenty-two major cities, including Havana, each associated with a zodiac sign. The grid of columns and rows identifies the distance between the cities and towns.

To the right of this table, referencing the founding symbol of the Aztec-Mexica empire, an eagle with a snake in its talons alights on a cactus; it holds a banner in its beak printed with the words "Me me felicemque! tuis quae pressatri imphis" (And it makes me happy to show you the impressive extent of your territories!). Above the bird's head, drums and weapons rest, symbolizing the Spanish conquest, while the new viceroy's coat of arms surrounded by flags hovers above. Surrounding this depiction is a vista of the Atlantic Ocean with a rising sun and ships sailing along with chorographic views of a port, probably Veracruz, where Ahumada would have landed. Mexico City, the capital of the viceroyalty, appears as an island floating in a lake. Above the city, another banner floats stating, "Hic tuus hinc nostras exturbat Cynthius umbras" (From here your island [probably a reference to the capital], this Cynthius, thrusts out into our shadows).[23]

In the upper center, the fifth section of the *Mapa y Tabla Geografica* presents a comprehensive view of New Spain and North America, tracing the coastlines of the Pacific Ocean from present-day Central America to northern California, the Gulf of Mexico, and the Atlantic Ocean from Florida to Virginia. Finally, the inscription at the bottom of the print includes the title of the document, a simple legend, and a summary of the information provided as well as instructions on how to use the distance chart. This text concludes with the name of Joseph Nava (the engraver), city of Puebla de los Angeles, and the date 1755.

The *Mapa y Tabla Geografica*, then, presented Viceroy Ahumada with a comprehensive picture of the whole of his administrative responsibility, New Spain, through visual metaphor as well as the geodetic map. The female figure of New Spain and the emblematic eagle with weapons and coat of arms refer to the imperial origins of the map, while the data presented in the table and map make New Spanish territory measurable and tangible, locating interior towns, cities, and ports and listing the distances between them. The *Mapa y Tabla* also conceptualizes territory as essentially a technological and economic enterprise, as García Cubas would in the next century.[24] This information links the space and commercial resources of New Spain to Spain's own internal colonization efforts, such as the roads project. In Fernando VI's extensive *Instrucciones* to Viceroy Ahumada, he outlines the incoming viceroy's numerous administrative duties, reminding him to pay special attention to the repair and security of the roads.[25] The *Mapa y Tabla* reflected the Bourbon regime's move toward a pragmatic philosophy of governance based on domination through empirical information for comprehensive management and administration. Metaphorical New Spain was made concrete by the delineation of territory, measurement of distances between places, and visualizations of towns and ports. Here, *by* metaphor, *by* measurement, and *by* map, this former city governor and war hero was prepared to administer the wealthiest of Spain's American territories in the name of the king.

But perhaps not, because, as discussed above, imperial "power is entangled spatially, since state power emerges in part from its territoriality, but this territory is never a homogeneous space." Consequently, throughout the eighteenth century, viceregal cartographic images of New Spain were entangled with other mappings made or commissioned by its inhabitants that both encompassed and resisted the *Mapa*'s reference to the ordered vision of Spanish administration. This is to say that there was an ongoing discourse about space among various voices of colonial Mexico, who had differing access to cartographic and technical knowledge, resulting in heterogeneous views of colonial spaces. These perspectives would reverberate in the works of García Cubas.

FOOTPRINTS AND TOPONYMS

In New Spain, the mapping of towns and surrounding areas was an important aspect of viceregal administration, tribute collection, and land granting.[26] Since the sixteenth century, indigenous groups had produced maps that visualized their territory for Spanish authorities. For example, a late sixteenth-century depiction of the pueblo of Amoltepec graphically displays indigenous

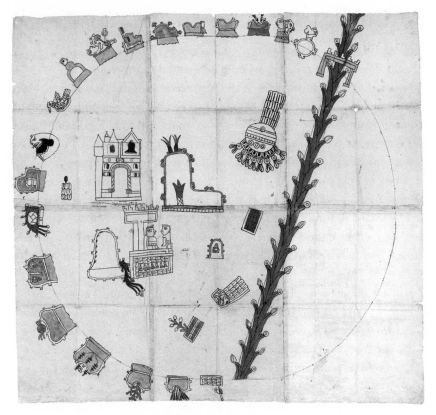

FIGURE 2.4. *Relación geográfica* map of Amoltepec (1580). Nettie Lee Benson Latin American Collection, University of Texas Libraries, University of Texas Austin.

spatial priorities (fig. 2.4). A church, denoted by the architectural elements of the arch and bell, appears above the ruler's palace, shown as two figures seated on a raised platform. A three-quarters circle of toponyms identifies the *atepetls*, ethnically based political entities that constitute Amoltepec territory. A local river is identified by the symbol for water showing diagonal blue waves ornamented with shells and circular forms.[27] Using an indigenous visual vocabulary, Amoltepec is depicted as a community embedded in networks of historical, kinship, and environmental relationships to the land. Mappings such as those of Amoltepec could not easily be incorporated into a comprehensive map of New Spain using European standards; indigenous maps described space within cultural idioms rather different from the mathematical structures of European maps.

By the end of the sixteenth century, colonial authorities had introduced land use principles based on the communal land tradition of early Spain, which specified that all land formally belonged to the king and was available for the "common use of Nature's fruits." Land could not be privately owned; however, usufruct allowed individuals to use a piece of land without owning it. As a result, formal land possession could be granted to individuals for specified agrarian or manufacturing uses; if the land was abandoned, it was available to other individuals. This tenurial system, however, did not function as conceptualized in Spain, because of the necessary accommodation to local conditions.[28]

Likewise, in New Spain, the Spanish system was acculturated to existing conditions of indigenous understanding of land, its use and possession. In his study of land use and land tenure in northeastern New Spain, geographer Miguel Aguilar-Robledo describes the parallels between Iberian and Meso-american landholding systems that facilitated the process of transfer and consolidation of the imposed model in New Spain. This partially transformed the Mesoamerican system to the Iberian code.[29] He writes,

> In a sense, the process of conversion [to the Spanish tenurial system], an enterprise of "cultural translation," encoded tenurial Mesoamerican forms, based on cadastral records and oral traditions, into the new, written, legal tenurial system. This allowed the Indians to keep some of their ancient lands, and even to reclaim usurped ones. This tacit compromise allowed some parts of the Mesoamerican model (e.g., communal landholding) to survive well into the colonial times or, in remote places, they even outlived the colonial regime.[30]

As a result of this cultural translation, diverse localist practices appear in maps that were included in case documents associated with the processes of land granting and litigation found in the Archivo General de la Nación (AGN), in Mexico City. These cases consist of handwritten documents, generally including a series of *autos*—notarized depositions from witnesses in a case—as well as summaries and pronouncements from viceregal officials. Numerous scholars have studied the legal documents extensively.[31]

Many land maps, which were not standard documents in these cases, have been removed from the original documents and placed in the AGN's Mapoteca for conservation and security purposes. The makers of these maps were highly diverse but generally fall into four categories: (1) anonymous individuals who may have been local artisans; (2) named individuals who are local officials; (3) named individuals who identify themselves as a *perito*, expert, but are clearly not trained in surveying; and (4) named individuals who identify themselves as *agrimensores*, trained surveyors. In general, maps made by the first

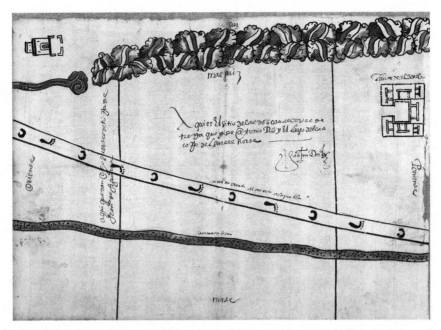

FIGURE 2.5. Huejotzingo (Puebla), 1591. Productor: Gaspar Derbés; alcalde. Archivo General de la Nación, Instituciones Coloniales, Collecciones: Mapas, planos e ilustraciones. Source: *Tierras*, vol. 1876, exp. 8, fol. 3.

three groups vary from simple line sketches to elaborately painted drawings of local landscapes. *Peritos* may annotate dimensions of the disputed land in their written descriptions but not on the drawings. In contrast, agrimensores present land in cartographic terms using geometric triangulation, sometimes to the exclusion of any natural features or reference to inhabitants.

Land maps have not been visually examined systematically or extensively by scholars, especially in terms of how they change over time in relationship to cartographic modes that were in use and evolving in Spain and New Spain. In a 1591 map of the pueblo of Huejotzingo in present-day Puebla produced by Gaspar Derbés, *alcalde* (mayor), for example, the road to the city of Puebla, marked by both footprints and horse (or donkey) hooves, cuts across a landscape and is surrounded by *milpas*, tilled land (fig. 2.5). To the north (noted at the bottom of the page), a grey line marks a *barranca* (ravine); in the south, the water source and a stone fence—depicted as bread-loaf-like shapes—are located. In the center, two straight lines delimit the land of Antonio Rodríguez, separating it from the land of Francisco Figueroa to the east and a U-shaped house in the west.[32] The map was made by mandate of the Real Audiencia, a governing body of New Spain, and demonstrates the use of indigenous figurative vision

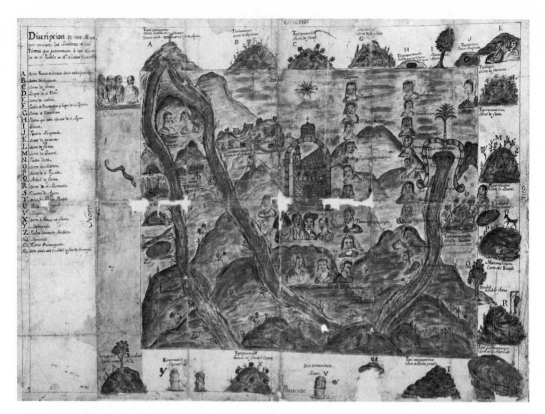

FIGURE 2.6. San Nicolás Tenazcalco (Edo. de Mexico), 1715. Productor: Carlos Romero de la Vega. Archivo General de la Nación, Instituciones Coloniales, Collecciones: Mapas, planos e ilustraciones. Source: *Tierras,* vol. 2999, exp. 15, fol. 6.

of land associated with sixteenth-century *relaciones geográficas* traditions.[33] Late sixteenth-century and early seventeenth-century land maps deploy indigenous visual vocabulary, such as footprints, for cases regarding the land and its uses. Throughout the seventeenth century, such maps continued to mix indigenous formats with European elements, such as a stylized drawing of a church with paths designated by footprints.

Moving to the eighteenth century, traditional indigenous elements in land maps and sketches are visible to varying degrees. A 1715 map of the pueblo of San Nicolás Tenazcalco, in the State of Mexico, made by Carlos Romero de la Vega, provides more detailed information (fig. 2.6). The associated document provides information about a request of a nearby town, San Francisco Sayaniquilpan, to purchase a piece of land designated as *real patrimonio*, that is, belonging to the king, for twenty-five pesos. A series of individuals, including

Juan de Santiago, an "Indio Principal," provide sworn testimony about the location of the land to officials.[34] The boundary of the land is described by metes and bounds, that is, by measuring distance and direction to and from a specified point.

The map of the pueblo depicts local physical features and monuments that are alphabetically indexed. The water source that passes through the village is marked by the letter A. Surrounding pueblos are named and marked by letters such as B, Thehuixtepetl (Serro de espinas—hill of thorns), and C, Tepequaxochtli (Serro de flores—hill of flowers), and so on. Below B and C, the village of San Nicolás appears as a grouping of buildings behind an arch framing a church facade. Above the arch, a compass rose identifies north. In the center panel and to the right of the buildings are the pictures of twenty-six local leaders, four of whom are unnamed while the others are identified by either indigenous or Hispanic names. Here, the visual vocabulary of the image of San Nicolás has incorporated European elements: representational depictions of natural features rather than logographs, indexing of landmarks, and the compass rose.

The Amoltepec, Huejotzingo, and San Nicolás images use cultural symbols and coordinates to describe the pueblo. They retain indigenous elements, especially the intentionality of identifying the land through historical connections and relationships. Notations are simplified within this highly graphic tradition, but it is clear that land never signified a purely territorial phenomenon; rather, territorial description was an inventory of physical markers indicating community boundaries. Thus, references to the indigenous community's relationships to the land through people and natural features remain consistent in various eighteenth-century maps of the interior spaces of Mexico.

THE MEASURING LINE

The 1728 map of San Mateo Aticpac (Tlaxcala) was employed in a dispute over boundary lines between the *naturales*, indigenous inhabitants, of the town San Mateo Aticpac and Santa Catharina, a *barrio*, or dependency, of San Mateo (fig. 2.7). Set in a sparsely forested landscape, a church is placed at the center with the pueblo name, San Mateo Aticpac, above it. Santa Catharina is located in a southern (left) section of the map, which at some point in the archiving of the case became separated from the section of the map now preserved in the AGN's Mapoteca but remains in the volume containing the *tierras* documents.

FIGURE 2.7. San Mateo Aticpac (Tlaxcala), 1728. Productor: Francisco Hualcoyotzien. Archivo General de la Nación, Institutiones Coloniales, Collecciones: Mapas, planos e ilustraciones. Source: *Tierras*, vol. 1470, exp. 2, fol. 45.

Four men with hats and capes are placed to the right or north of the villages. Three of these individuals stand holding various objects, while a fourth kneels to mark a line that radiates from the church and extends to the north.

The 194 pages of the case include detailed *autos* from the leaders of the two pueblos as well as surrounding villages. Both sides describe the justification of their claims as based in ancient *títulos*, primordial land titles, which were asserted to date to the pre-Spanish period.[35] The representatives from San Mateo present *títulos* described as "testaments from the time of Gentilidad" (paganism, that is, pre-Christianization), stating that the papers outline their land rights.[36]

According to case documents, at approximately eight o'clock in the morning on August 10, 1728, in the pueblo of San Mateo Aticpac, Don Diego de

FIGURE 2.8. San Mateo Aticpac (detail).

Santiago, Don Pedro de la Cruz, and Don Pablo de los Santos, local leaders, met with Don Juan Esguissuchitl, a scribe, in an area to the north of the church known as La Concepcion. With documents in hand, they walked off the boundaries according to the *títulos*, which included a nopal plant, a dry *barranca*, or water channel, and a small bridge as boundary markers. The documents state that they made a map of all the lands and have another map dated back to 1597.[37]

Above and to the left of the figures depicted in the image, the word "Conspsion" (*sic*) is written above them, locating them near La Concepcion (fig. 2.8). The image, however, probably does not depict the August 10 survey, because the map is contained in a forty-eight-page document included in the case, which is dated May 26 (1728)—more than two months before the notarized August boundary marking occurred—and annotated on the title page as "se declaron por falsos y se mandaron separar" (declared to be false and ordered to be separated).[38] The almost illegible and unnotarized testimony, with each recto page marked "son falsos" (these are false) across the top in a different handwriting, seems to have been presented by the community of Santa Catharina. Further, the handwriting of these pages is very similar to the handwriting on the Aticpac image. Rather than just indicating the bounded land as described in the *títulos*, however, the figures of the map may be marking the distance of the land from the church, as the official documents indicate that demarcations are to be made from the church.[39]

Ultimately the case was sent to the Real Audencia in Mexico City for final determination. Overall, this case elucidates how indigenous groups would make land claims against each other in the early eighteenth century. While the case is challenging to follow with all its claims and counterclaims, it is clear, nevertheless, that the litigants argued their cases using documents that they considered to be historical, which defined boundaries using physical markers.

At the same time, the Aticpac map, with the image's detail demonstrating visual reference to land-measuring methodologies, opens a discussion about how the measuring line would make its way into late colonial land maps. In his study of the rise and decline of colonial surveying, Aguilar-Robledo traces the problematic development of surveying New Spain. By the eighteenth century, surveying remained underdeveloped due to the ambiguity of royal decrees that defined land grants, a flawed system of weights and measures, and poor development of surveying tools and techniques.[40] The resulting landholding problems, along with the high costs of surveys, allowed landowners to obtain grants for their land without mandatory fieldwork, instead relying on empiricism or observed local knowledge in these land descriptions. The documents are filled with pages and pages of *autos* from witnesses brought by the litigants, who explain their understanding and/or knowledge of the disputed property. As a result, highly varied maps were drawn for land cases where boundary and other disputes were brought to authorities for resolution.

The 1723 map produced by Julio García Morón, in another example, depicts the hacienda of Don Juan Primo as well as his ranch, San Miguel el Grande (fig. 2.9). The bird's-eye view of the space shows clusters of buildings, a creek, and the sources of water. Roads—marked in red umber—cut across the map page, through the town, and continue to Queretaro and Mexico City. The whole landscape is set in eight lines that radiate from a compass rose; two additional lines run roughly parallel to the north-south line established by the compass rose. In the maps of San Mateo and Hacienda Primo then, landscape and built space remains the main descriptors overlaid with divergent lines.

In contrast, the 1732 survey map of the area around the town of Tepeaca produced by Maximiliano Gómez Daza is part of a set of documents that explicates a request to rectify earlier measurements made by a certain Licenciado Cabrera (fig. 2.10). Using lines and letters, the diagram depicts the jurisdiction of Tepeaca, its royal lands, the roads to Tlacotepeque and Tlaco, the ranch of a *cofradía*, or guild, and the pueblos of San Luis and Andrés. In his signed description of the diagram, Gómez Daza, who identifies himself as Agrimensor General de Nueva España, annotates the boundaries beginning at A, shown in the upper right corner.[41] On subsequent pages, he presents the calculations for the area of each of the separate sections of the land, identified by numbers.[42] Gómez Daza also includes a highly critical analysis of the previous work of

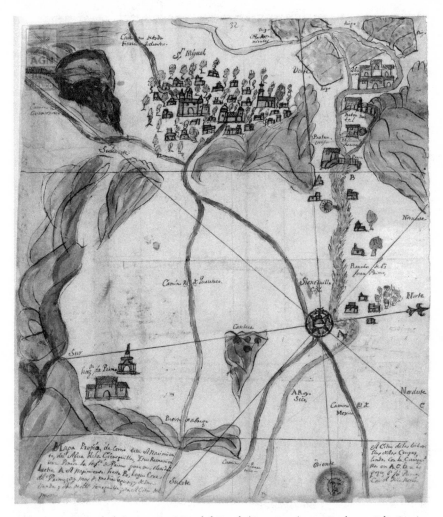

FIGURE 2.9. Hacienda de Primo y San Miguel el Grande (Guanajuato), 1723. Productor: Julio García Morón. Archivo General de la Nación, Institutiones Coloniales, Collecciones: Mapas, planos e ilustraciones. Source: *Tierras,* vol. 258, exp. 4, fol. 90.

Licenciado Cabrera, explaining that "some who are not proficient at this science [surveying]" make errors due to their lack of direct experience.[43]

In the center are the measurements of the land of General Don Nicolás de Villanueva, showing the location of his livestock at the boundaries of the pueblos. Geometric lines demarcate the area of the various towns and pueblos, while letters do not locate landmarks, as in the previous maps, but calculation points. People and landscape have vanished. Bearing and distance describe the

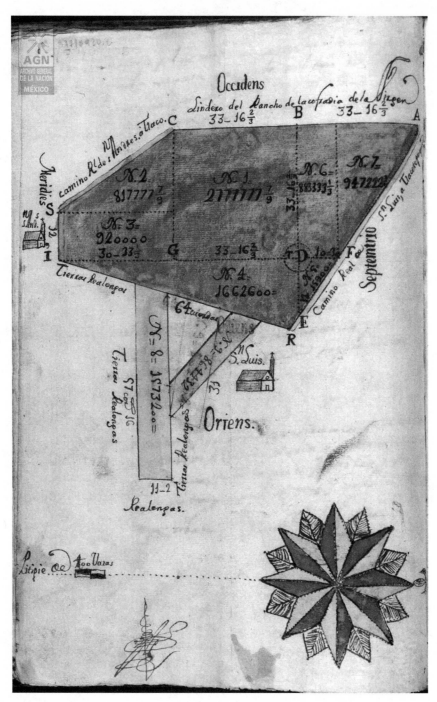

FIGURE 2.10. San Luis de los Chochos (Tepeaca, Puebla), 1732. Productor: Maximiliano Gómez Daza. Archivo General de la Nación, Instituciones Coloniales, Collecciones: Mapas, planos e ilustraciones. Source: *Tierras,* vol. 487, exp. 1, fol. 96v.

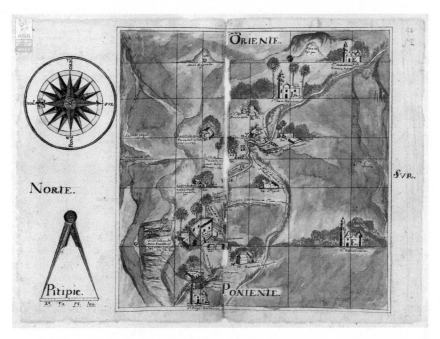

FIGURE 2.11. San Juan Axalpan, San Sebastián Zinacantepec, San Gabriel Chila y San Diego Chalma y hacienda Santísima Trinidad y San José Buenavista (Tehuacan, Puebla), 1780. Productor: Jacinto de Espinoza, agrimensor. Archivo General de la Nación, Institutiones Coloniales, Collecciones: Mapas, planos e ilustraciones. Source: *Tierras,* vol. 1058, exp. 2, fol. 52.

land: the compass rose in the lower right corner locates direction, and the small bar scale in the lower left corner indicates distance.

Moving to the late century, in the 1780 map of the Tehuacan area (Puebla), five towns and a hacienda are shown in chorographic view, tucked into the rolling landscape (fig. 2.11). Signed by Jacinto de Espinoza, *agrimensor*, this map is brought as evidence by Don Antonio Martín, governor of the community of Axalpan, in a lawsuit over the allocation of water from the *barranca* (gully) Chalma. Rather than filling the space of the paper, as in the previous examples, a frame contains the verdant landscape within a gridded space. The compass rose and *pitipié*, a compass spanning a distance scale, are not integrated into the image but placed in the left margin, taking up almost a quarter of the page. The gridding, in fact, does not emphasize mathematical description, because bearing and distance are not critical to the purpose of this map. Rather, Espinoza provides a ten-page detailed description that uses alphabetic letters to locate sites and roman numerals to trace the movement of the water.[44] On behalf of the indigenous inhabitants, lawyers also present a one-hundred-page brief,

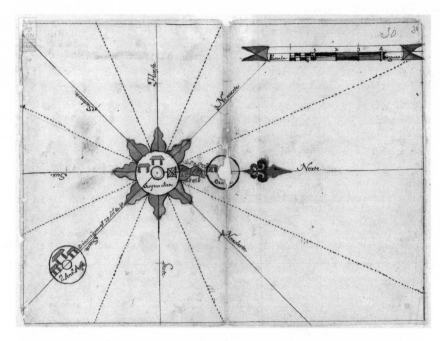

FIGURE 2.12. Haciendas San Juan Bautista Suytunchin y San José Occhac (Yucatán), 1787. Productor: Santiago Arosteguidurán, agrimensor. Archivo General de la Nación, Institutiones Coloniales, Collecciones: Mapas, planos e ilustraciones. Source: *Tierras,* vol. 1061, exp. 1, fol. 30.

which claims that the *naturales* (indigenous inhabitants) of the pueblos have had water rights since the time of their *gentilidad* (understood as heathenism) and includes numerous *autos* testifying to these rights.

In a 1787 example, two haciendas in present-day Yucatán become integrated into a compass rose with Suytuncten (also identified as San Juan Bautista Suytunchin) in the center and Occhac (San José Occhac) placed on the radiating north line (fig. 2.12).[45] The bar scale in the upper right corner indicates distances. Notably, small icons of buildings within a circle identify the hacienda with no reference to physical landscape; rather, the distance between Suytuncten and Occhac along the north axis describes the space and form of the compass. This map, produced by Santiago Arosteguidurán, is part of a demand by Don Gabriel Bautista that Don Joseph Fajardo suspend the population expansion that had begun at Occhac. Arosteguidurán presents a ten-page description of the measurements of distance, carefully documenting how they were obtained.[46] And, in figure 2.13, a 1791 image of Santa Isabél Chalma, in the state of Mexico, depicts land as geometric units and associated text, without detailed landscape imagery. Signed by Diego Muñoz, *agrimensor*, the map was part of a land dispute over land allocation to the town of Santa Isabél Chalma, which

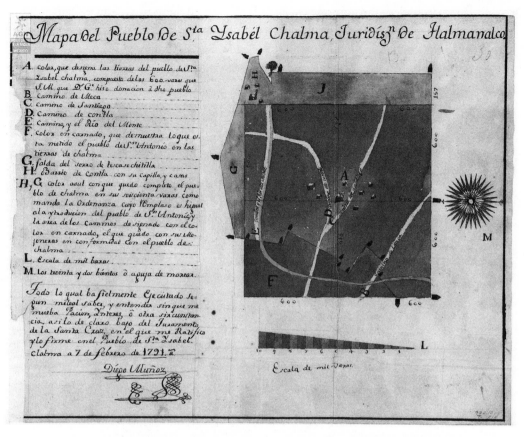

FIGURE 2.13. Sta. Isabél Chalma (Edo. de México), 1791. Producer: Diego Muñoz, agrimensor. Archivo General de la Nación, Institutiones Coloniales, Collecciones: Mapas, planos e ilustraciones. Source: *Tierras*, vol. 1518, exp. 5, fol. 30.

was opposed by Don Luis Paez de Mendoza. The map shows Santa Isabél at the center, marked A, of four quadrants of six hundred *varas* (approximately five hundred meters) per side depicting the allocation.[47] On the map, Muñoz also marks an area F, which overlaps with the land of another town, San Antonio.

In this series of examples, then, mathematical survey methodology with its attendant lines, scales, and compass roses is part of the descriptive vocabulary of land maps—but they are incorporated in divergent ways according to local views and needs, and doubtlessly the skill and knowledge of the mapmakers. At the same time, a multiplicity of local mapping practices continues through the century as local domination through and resistance to land claims deploy diverse lived descriptions of property.

The land cases discussed above add insights as well as demonstrate the need to better understand how ownership of land and land rights played out visually as a struggle between the local understanding of historical place, differing uses of measuring methodologies, and the legal framework imposed by the Spaniards. In the documents reviewed, their associated maps were not considered primary evidence in a case but were included as visual testimony presented along with the written *autos* by various individuals. The primary evidence consisted of the testimony of the witnesses who articulated the historical context of the land, and the community's or individual's connections to it.

NARRATIVE MAPPINGS

A scene of verdant and productive landscape fills the page of a 1767 map depicting the lands around the Rancho Apetlanca, in present-day Morelos, which was a part of a dispute over a certain tract of land between Don Isidro Roman and Don Juan de Pilar and the *naturales* of the pueblos of Quezala y Chilachia (fig. 2.14). The ranch of Apetlanca occupies the northerly part of the map, and Quezala is located to the south. Numbers from 1 to 21 index various locales, including natural features, buildings, monuments, the Rancho Apetlanca, and indigenous pueblos. A document dated to January 19, 1777, states that at eight o'clock in the morning, Don Joseph Antonio de Mendivil, a local official and judge, went into the countryside accompanied by Don Pedro Baena, *perito*, along with four other witnesses to mark off the disputed land. Five pages of testimony carefully describe the natural features that were used as boundary markers and their compass directions.[48] In the upper part of the map, made by Baena, two men on horseback oversee three other men, who hold a rope, showing the taking of a measurement in *cordeles* (one *cordel* equals approximately fifty *varas*, or 139 feet). This measuring activity, however, is secondary in the visual image, which emphasizes human activity in the countryside.

For this discussion, the map visualizes an intensely active space. The Apetlanca landscape visually references another mapping practice appearing in New Spain. Identified as narrative mapping, the lands and landscapes of New Spain are described through written texts that elucidate the intense personal spaces that form around geographic facts and intricate networks of social relations associated with notions of *patria*, or homeland. These were often written by elite criollos, who felt their privileges in jeopardy as a result of Bourbon curbing of local authority. New Spain–born Spaniards found themselves in a problematic position. On the one hand, they could not assert themselves to be original to the land because of indigenous peoples' primordial claims to land.

FIGURE 2.14. Rancho Apetlanca, Zacualpan (Morelos), 1767. Productor: Pedro Baena, perito nombrado. Archivo General de la Nación, Institutiones Coloniales, Collecciones: Mapas, planos e ilustraciones. Source: *Tierras,* vol. 3600, exp. 6, fol. 31.

On the other hand, European-born Spaniards did not accept criollos as real Spaniards, because of their place of birth. These elites existed in ambiguous space, but through narrative mapping, criollos would create conceptual space that certified their distinctive relationship with the land of New Spain.

A midcentury example of such practices is found in the *Theatro Americano, descripcion general de los reynos y provincias de la Nueva España* written by José Antonio de Villaseñor y Sánchez (ca. 1700/5–1759), a criollo Contador General de la Real Azogues. The 1741 edict by Felipe V ordered that the viceroy of New Spain produce up-to-date and "detailed information about the true state of these provinces." In 1743, Villaseñor was asked to respond to the king's mandate. To this end, local administrators across viceregal Mexico were required to provide detailed written summaries about the climate, agricultural products, and inhabitants of the territories they managed. It would be Villaseñor's task to collate the diverse data into a two-volume compendium titled *Theatro*

Americano. Thirty copies were printed in Mexico in 1746 and 1748. Villaseñor's volumes synthesized for the king a catalog of the spatial distribution of demographic and economic data about midcentury New Spain.

Villaseñor, however, envisioned the data within a distinct New Spanish viewpoint by locating them within historical networks. The introductory chapters 1 and 2 synthesize a history of New Spain from biblical antediluvian times, to the arrival of indigenous people, to the Spanish conquest and rule. Having set the historical context, Villaseñor argued in chapter 3 that New Spain must be seen in its historical and physical wholeness.[49] This is a critical move that visualizes a New Spain that could not be measured exclusively by its physical resources or distances between towns and cities. As a result, Villaseñor situates the reader in the physical space *and* history of New Spain, and does not describe viceregal Mexico as just a physical extension of Spain or within Spanish history. Throughout the *Theatro Americano*, there emerges a self-referential space—a holistic space from which to view *and* consume New Spain from New Spain, not from Old Spain.

Villaseñor's notion of geography as an observational practice from which to locate a site for knowledge production would appear in other texts from the second half of the eighteenth century. In the *Bibliotheca Mexicana* of 1755, Juan José de Eguiara y Eguren (ca. 1696–1763), a cleric and educator born in New Spain, reacts to contemporary European authors who disparage the Americas as a desolate place unable to support intellectual activity. The *Bibliotheca Mexicana*, which was not completed, catalogs the history of New Spain's institutions of higher education and the lives and scholarly works of individuals from pre-Hispanic times through the viceregal present. Deeply attached to the land and culture, Eguiara y Eguren begins his work with a prologue that sets the geographic stage for the study. He explains, "Among the varied climates of the globe, none is more suitable to the inspiration of people of talent than the sky of Mexico; so that those who have become familiar with ancient Athens and now contemplate Mexico City consider them to be very close in similarity for the gentleness of their skies and their airs with which they [people] sustain and refine the shrewdness, ability and grandeur of their wits."[50] Here, he associates the climate and environment of Mexico City with Athens in order to represent America not as a hostile natural environment but as "a space amenable to the cultivation of eloquence and knowledge."[51] Confirming his membership in a cultured New Spanish community, Eguiara y Eguren annotates the physical space of New Spain as a place from which to produce and consume the achievements of New Spain. For our discussion, however, we may observe that, much as in the *Theatro Americano*, in what appears initially to be the marginal introductory sections in the *Bibliotheca Mexicana*, Eguiara y Eguren situates the reader as a viewer of the physical space of New Spain from which to produce and consume the intellectual achievements.

The reasons for these geographic descriptions and references are social and political and suggest that eighteenth-century criollo mappings utilize narratives that visualize space. These descriptive writings were possibly embedded in concepts of discourse as mappings that may have been transferred from Spain and adapted to local situations. Specifically, a conceptualization of space in and through travel comes out of early medieval Iberian traditions in which geography was organized as a journey, as a linear movement through space organized as a route of travel. Ricardo Padrón explains this discourse of travel:

> By reminding us, at this point, that we need to backtrack to a particular node in a network if we wish to move in a certain direction, the discourse invites us to relate to the territory from a particular point of view. It interpelates [*sic*] the reader, not as an onlooker looking down on the territory from a height, as in a map, but as a traveler, moving through the territory, place by place, along routes. . . . Territory here is conceptualized in and through travel, in ways that resist their representation in the new cartographic language of the gridded map."[52]

Likewise, in the eighteenth century, then, territory was also conceptualized by certain writers in and through verbal self-mapping practices that shaped a criollo spatial stage.

The *Rusticatio Mexicana*, a bucolic poem by Rafael Landívar (1731–93), a Jesuit educator born in Guatemala, published in 1781–82, also deploys narrative practices that depict the land and environment. Landívar writes, "But I, through love for my native land, enjoy most of all to visit the ever verdant fields of my country."[53] He continues by describing physical phenomena—the lakes, springs, a waterfall, and volcanoes; resources—dyes [cochineal, indigo], sugar, gold, and silver; domestic and wild animals; and sports. In his descriptions of the lake around Mexico City, for example, Landívar describes in detail the lake's crystal-clear fresh water and how indigenous people use the water as a resource and for recreation. He also provides the reader with details about the devastating destruction caused by a volcanic eruption in which "everything was destined rather to perish in the approaching flames."[54] Traveling to southern New Spain, Landívar moves to the almost microscopic; he describes the natural history of the cochineal insect—used to obtain a valuable dye: "It is gifted with a gentle nature, abhorring murder among its fellows, abhorring civil disorder."[55] Then Landívar moves on to discuss the beaver, "a wary and intelligent animal, possessing great gifts beneath a shaggy exterior. . . . It has inherited noble traits." Specifically, due to their alert disposition and industrious nature, these animals "establish along river banks living quarters for their people, to construct dams cross the streams, and to rule their great city with uninterrupted peace."[56] Landívar, as an eyewitness, undertakes an itinerary for the reader across diverse environments, showing that they are not desolate or bereft of productive activity but that nature is active at all levels—from lake to volcano to beaver to insect—and assuring that its inhabitants developed modes of knowledge that enabled them to know, as well as to "intervene upon and assert authority over parts of the environment."[57] These descriptions emphasize spatial wholeness in contrast to the Spanish administrative mapping of fragmented territory—that of ports and intendancies—as well as the localist views of some of the land maps.

Eguiara y Eguren's and Landívar's writings may also manifest the second Iberian concept that underlies criollo mapping practices: membership in a community. Recall that Eguiara y Eguren claims membership in a community of intellectuals in New Spain. Landívar further reiterates this notion in his natural history descriptions through analogies with communities in nature. He cites the cochineal insects' "abhorrence of murder among its fellows" and the beavers who "rule their great city with uninterrupted peace."

The description of unique physical environments and the focus on community in narrative mapping are intertwined and strongly affirmed in 1771 in a presentation to the king entitled "Representación humilde que hizo la imperial, nobilíssima y muy leal ciudad de México en favor de sus naturales," written on behalf of the City Council by Antonio Joaquín de Rivadeneira y Barrientos, a criollo judge in Mexico City.[58] In reaction to ongoing Spanish attempts to increase administrative control over New Spain, resulting in diminished criollo access to public and ecclesiastic offices, Rivadeneira urges the king to select *españoles americanos*, American-born Spaniards, over foreigners, that is, European-born Spaniards. He reasons that it is the affective relationship constituted by place that distinguishes a native from just an inhabitant because, by nature, humans have inherent affections for the soil on which they are born and disaffection for all others.[59] While Rivadeneira's text does not describe geography, it does map geography's effects. He observes that native-born individuals understand their land, and, when European Spaniards try to impose foreign ideas, they fail because these foreigners are unable to comprehend the land and its people. Rivadeneira respectfully but emphatically argues that foreigners can neither produce nor consume knowledge about New Spain because they are not of the soil—that is, native born. He counsels the king to appoint administrators with origins and experience in New Spain, because their knowledge results in the most effective and productive government. Rivadeneira's discourse on love of land and country reflects the experiential, self-referential premises of narrative mapping practices.

Eguiara y Eguren, Landívar, and Rivadeneira each frame a discourse that emphasizes the idea of "belonging," certified through personal lived experience of cultural achievements and natural environment. Their references imply the existence of a community of people who belong in and to the spaces of New Spain. Each author maps the contours and boundaries of this community differently: Eguiara y Eguren—through its cultural achievements, Landívar—through experienced environments, and Rivadeneira—through innate affection for and knowledge of place. The writers describe what they might expect their readers to comprehend implicitly: natural membership in a community understood as based on affinity and integration for the common good. Their references to community as natural ties emerge from two concepts—*naturaleza*—nativeness and *vecindad*—citizenship—found in Spanish discourses and transferred and adapted in New Spain.

In her study of these concepts, historian Tamar Herzog explains that in Spain these designations of community membership were used to distinguish natives from foreigners. Such membership was socially recognized and negotiated; citizenship was identified by performance, behavior, and reputation. In analyzing the transfer and adaptations of these concepts to the Americas, Herzog concludes that criollo discourse sought to ignore formal definitions and boundaries and "to place emphasis instead on 'natural ties' that united people who loved one another . . . [and were integrated] into a community." "Love of the local community and citizenship could thus lead to nativeness."[60]

Criollo narrative mapping practices, then, trace itineraries through the physical environment and demarcated communities that did not necessarily have measurable boundaries but were marked by boundaries of affinity and commonality that certified inclusion and, concomitantly, exclusion. Narrative mapping addressed the liminal identity of elites by specifying the existence of the spaces of *españoles americanos*.

Narrative and cartographic mapping interweave in the work of José Antonio de Alzate y Ramírez (1737–99), an internationally known and respected criollo scientist born in New Spain, who produced both geographic as well as cartographic materials. In his publications, he, too, undertook geographic narratives that describe and discuss various activities, resources, and conditions of New Spain. For example, he explains and illustrates the natural history of the cochineal insect and illustrates the production of a valuable dye made with this insect.[61] Further, like Eguiara y Eguren, his writings address the problematic assertions of foreigners: "In the [works] of most of the authors who have written of this America [New Spain] one finds some very crass mistakes, and so I propose to provide some emending pieces that serve as a corrective."[62]

Alzate also attempted to address the lack of maps through a series of cartographic productions. In 1763, he produced *Plano de la Nueva España*, published in the *Historia de Nueva España* of Antonio Lorenzana, the archbishop of Mexico.[63] Identifying the divisions of the New Spain landmass according to bishoprics and provinces within the grid of longitude and latitude, the map traces Hernán Cortes's travels from the Veracruz coast to the Aztec capital, Tenochtitlan, that is, Mexico City. In 1767, Alzate also produced an atlas for the archbishop, the *Atlas eclesiástica del Arzbispado de México*, as part of a project to reorganize the archbishopric. This atlas includes a map of the whole archbishopric as well as eighty-seven maps and plans showing the location of missions and churches. Both of these projects function to describe New Spain in the context of Spanish interests—the history of the conquest and ecclesiastically managed spaces.

In 1767, Alzate would produce a map depicting the whole of North America as well as archbishoprics and various towns and cities of New Spain.[64] The audience for this map was not Spanish, however, but the scientists of the Academy of Science in Paris. He dedicated a copy of the map to members of the Paris Royal Academy of Science and sent it to France. The academy published and disseminated the 1767 map, but the print copies did not arrive in New Spain until 1792 (fig. 2.15).[65] *Nuevo mapa geographico de la America septentrional*, a version of Alzate's original map printed in France, displays scale, latitude, and longitude; Alzate locates New Spain's territory in relationship to other continents. His cartographic work and geographic perspectives provide viewers and readers with both an overview of a physical space and observed information, emphasizing visual description of self-referential geography. Distinct from Lorenzana, Alzate attempts to understand New Spain through cartography as well as the visualization of experiential geographic information about New Spain's people and natural history.

CONCLUSION

By the end of the eighteenth century, various understandings of colonial land coexisted uneasily; these are manifested in diverse mapping practices. Pres-

FIGURE 2.15. José Antonio de Alzate y Ramírez. *Nuevo mapa geographico de la America septentrional, perteneciente al virreynato de Mexico dedicado à los sabios miembros de la Academia real de las Ciencias de Paris por su muy rendido servidor, y capellan, Don Joseph Antonio de Alzate, y Ramirez, Año de 1769.* Courtesy of the John Carter Brown Library at Brown University.

sured by imperial demands to secure administrative reform and commercial expansion for Spain, Bourbon officials mapped eighteenth-century New Spain as empirical space, a place where resources and inhabitants were to be located and measured—ordered and embedded in geodesic knowledge. Rooted in Spain's medieval land tenure system, this mapping insisted that imperial power emerged from claimed territoriality. In contrast, exemplifying that "territory is never a homogeneous space," indigenous populations envisioned the land as an originating space, where inhabitants were not just located but dominated the space through primordial connections and thorough integration into the landscape. These mappings located territory in the network of culture and history rather than in the grid of geometry.

Concurrently, as a result of the cultural adaptation and transition to the tenurial legal system, landowners, indigenous and nonindigenous, regularly disputed land and water rights and deployed diverse lived descriptions of their property, including maps and sketches. And criollo intellectuals understood the shifts in Spain's governmental practices and the resulting alteration of the relationship between peninsula and American territories as threatening to their existing positions of authority in New Spanish society. They resisted the notion of the spaces of New Spain as exclusively defined empirically, primordially, or legally. Instead, through their writings, they envisioned the land as a unique and complex physical environment that elite Spaniards born in New Spain interacted with to construct an amalgam identification and space for *españoles americanos*.

Although the line and the boundary marker are critical in all of the late colonial mapping practices of New Spain, neither a definitive line nor a fixed boundary separates colonization from resistance to colonization. Distinct in intention, format, and production, diverse eighteenth-century mapping practices utilize common constructs and cross-references that slide along a continuum of domination and resistance. Criollo narrative mapping practices adapt Iberian constructs of itinerary travel and community and, for José Antonio de Alzate y Ramírez, gridded space. Landowners emphasize their lived experience of a territory in the context of the tenurial system. Indigenous maps, while emphasizing cultural markers, nevertheless incorporate Spanish instrumentations of indexing and the compass rose. Concomitantly, loud silences mark Spanish mappings—meaning that, in claiming imperial space through metaphor, measurement, and map, these practices resist indigenous and criollo perceptions of land.

Thus, viceregal cartography and mapping may be understood as technological and classificatory systems deployed in the service of Bourbon imperial and scientific projects. Cartography and mapping were also means deployed to resist those very same projects. Marked by conditions of contingency and contradiction, then, the mapping practices of New Spain were irrevocably entangled. This entanglement continued to be manifested in nationalist mapping of nineteenth-century Mexico, as seen in Antonio García Cubas's atlas pages, which display the continuing contradictions of colonial perspectives that mark the problematics of the move from colonization to decolonization.

NOTES

1. See Rabiela Hira de Gortari, "Nueva España y México: Intendencias, modelos constitucionales y categorías territoriales, 1786–1835," *Scripta Nova: Revista Electrónica de Geografía y Ciencias*

Sociales, vol. X, no. 218 (72) (2006), http://www.ub.edu/geocrit/sn/sn-218-72.htm, viewed 2013; Áurea Commons, "La organización territorial de España y sus posesiones en América durante e siglo de las luces," in *La geografía de la ilustración*, ed. José Omar Moncada Maya (México, D.F.: Instituto de Geografía, Universidad Nacional Autónoma de México, 2004), 41–81; and Horst Pietschmann, *Las reformas borbónicas y el sistema de intendencias en Nueva España* (México, D.F.: Fondo de Cultura Económico, 1996).

2. See Raymond Craib, "Cartography and Power in the Conquest and Creation of New Spain," *Latin American Research Review* 35, no. 1 (2000): 7–36.

3. Joanne P. Sharp, Paul Routledge, Chris Philo, and Ronan Paddison, eds., *Entanglements of Power: Geographies of Domination/Resistance* (New York: Routledge, 2000), 21.

4. Ibid., 7.

5. Ibid., 27.

6. The concluding chapter of Jaime E. Rodríguez O., *"We Are Now the True Spaniards": Sovereignty, Revolution, Independence and the Emergence of the Federal Republic of Mexico, 1808–1824* (Stanford: Stanford University Press, 2012), provides a comprehensive overview of the struggle to balance forces of domination and resistance, which were linked back to political upheavals in Spain, as Mexico was forming itself as a nation in the early nineteenth century.

7. See Magali Carrera, *Traveling from New Spain to Mexico: Mapping Practices of Nineteenth Century Mexico* (Durham: Duke University Press, 2011), chaps. 5–7, for a detailed analysis of Antonio García Cubas's work.

8. See Beatriz Berndt León Mariscal, "Discursos de poder en un nuevo Dominio," *Relaciones* 26, no. 101 (2005): 227–59.

9. Ralph Bauer, *The Cultural Geography of Colonial American Literatures: Empire, Travel, Modernity* (Cambridge: Cambridge University Press, 2003), 19.

10. On the standardization of map scales, see Maria de Carmen León García, "Cartografía de los ingenieros militares en Nueva España, segunda mitad del siglo XIII," in *Historias de la Cartografía Iberamérica: Nuevos caminos, viejos problemas*, ed. Héctor Mendoza Vargas and Carla Lois (México, D.F.: Instituto de Geografía, Universidad Nacional Autónoma de México, 2009), 444–48 and 451n4.

11. Michael Crozier Shaw, *"El siglo de hazer caminos*: Spanish Road Reforms during the Eighteenth Century; A Survey and Assessment," *Dieciocho* 32, no. 2 (2009): 413–415 and 444, 447.

12. Ibid., 431.

13. Antonio López Gómez and Carmen Manso Porto, *Cartografía del siglo XVIII: Tómas López en la Real Academica de la Historia* (Madrid: Real Academica de la Historia, 2006) provides a comprehensive overview of the production of the *Diccionario*. For a succinct overview see Horacio Capel, "Los diccionarios geográficos de la ilustración española," in Moncada, *La geografía de la ilustración*, 83–156.

14. See David Turnbull, "Cartography and Science in Early Modern Europe: Mapping the Construction of Knowledge Space," *Imago Mundi: The International Journal for the History of Cartography* 48, no. 1 (1996): 5–24.

15. *Real cédula ordenando se envien completos informes sobre nucleos urbanos, demograficos, economicos y eclesiasticos de todos los territories de Indias: Madrid, 19 julio 1741*. Reproduced in *Cuestionarios para la formación*, ed. Francisco de Solano and Pilar Ponce (Madrid: Consejo Superior de Investigaciones Científicas, 1988), 141.

16. See Juan Pimental, "The Iberian Vision: Science and Empire in the Frameworks of a

Universal Monarchy 1500–1800," *Osiris* 15 (2000): 17–30; and Jorge Cañizares-Esguerra, "Iberian Colonial Science," *Isis* 96 (2005): 64–70.

17. Daniela Bleichmar, *Visible Empire: Botanical Expeditions and Visual Culture in the Hispanic Enlightenment* (Chicago: University of Chicago Press, 2012), 29–33. See also Gabriel Paquette, *Enlightenment, Governance, and Reform in Spain and Its Empire, 1759–1808* (New York: Palgrave Macmillan, 2008), 93–97.

18. For discussion of intensified Spanish coastal mapping of New Spain in the late eighteenth century, see Louisa Martín-Merás, "La Expedición Hidrográfica de Atlas de la América sepentrional, 1792–1805," *Journal of Latin American Geography* 7, no. 1 (2008): 203–18; José Omar Moncada Maya, "Los ingenieros militares en la Nueva España del siglo XVIII. Promotores de la illustración," in Moncada, *La geografía de la ilustración*, 199–226; and *El ingeniero Miguel Constanzó: Un militar ilustrado en la Nueva España de siglo XVIII* (México, D.F.: Universidad Nacional Autónoma de México, 1994); and John Lieby, "Miguel Contansó and His 1799 Report on the Defenses of Veracruz," *Jahrbuch für Geschichte Latinamerikas* 42 (2005): 33–46.

19. The *Atlas de América* is discussed and illustrated in López Gómez and Manso Porto, *Cartografía del siglo XVIII*, 401–47.

20. Nuria Valverde and Antonio Lafuente, "Space Production and Spanish Imperial Geopolitics," in *Science in the Spanish and Portuguese Empires (1500–1800)*, ed. Daniela Bleichmar et.al. (Stanford: Stanford University Press, 2009), 198.

21. See Guadalupe Pinzón Ríos, "Los mapas del Pacifico novohispano: Apropiación y defensa de los litorales durante el siglo XVIII," in Mendoza Vargas and Lois, *Historias de la Cartografía Iberamérica*, 183–210.

22. Ernest J. Burrus, *La obra cartográfica de la Provincia Mexicana de la Compañía de Jesús (1567–1967)* (Madrid: Ediciones José Porrúa, 1967), 1:82. In footnote 6, Burrus traces the origin of this attribution, mentioning that Humboldt cites this map made by the "Jesuits of Puebla de los Angeles." In a subsequent footnote, Burrus also questions how the initials P.J.E. found in the dedications correlate with this attribution. He does not, however, offer an alternative designation.

23. In Greek mythology, Cynthius is the surname of Apollo, which was derived from Mount Cynthus on the island of Delos, his birthplace. This island reference probably links to Mexico City, which was founded on an island located in Lake Texcoco.

24. Valverde and Lafuente, "Space Production," 208.

25. [Item] "37. Pondreis especial atencion en el reparo y seguridad de los caminos y de todas las obrás públicas que consideráreis necesarias y convenientes en todas las ciudades y pueblos principales." (37. Special attention shall be put on the repair and safety of the roads and all public works considered necessary and proper in all major cities and towns.) "Instrucciones general que trajo de la corte El Marqués de las Amarillas: Expedida por la via del consejo," in *Instrucciones que los Virreyes de la Nueva España dejaron a sus sucesores* (México, D.F.: México, Imprenta Imperial, 1867), 71.

26. Miguel Aguilar-Robledo, "Land Use, Land Tenure, and Environmental Change in the Colonial Jurisdiction of Santiago de los Valles de Oxitipa, Northeastern New Spain" (Ph.D. dissertation, University of Texas, Austin, 1999), 229–32. Aguilar-Robledo adds that such mapping provided visual renditions of areas; recorded the indigenous towns and toponyms next to the granting areas; "accurately" located the requested and granted lands, based on permanent and distinguishable features of the landscape (rivers, arroyos, hills, indigenous mounds); portrayed

the process of "filling in" an "empty" space; denoted "useful" recourse for the settlement and land development purposes; encoded family symbols (scales, directions, cherubs, stylized natural features) on the unknown landscapes of New Spain; and symbolically apprehended, re-created, and took over, as an act of rule, the newly conquered lands. My thanks to Ray Craib for pointing out Miguel Aguilar-Robledo's work on land tenure in Mexico to me.

27. Barbara Mundy, *The Mapping of New Spain: Indigenous Cartography and the Maps of the Relaciones Geográficas* (Chicago: University of Chicago Press, 1996), 112–13. A broad overview of local and regional mapping in New Spain is provided by Víctor Manuel Ruiz Naufal in his essay "La faz del terruño: Planos locales y regionales siglos XVI–XVIII," in Mendoza Vargas and Lois, *Historias de la Cartografía Iberamérica* (n. 10 above), 33–69.

28. Aguilar-Robledo, "Land Use," 152–53. In chapter 4 of this dissertation, Aguilar-Robledo provides a more detailed description of the Iberian landholding system and its acculturation into New Spain.

29. ibid., 339–40.

30. Ibid., 159.

31. See Georgina Endfield, "'Pinturas,' Land and Lawsuits: Maps in Colonial Mexican Legal Documents," *Imago Mundi* 53 (2001): 7–9; Peter Gerhard, *A Guide to the Historical Geography of New Spain* (Norman: University of Oklahoma Press, 1993); and Francisco de Solano, *Cedulario de tierras: Compilación de legislación agraria colonial (1497–1820)* (Madrid: Consejo Superior de Investigaciones Científicas, 1988). Dorothy Tanck de Estrada, *Atlas ilustrado de los pueblos de indios Nueva España, 1800* (México, D.F.: Colegio de México, 2005), provides an excellent historical and cartographic study of nearly five thousand *pueblos de indios* that existed in late viceregal New Spain.

32. Huejotzingo (Puebla), 1591. AGN catalog no.: 1285. AGN source: *Tierras*, vol. 1876, exp. 8, fol. 3r.

33. See Mundy, *The Mapping of New Spain*, for an overview of this tradition.

34. San Nicolás Tenezcalco (Edo. de Mexico), 1715. Producer: Carlos Romero de la Vega. AGN catalog no.: 2288. AGN source: *Tierras*, vol. 2999, exp. 15, fol. 80r.

35. "Primordial titles are indigenous-language, municipal histories containing extensive descriptions of the communities' territorial boundaries and land-holdings," from Stephanie Wood, "The Social vs. Legal Context of Nahuatl Títulos," in *Native Traditions in the Postconquest World*, ed. Elizabeth Hill Boone and Tom Cummins (Washington, DC: Dumbarton Oaks, 1998), 201–28.

36. San Mateo Aticpac (Tlaxcala), 1728. Producer: Francisco Hualcoyotzien. AGN catalog no.: 1042.1. AGN source: *Tierras*, vol. 1470, exp. 2, fol. 8v.

37. Ibid., fols. 9r and 9v.

38. Ibid., fol. 73r.

39. See n. 47.

40. Miguel Aguilar-Robledo, "Contested Terrain: The Rise and Decline of Surveying in New Spain, 1500–1800," *Journal of Latin American Geography* 8, no. 2 (2009): 23. This article provides excellent information on the specific use of tools as well.

41. San Luis de los Chochos (Tepeaca, Puebla), 1732. Producer: Maximiliano Gómez Daza. AGN catalog no.: 0722. AGN source: *Tierras*, vol. 487, exp. 1, fols. 97r–98v.

42. Ibid., fol. 99r.

43. Ibid., fol. 100r.

44. San Juan Axalpan, San Sebastián Zinacantepec, San Gabriel Chila y San Diego Chalma y

hacienda Santísima Trinidad y San José Buenavista, Tehuacan (Puebla), 1780. Productor: Jacinto de Espinoza, agrimensor. AGN catalog no.: 0915. AGN source: *Tierras*, vol. 1058, exp. 2, fols. 53v–57r.

45. A possible origin for the format of this Yucatec map is discussed by Amara L. Solari in "Circles of Creation: The Invention of Maya Cartography in Early Colonial Yucatán," *Art Bulletin* 92, no. 3 (2010): 154–68.

46. Haciendas San Juan Bautista Suytunchin y San José Occhac (Yucatán), 1787. Productor: Santiago Arosteguidurán, agrimensor. AGN catalog no.: 0916. AGN source: *Tierras*, vol. 1061, exp. 1, fols. 26v–29r.

47. In 1567, landless Indian pueblos were granted five hundred *varas* of land and Spaniards were prohibited from raising stock within one thousand *varas* of this land. Known as *tierras por razón de pueblo* (by right of township, which became know as *fundo legal* [legal allotment] in the nineteenth century), the fifteen hundred *varas* were to be measured from the town's church to the four "winds," that is, all directions. In 1687, the king increased the *tierras por razón* by one hundred *varas* to six hundred *varas*, to be measured from the last house of the pueblo. Landlords and stockraisers contested this, so that in 1695, the king had to revert to the previous format of measuring from the pueblo's church. Aguilar-Robledo, "Land Use" (n. 26 above), 215–16. See also Stephanie Wood, "The *Fundo Legal* or Lands *Por Razón de Pueblo*: New Evidence from Central New Spain," in *The Indian Community of Colonial Mexico: Fifteen Essays on Land Tenure, Corporate Organizations, Ideology and Village Politics*, ed. A. Ouweneel and Simon Miller (Amsterdam: CEDLA [Center for Latin American Research and Documentation], 1990), 118–19.

48. Rancho Apetlanca, Zacualpan, Morelos (Morelos) 1767. AGN catalog no.: 2489. AGN source: *Tierras Fuente*: *Tierras*, vol. 3600, exp. 6, fols. 28v–30r. The follow notation appears below the figure on the right end of the rope: 102 crds [cordeles] de 50 v [varas].

49. José Antonio de Villaseñor y Sánchez, *Theatro Americano: Descripcion general de los reynos y provincias de la Nueva España y sus juridicciones,* libro primero, chap. 3, 20. Further, aligning with the idea of holistic viewing, Villaseñor y Sánchez produced a manuscript map of New Spain as part of the *Theatro Americano* project, which was never printed or circulated. See Michel Antochiw, "La visión total de la Nueva España: Las mapas generales del siglo XVIII," in *México a través de los mapas,* ed. Héctor Mendoza Vargas (México, D.F.: Instituto de Geografía, Universidad Nacional Autónoma de México, 2000), 73–77.

50. Anthony Higgins, *Constructing the* Criollo *Archive: Subjects of Knowledge in the* Bibliotheca Mexicana *and the* Rusticatio Mexicana (West Lafayette: Purdue University Press, 2000), 39.

51. Ibid., 33 and 67.

52. Ricardo Padrón, "Mapping Plus Ultra: Cartography, Space and Hispanic Modernity," *Representations* 79 (2002): 45, 48.

53. Andrew Laird, *The Epic of America: An Introduction to Rafael Landívar and the* Rusticatio Mexicana (London: Gerald Duckwork & Co., 2006), 124.

54. Ibid., 135.

55. Ibid., 150.

56. Ibid., 162–63.

57. Higgins, *Constructing the* Criollo *Archive*, 167.

58. From "Representación humilde que hizo la imperial, nobilíssima y muy leal ciudad de México en favor de sus naturales," in *Colección de documentos para la guerra de independencia de México de 1808 a 1821,* ed. J. E. Hernández y Dávalos (México, D.F.: Comisión Nacional para las Cele-

braciones del 175 Aniversario de la Independencia Nacional, 1985), vol. I, 427. For an overview of Rivadeneira y Barrientos's contributions, see D. A. Brading, *The First America: The Spanish Monarchy, Creole Patriots and the Liberal State, 1492–1867* (Cambridge: Cambridge University Press, 1991), 479–83.

59. "Representación," 430. "Entre los efectos naturales se cuenta con mucha razón el amor, que tienen los hombres a aquel suelo, en que nacieron; y el desafecto a todo otro; siendo estos dos motivos los más sólidos principios, que persuaden la colocación de el natural, y resisten la de el extraño."

60. Tamar Herzog, *Defining Nations: Immigrants and Citizens in Early Modern Spain and Spanish America* (Durham: Duke University Press, 2005), 151.

61. Found in José Antonio Alzate y Ramírez, Plancha 7, fig. 1: Indio que recoge la cochinilla con una colita de venado, 1777. AGN catalog no.: 0126. AGN source: *Correspondencia de Virreyes*, 1a. serie, vol. 90, exp. 56, fol. 262.

62. "En los más de los autores que han escrito de esta América se hallan algunos erroes crasísimos, y así me propongo ir dando algunos pedazos enmenedados, para que les sirvan de correctivo." *Diario Literario de México*, 12 marzo 1768, n.p. Quoted in Alberto Saladino García, *Dos científicos de la ilustración hispanoamericana: J. A. Alzate y F. J. De Caldas* (México: Universidad Autónoma del Estado de México, 1990), 283.

63. See Hernan Cortés, *Historia de Nueva-España/escrita por su esclarecido conquistador Hernan Cortes* (México: en la imprenta del Superior Gobierno, del Br. D Joseph Antonio de Hogal, 1770).

64. For an overview of Alzate's cartographic work see Michel Antochiw, "La visión total de la Nueva España: Las mapas generales del siglo XVIII," Mendoza Vargas, *México a través de los mapas* (n. 49 above), 71–88.

65. Alzate explains the history of the map's production and reproduction as part of discussion of New Spain's geography in the *Gazeta de Literatura* 3, no. 6 (January 22, 1793): 46. See Fiona Clark, "The *Gazeta de Literatura de México* (1788–1795): The Formation of a Literary-Scientific Periodical in Late-Viceregal Mexico," *Dieciocho* 28, no. 1 (2005): 7–30.

REFERENCES

AGN catalog no.: Archivo General de la Nación, *Catálogo de ilustraciones*.
AGN source: Archivo General de la Nación, document group.

CHAPTER THREE

CARTOGRAPHY IN THE PRODUCTION (AND SILENCING) OF COLOMBIAN INDEPENDENCE HISTORY, 1807–1827

Lina del Castillo

INTRODUCTION

The two maps with which we begin show different parts of South America. A closer look suggests a family resemblance: they both bear the name "Colombia." The continental vision (fig. 3.1), printed in London in 1807 and titled "Colombia Prima or South America," audaciously argues that "Colombia" was all of South America. The other map (fig. 3.2) is taken from the first national atlas of the independent Colombian Republic, printed in Paris in 1827. Although it did not claim all of South America, the second image did claim territories for the Colombia that we now recognize as Panama, Ecuador, Venezuela, and Colombia, as well as parts of Costa Rica, Nicaragua, Honduras, Peru, Guyana, and Brazil. The result: a generously large version of a country scholars now remember as Gran Colombia, the political body that unified the Captaincy General of Venezuela with the Viceroyalty of New Granada (including Panama and Quito) against Spanish royalist forces from 1819 to 1830.

The makers of the "República de Colombia" image needed to convince international and national audiences that Colombia existed as an independent,

FIGURE 3.1. William Faden and Louis Stanislas d'Arcy de la Rochette, composite of sheets 1–8 of "Colombia Prima or South America, In which it has been attempted to delineate the Extent of our Knowledge of that Continent Extracted Chiefly from the Original Manuscript Maps of His Excellency the late Chevalier Pinto Likewise from those of Joao Joaquin da Rocha, Joao da Costa Ferreira, El Padre Francisco Manuel Sobrevielo &c. And From the most Authentic Edited Accounts of Those Countries" (London, June 4, 1807). 256 × 174 cm. Scale 1:3,000,000. David Rumsey Map Collection.

sovereign entity. Therefore, in addition to this image of the nation, the atlas included larger-scale images for each of the twelve departments that made up Colombia, several of which identified the location of independence battles in order to help illustrate the multivolume narrative history of the revolution that this atlas formed a part of. But for this image to do the work that its makers wanted it to do, the atlas needed to engage in a kind of cartographic

FIGURE 3.2. José Manuel Restrepo, "Carta de la República de Colombia," engraved in Paris by Darmet, calle du Battoir, no. 3, written by Hacq, 1827. 49 × 60 cm. Scale: 1:5,500,000. Map forms part of "Historia de la revolucion de la Republica de Colombia, por Jose Manuel Restrepo, Secretario del Interior del poder ejecutivo de la misma Republica. Atlas" (Paris: Libreria Americana, Calle del Temple, no. 69, 1827) (verso of title page). Imprenta de David, Calle del arrabal Poissonniere, no. 6 en Paris. David Rumsey Map Collection.

and historical cannibalism. The 1807 "Colombia Prima," a rare map, whose significance in the process of Spanish American independence has largely been overlooked, was only one of the atlas's many victims. This chapter traces out the circumstances behind the ideation, creation, and circulation of these two interconnected "Colombias," highlighting how each was a product of different transatlantic social and political networks that were dedicated to the independence of Spanish America, albeit in different ways. In both cases, timing mattered. To best understand the historical context and meaning of the content displayed on the two maps, we must situate them within the fast-paced geopolitical changes that occurred in the early nineteenth century as the Spanish Atlantic monarchy began to dissolve. These little-known, understudied maps offer a visual archive of information that helps us better see, on the one hand, the grand imperial designs and republican territorial desires that sprung from the Spanish imperial crisis and, on the other, just how contingent those purportedly permanent visions actually were.

In this sense, the chapter underscores how cartography attempted to cleanse imagined territories of inconvenient pasts at the behest of the grand ideological projects that emerged from the shocks of transatlantic revolutions.[1] Deliberate acts of forgetting on maps and through histories yielded in time master metanarratives of independence that have triumphed into the current day.[2] The visual quality of printed maps and their claims to scientific objectivity offered Spanish American elites, and the British imperial agents and French cartographers they worked with, a powerful way of excluding troublesome facts from the kind of territorial history they needed to promote on both sides of the Atlantic.[3] This was true for Francisco de Miranda (1750–1816), the Venezuelan cosmopolitan creole who built his life around Spanish American independence well before Napoleon invaded Iberia, and who, as this chapter demonstrates, led the transatlantic charge to produce a continental "Colombia Prima" kind of vision by 1807. This was also true for the *partido de los Libertadores*, or the Liberator Party, the select group of men that had joined the independence cause, and who governed with—and supported—Bolívar from roughly 1819 until Gran Colombia's dissolution in 1830.[4] Nevertheless, while examining what is included and erased from a map's face may reveal some deliberate acts of forgetting, other forms of forgetting are less deliberate. Those who had intimate connections with the processes of map production and dissemination often did not leave a direct physical trace of their interactions with these sources. This missing part of the cartographic historical record is demonstrative of the way maps as material objects were made and circulated. Delving into contemporary carto-bibliographies, newspaper reports, letters, and memoirs help make visible, at least to an extent, the men who were most invested in influencing how a shape-shifting "Colombia" needed to be imagined cartographically by the different scientific societies, literary communities, and diplomatic circles that these men plugged in to, depended on, and were embedded in.

Printed cartography marshaled for Spanish American independence was not meant to decolonize minds. As José Moya has recently argued, postcolonial studies emerged from, and responded to, the post–World War II independence of former British and French colonies in Africa and Asia. This was the case even though European cultural penetration was relatively superficial when compared to the depth of Iberian colonialism in the Americas.[5] Proindependence leaders in Spanish America did not "hijack" cartographic language in order to subvert a colonial order that had placed unwelcome foreigners in a position of control and dominance over the lives and labor of indigenes. Neither did independence deliver a radical break with European colonialism by the end of 1820s.[6] Instead, insurgent leaders in America saw themselves both as indigenes and as Europeans; they were less interested in overturning the existing social order than they were in maintaining what they believed was their place as the "nobility" of the "New Continent" with legitimate authority to rule over a territory and a people.[7]

In two major sections, this chapter contextualizes the conceptualization, production, content, and circulation of the two Colombia maps signaled above. The first section explains the origins of "Colombia Prima" by turning to the transatlantic moves made by Miranda during the late eighteenth and early nineteenth centuries. The chapter builds on the work of Karen Racine and others by highlighting cartographic evidence that suggests how Miranda's travels allowed him to develop close relationships with US and British politicians, military men, diplomats, and mapmakers. Miranda's relationship with the London-based cartographer William Faden and his renowned geographer Louis Stanislas D'Arcy de la Rochette facilitated transportation of "Colombia Prima" from London across the Atlantic, where it became a gift intended to cement diplomatic relations between Caracas and Santafé (Bogotá). Circulation of "Colombia Prima" in Spanish America may have aimed at shaping proindependence geographic imaginations, but it backfired. On the one hand, tensions between the rival juntas that had sprouted up throughout the Spanish Empire after the abdications of the Spanish kings at Bayonne in 1808 had made the centralized continental vision depicted by this map an impossible ideal by 1812 with the fall of the First Republic. On the other, "Colombia Prima" had joined other British mapping efforts to put Great Britain's imperial designs ahead of Spanish American territorial claims.

Not until 1819, when the term "Colombia" emerged in Spanish America to refer to an officially founded republic, did Spanish Americans scramble to combat the cartographic territorial limits that a British "Colombia Prima" tried to impose. The second section considers the conception, drawing, and printing of the 1827 *Historia de la revolución de la República de Colombia: Atlas*, demonstrating how changes from the manuscript version to the print edition formed part of a critical diplomatic effort on the part of Colombian representatives abroad. One way was through the atlas's ultimate selling point. What made the atlas "preferable to any other printed map" of Colombia was how it identified the places where major independence battles were fought.[8] Which battles and cities made it onto the final printed version of the atlas, and which did not, when viewed in light of competing contemporary narratives of independence, reveals how the international network of the Liberator Party worked to place Bolívar at the center of the revolution at the expense of several other key players. Centering Bolívar formed part of a larger political strategy: it helped consolidate el Libertador's political legitimacy as well as that of the men allied to his cause at a time when European powers, especially France, were reluctant to recognize Colombian independence in the first place. Reading the 1827 atlas in this way helps us better remember how fragile Bolívar's control over the patriot forces was, the crucial turning points that contributed to his ascendency, and the role of international diplomacy and cartography in the construction of Bolívar's authority.

Both "República de Colombia" and "Colombia Prima" were maps that clamored to be seen, recognized, and revered as true and scientific.[9] Both were cartographies of Colombian independence. Neither was successful in creating a stable Colombian territorial geo-body. It is precisely the ephemeral nature of these representations that makes them interesting. Tracing the discursive strategies and transatlantic intrigues involved in the mapping out of these early (and short-lived) visions of Colombian territory allows us to better see how proindependence leaders turned to the "objective" language of cartography in their efforts to transubstantiate contentious political realities on the ground into seamless images intended to shape the mental maps and historical imaginations of audiences abroad. The kinds of geopolitical entities that these images championed reveal some of the paths considered, but ultimately not taken, during the tumultuous, uncertain moments when Spain's Atlantic empire was in the process of disintegrating.[10]

PART 1. COLOMBIA PRIMA

FRANCISCO MIRANDA IMAGINES A "PUEBLO COLOMBIANO"

José Manuel Restrepo's *History of the Revolution* and its illustrative atlas placed Bolívar at the center of Colombia's creation and liberation. Although this perspective informs the entire text, Restrepo's dedication was most explicit: "Ever since I decided to devote part of my leisure time to the daring enterprise of writing the History of the Revolution of the Republic of Colombia, the idea naturally occurred to me that I should dedicate the work to you [Bolívar], its creator and liberator."[11] Despite what Restrepo would have us believe, el Libertador was not the first person to imagine a Pueblo Colombiano. Rather, it was the *generalísimo* of Venezuela's First Republic, Francisco de Miranda.

Karen Racine has highlighted Miranda's friendships with many of his generation's most important political figures, including George Washington, Alexander Hamilton, Thomas Jefferson, Catherine the Great, British prime minister William Pitt, and, of course, Simón Bolívar.[12] Miranda's cosmopolitan friendships reflected his ability to connect with various mercantile, political, and military networks interested in the same goal: opening Spanish America's markets. They also suggest Miranda's extensive travels. In 1770, when barely twenty years old, Miranda left Caracas, but he still enjoyed generous economic support from his father, whose contribution to the Spanish royal government allowed Miranda to tour Spain, North Africa, Havana, and Florida as a navy captain, where he learned new languages, forged new and lasting friendships, and augmented his wealth. By August of 1782, Miranda's interactions with other Spanish navy officers were proving less than smooth sailing; he was charged with espionage, contraband trade, and possession of books prohibited by the Inquisition. Worried about the consequences his trial would bring, Miranda went into exile. He left for Spain via North America, a place that caught his interest, given its recent independence. Miranda landed in North Carolina in 1783 and began a two-year tour of the United States, which included stops in Charleston, Philadelphia, New York, and Boston.[13] Throughout his travels his senses were inundated with a new way of referring to the Americas: Columbia.

By the 1780s, calling America "Columbia" had become thoroughly conventional.[14] The argument that the New World should bear some form of Christopher Columbus's name instead of Amerigo Vespucci's had circulated in the Americas and Europe as early as the sixteenth century. "Columbia," however, did not emerge as a signifier for North America until the early eighteenth century, when *Gentleman's Magazine* mockingly satirized how the "conquests and acquisitions in Columbia," carried out by the British contributed little to its power as an empire.[15] Satire gave way to poetry as "Columbia" inspired songwriters and poets on the other side of the Atlantic seeking an evocative identity for America that could help sever it from British ties. The end of the American Revolution in 1783 inspired poetic transformations of places. The premier institution for higher learning in New York City could no longer remain King's College; it reopened as Columbia College in 1784. Coins minted between 1785 and 1786 erased the Indian maiden that had passed from hand to hand as legal tender, and the female allegory named "Columbia" took her place.[16] In 1786 South Carolina decreed "Columbia" would be the name for its new capital city. The United States Congress followed suit, locating the national capital on the shores of the Potomac in the newly named District of Columbia. In short, from 1783 to 1784, when Miranda toured the United

States, he also was touring a newly independent Columbian nation. Miranda did not forget Columbia on his way to Europe in 1784.

As early as 1788, we find Miranda writing letters to sympathetic European royalty such as Prince C. Landgrave de Hesse, thanking him for supporting the ideal of independence for a "disgraced Colombia."[17] It seems Miranda, a careful student of languages, decided to Latinize the Anglicized version of Columbus's name. As self-appointed general commander of a "Colombian" army, he announced his arrival in Caracas in 1806 with a proclamation "to the inhabitants of the Americo-Colombian continent," inciting all sixteen million people to rise up against the Spanish, assuring them that, "just as desire indubitably constitutes our independence, union will assure us permanence and perpetual happiness."[18] In between his informal pleas for European support of an independent "Pueblo Colombiano" in the late 1780s and his (ultimately unsuccessful) Venezuelan proclamation against the Spanish in 1806, Miranda found the time, friendships, and resources in London that allowed for a literal mapping-out of a Colombian community.[19]

Miranda was well aware of the power of maps to convince allies of the geopolitical and economic value of South America. In 1790, Miranda unfurled maps of America by Jean Baptiste Bourguignon d'Anville for William Pitt. Miranda remembered how he taught the British statesman about "the geography of Chile, Peru, etc. Like a good schoolboy, he [Pitt] went on all fours to understand the map, which lay across the carpeted floor."[20] At the end of his physical interaction with the map, Pitt was impressed. He begged Miranda to send him more information on the Tupac Amaru and Comunero rebellions in South America. D'anville's maps were evidently useful, but Miranda needed more maps of the region if he wished to convince his British allies and potential backers in the United States to participate in a bold endeavor: a three-pronged invasion and liberation of "Colombia." Miranda strategically used the growing threat of French designs on Spanish territories in the New World as a foil: "What measure would be easier and at the same time more effective than to detach from Spain an immense dominion with a population and richness which constitute a mass of resources that by a counterstroke could be turned to the advantage of France, in whose interests Spain is so blindly involved?"[21] Miranda proposed that the United States attack Spanish America from Panama and the British forces invade from Buenos Aires, while he would lead forces into his home territory of Coro, Venezuela. Maps of South America proved essential for this coordinated invasion, and the leading British map publisher of the period, William Faden (1750?–1836), proved critical for these cartographic ends.

We first have evidence of Miranda's working relationship with Faden in 1792, when the mapmaker billed Miranda for a collection of maps delivered to his home for the sum of about sixty pounds.[22] Miranda again called on Faden in August of 1804 in order to build the cartographic arsenal he needed to illustrate his plan of attack.[23] By then the United States had signed a treaty with France, precluding any inclination to invade Panama. Miranda was undaunted. In October of that year, Miranda met with Commodore Sir Home Riggs Popham (1762–1820), William Pitt (1759–1806), and Henry Dundas, Lord Melville (1742–1811).[24] Miranda later recalled that after breakfast, "the table was cleared, and the maps were unrolled."[25] It was on these maps that these men negotiated the future of Spanish America, determining which points were the most advantageous for the operation of expeditionary forces. These plans culminated in the controversial and failed British invasion of Rio de la Plata in 1806, the same year that Miranda landed at La Vela de Coro in Venezuela, proclaiming Colombian independence, also a disastrous expedition.[26] The dismal failure of both missions had badly damaged Miranda's credibility.[27] He returned to England at the end of 1807, chastened. The radically changing geopolitical landscape in Europe from 1807 to 1810 forced a London-based Miranda to adjust his vision of what an independent Spanish America might look like.

BRITISH IMPERIAL DESIGNS SEEN THROUGH "COLOMBIA PRIMA"

While it is possible to piece together Miranda's role in the production of Faden's maps and the uses Miranda put Faden's maps to, finding evidence demonstrating the extent to which Miranda influenced the actual content of Faden's cartography is challenging, to say the least. Bookkeeping records and letters tie Miranda to Faden through the production of "Mapa Geográfico de América Meridional," originally engraved by Juan de la Cruz Cano y Olmedilla in 1775.[28] These records do not reveal much more than the fact of their working relationship on a map that was, ultimately, an exact replica of an already existing map. It is by turning to the social history of Faden's 1807 edition of "Colombia Prima" that the extent of Miranda's ideological influence on Faden's cartographic production becomes clearer.

The first and most obvious clue that this map might reflect Miranda's vision is its title (fig. 3.3). Although printed in London in 1807, almost a century after the term "Columbia," with a *u*, had been circulating in London, and several years after it had already been popularized in the United States, the spelling of the word on this version of Faden's map reflected Miranda's Latinized version

FIGURE 3.3. Faden and De la Rochette, "Colombia Prima, or South America" (detail). David Rumsey Map Collection.

with two *o*'s. Perhaps this may have been due to a British reaction to a sore subject: Columbia as an independent North America. Even if this is the case, this printed map is the first to use "Colombia" as a name for South America, not North America. "Colombia Prima" may be linked to Miranda beyond the mere fact of its title, however. Although Miranda himself is nowhere mentioned on the map, a close friend of Miranda's is: "the late eminent and learned Geographer, Louis Stanislas Darcy de la Rochette" (1731–1802) (fig. 3.3). De la Rochette started working for Faden in 1780 and quickly become Faden's most respected geographer. When in June of 1791 Miranda ordered several maps from Faden to complement his own collection, De la Rochette was the one responsible for presenting Miranda with Faden's catalogs.[29]

Contact between Miranda and De la Rochette over map selection deepened into a working friendship committed to Spanish American independence. By 1798, Miranda had asked De la Rochette to help him with a very sensitive endeavor: editing the writings of Juan Pablo Viscardo y Guzmán.[30] This Peruvian Jesuit priest had also promoted Spanish American independence during his exile in Great Britain during the late eighteenth century.[31] As he lay on his deathbed during the winter of 1797–98, Viscardo y Guzmán gave his close friend the United States ambassador, Rufus King, his books and various manuscripts. King, aware of Miranda's interests, turned these papers over to the Venezuelan. The most famous tract Miranda translated and circulated from this collection was "A letter to Spanish Americans."[32] Viscardo's papers also included a longer monograph that Miranda and De la Rochette jointly revised, edited, and translated. Although the edited version of book was never published in their lifetimes, drafts bearing both their handwritten commentary testifies to the friendship between the two men and their commitment to independence.[33] The intense working relationship Miranda and De la Rochette shared in order to further the cause of Spanish American independence helps explain why the 1807 engraving of South America bears the title "Colombia Prima." Faden honored the memory of De la Rochette, recently deceased in 1802, by engraving the political commitments of his esteemed geographer onto this map's face—commitments Miranda had helped inspire.

When one goes beyond the title of "Colombia Prima," and the dedication to De la Rochette, however, little else suggests significant contributions by Miranda to Faden's cartography. In fact, it is possible to see how this map actually runs counter to the territorial interests of an independent Spanish South America. Taking account of the other informants that contributed to the making of this map helps explain why "Colombia Prima" cuts short the extent of Miranda's grand vision. Much like other maps, "Colombia Prima" has a cartouche that reveals its composite nature, identifying the experts that helped shape its contours (fig.3.3). Beyond honoring a deceased De la Rochette, the cartouche also announces its use of the original manuscript maps of "his Excellency the late Chevalier Pinto, Likewise from those of Joao Joaquim da Rocha, Joao da Costa Ferreira, El Padre Francisco Manuel Sobreviela." Sobreviela, a Franciscan friar who participated in the relocation of Indian settlements near the Amazonian region of Peru, and had drawn maps of the region, was the only contributor who was not Portuguese.[34] The rest of the contributors were primarily Portuguese officials with expertise in the Amazonian region. Chevalier Luis Pinto was minister from Portugal to the court of Great Britain.[35] Joao Joaquim da Rocha, originally from Minas Gerais, was a young adviser to the Portuguese emperor.[36] Joao da Costa Ferreira was a Portuguese military engineer charged with demarcating the boundary between the Portuguese and Spanish Empires during the 1780s.[37] Why these informants?

The 1807 date of the first publication of "Colombia Prima" provides a partial explanation. During most of that year, Miranda had been busy extricating himself from a losing campaign in Coro, Venezuela, making it difficult for the criollo to have any input into the map itself. More broadly, however, 1807 also marked the beginning of a tumultuous time for the Portuguese and Spanish Crowns. Napoleon's invasion of the Iberian Peninsula threatened. The British, already closely allied with Portugal, aided in the evacuation of its court from Lisbon, shepherding thousands of people, their belongings, and archives across the Atlantic to Rio de Janeiro, maps included. Therefore, although the map may announce that De la Rochette, Miranda's close friend, was the chief geographer, his death in 1802 suggests other interests may have been placed in the cartographic foreground. Spanish-Portuguese claims on the Amazonian border, clearly, were of primary concern given the expertise and origins that Faden's "Colombia Prima" highlighted. Not surprisingly, the Portuguese Crown, which opened Brazil's ports to a kind of "free trade" that inevitably placed Britain, its naval shepherd and protector, at an unparalleled advantage, came out the clear territorial winner on this map. But other imperial powers competing with Spain for territory in South America benefited as well.

To best gauge the new claims that Faden's 1807 "Colombia Prima" made on supposedly Spanish territories, consider comparing it to Faden's 1799 replica of Cruz Cano's "America Meridional" (fig. 3.4 vs. fig. 3.1). Despite the fact that Faden produced both images, they differ significantly with respect to boundaries. While the Portuguese Crown's (and, by extension, Great Britain's) ample claims on Amazonia expand notably, Dutch Guyana deepens its reach south, well into what Cruz Cano had identified as French Guyana. French Guyana, in turn, grows inland to the west, claiming territory that Cruz Cano had deemed Spanish. Given British interests in Dutch Guyana by the turn of the nineteenth century, this transformation of the Guyanas benefited Great Britain in particular. In short, "Colombia Prima" shrinks the claims of Spain as it proposes a new kind of geopolitical distribution of South America among European empires in the frontier region between Spanish and Portuguese America, giving Britain the greatest advantage.

Such stark new claims required assertions of accuracy. The new boundaries in "Colombia Prima" are buoyed by the erasure of territorial history, at least from the map face. Guyana's contested territorial history was the object of description and explanation in Cruz Cano's original map in 1775. The exact replica by Faden in 1799 is faithful to that history and reprints it in its entirety (fig. 3.5). Dutch Suriname and French Guyana, according to Cruz Cano, bordered each other in the "Territory named Guayana discovered in 1499 by Alfonso de Ojeda, Americo Vespucio, and Juan de la Cosa, but because the land was abandoned by the Spanish, the French established themselves there in 1635. The Dutch arrived in 1663. Four years later, in 1667, the English took over, and in 1676, the Dutch returned to establish their presence."[38] "Colombia Prima," printed in 1807, less than ten years after Faden's 1799 replica, leaves an open blank space where Cruz Cano's dense text had been (fig. 3.6). The erasure of this contested history, together with the use of color and boldface typography

FIGURE 3.4. Juan de la Cruz Cano y Olmedilla, "Mapa Geográfico de América Meridional" (London: William Faden, 1799). 185 × 130 cm. Scale: 1:4,300,000. David Rumsey Map Collection.

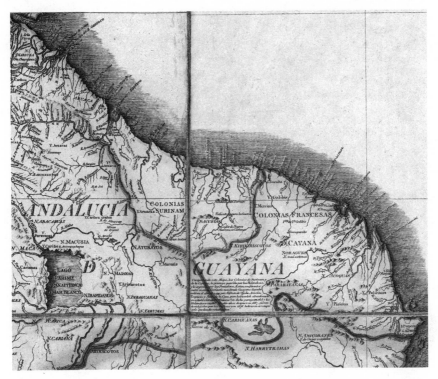

FIGURE 3.5. Cruz Cano y Olmedilla, "Mapa Geográfico de América Meridional" (Guayana detail). David Rumsey Map Collection.

to solidly demarcate "Dutch Guyana," elides how European imperial powers swapped this territory back and forth, not only during the early colonial period, but also from the late eighteenth into the early nineteenth century.[39] Such erasures helped solidify boundaries in ways that helped enhance the British Crown's potential claims on the region.

Weaving together the visible and less visible social webs that contributed to the production of "Colombia Prima"—and the ways in which this map attempted to reshape imperial boundaries of South America—points to the transatlantic alliances, hostilities, and territorial desires that led to the creation of this map in London in 1807. Miranda's own interest in procuring maps of the Americas starting in 1791 led to a friendship with Faden's revered geographer, De la Rochette. By 1798, United States ambassador Rufus King, friend to Miranda, had provided the funds, the incendiary proindependence manuscript texts, and the map-printing projects that Miranda (and by implication De la Rochette) could work on. Although the United States' disappointingly signed a treaty with France in 1800, foiling Miranda's plans for a three-pronged inva-

FIGURE 3.6. Faden and De la Rochette, "Colombia Prima," 1807 (detail). Note the erasure of text on French Guyana. Also note "Punta Barrima, or Cape Breme of the Dutch" in the upper left-hand corner of the image. David Rumsey Map Collection.

sion of South America, he was not dissuaded.[40] It seems Miranda, after so many years in London, had bought into a long-standing belief about the Spanish monarchy: that its holdings in the Americas were only weakly defended, and with just a few well-calculated attacks, the entire empire would come down like a house of cards. Miranda also believed that an independent "Colombia," poised between the North and South Seas, had extraordinary commercial potential, with or without the United States. He therefore employed his cartographic arsenal to plan an attack on Spanish South America with the aid of the British.[41] Miranda and his British counterparts in Buenos Aires were proven wrong. Miranda returned to London late in 1807, defeated.

A major problem Miranda faced while in Venezuela in 1806 was convincing Spanish Americans about the need for independence from Spain in the first place. Not until the Spanish Kings abdicated the throne to Napoleon at Bayonne in 1808 did stirrings start to ripple out. The resulting crisis in legitimacy prompted local juntas throughout the Spanish Empire to claim that sovereignty reverted to the people, not the French.[42] Disintegration came fast: starting in 1809 juntas in Spanish America, styled on those of the peninsula, declared their loyalty to Fernando VII, yet many sought autonomy from the more traditionally dominant neighboring urban centers like Santafé, Lima, and

Caracas. In the meantime, Miranda's Grafton Street home drew independence sympathizers from all over, and twenty-seven-year-old Simón Bolívar was one of them. The young Bolívar was awed by Miranda's impressive connections to high-ranking British and American officials. Realizing their utility, Bolívar invited Miranda to return to Caracas, disobeying direct orders from the city's junta; loyalty to Fernando VII, not independence as proposed by Miranda, was the junta's priority.[43]

In December 1810 Miranda landed in Caracas. Faden's "Colombia Prima" was folded neatly among his belongings. Miranda immediately set to work on a looming problem: the lack of unity and consensus around independence. On the ground, Miranda's imagined "Colombia" was by no means a united polity. The composite of dozens of city-states of the late eighteenth century had sprouted dozens more. Although loyalty to Fernando VII remained strong, developments in Spain began to make Miranda's push for independence easier. Early in 1810, as the Junta Central beat a hasty retreat to Cádiz, it dissolved and established the Regency Council, a move that called into question Fernando VII's ability (or desire) to return to the throne.[44] Juan Germán de Roscio, during the Caracas junta's debates, stated in no uncertain terms: "Upon seeing the act of installation by the Regency Council, we can see that if they remembered América, it was only to continue to make empty promises while declaring solemnly its slavery, and offer it a theory of liberty that would disappear given the calculus under which it subjected American representation in Practice."[45] Caracas became one of the first Spanish American juntas to declare independence outright, on July 5, 1811. But Caracas understood it could not go it alone. Neighboring juntas needed convincing that independence was possible, unity essential, and both geopolitically advantageous. As we will see, the ceremonious circulation of De la Rochette's map was intended to produce just that.

"COLOMBIA PRIMA" CIRCULATES IN SOUTH AMERICA

This aspect of the social history of "Colombia Prima" involves the Chilean canon Doctor José Joaquin Cortés de Madariaga, who had met Miranda during a London sojourn in 1803.[46] By early 1811, Miranda had appointed Madariaga official diplomat for the Caracas junta and sent him to negotiate a federated union with the Santafé (Bogotá) junta.[47] The mission: officially unite these two "Pueblos Colombianos" through an official treaty.[48] Once signed, the treaty created La Confederación de la Tierra Firme. Although several surrounding towns had yet to submit to the authority of either Caracas or Santafé, the juntas of both these cities considered that they, much like the peninsular juntas, had the right to represent the pueblos.

Canon Madariaga ceremoniously presented gifts to the Santafé junta, which included "eight extremely important and exact maps of our continent, South America, or Colombia Prima, ordered by the late eminent and wise geographer Luis Stanislao D'Arcy de la Rochette." The Santafé junta, in acknowledgment, praised both Francisco Miranda and Cortés de Madariaga, stating, "As long as America has Mirandas and Corteses, the civil liberty of its sons will not fall prey to ambitions like those of the French pueblos."[49] "Colombia Prima," for all its British imperial designs on South America, nevertheless remained valuable to Miranda and those interested in Spanish American independence. That is because it illustrated the imagined community Miranda had in mind: a continental federal polity, modeled on the kind of "Columbian" nation he had found in the United States, but, once Latinized, something new altogether. "Prima," meaning "the first" in Latin and "the force that triumphs" in Spanish, became a fitting title for this "Colombian" map of independence. This utopian community was not to be, however.

Despite the circulation of insurgent diplomats and proindependence maps, the *confederación*, based on an imagined "Colombian Pueblo" never materialized. Caraqueño authorities insisted that until the Santafereños also declared their absolute independence from Spain, the *confederación* itself was impossible. Santafereños not only maintained their loyalty to Fernando VII; they also were embroiled in struggles over regional power with towns that refused to accept Santafé's authority to lead in the absence of the monarch. By July of 1812, after a devastating earthquake in Caracas and a key royalist military victory at Puerto Cabello in Venezuela, Miranda understood that the patriot cause was lost altogether. He signed an armistice with Spanish captain Monteverde, bringing an end to the First Republic.

PART 2. REPÚBLICA DE COLOMBIA

REFORMING DE LA ROCHETTE AT ALL POINTS NECESSARY

The fall of the First Republic suggests the circulation of "Colombia Prima" did little to convince the many fragmented juntas of the value of a unified, continent-wide imagined community. It was the return of Fernando VII in 1814, and the escalating violence he brought to Spanish America, that proved the need for unity against a common enemy. This was especially true for the region hardest hit by the Reconquista armies: the Captaincy General of Venezuela and the New Granada Viceroyalty. The Gran Colombian Republic that emerged in 1819 was, ultimately, held together by its military's ability to fend off the royalist threat. By 1826, the military victories of Spanish American armies, together with the Del Riego Rebellion of 1820, had made it impossible for royalist forces to pose any significant threat to South America.[50]

Much to the chagrin of Spanish American elites, the powers of the earth outside the Spanish Empire nevertheless understood "independence" as more of a "civil war" among Spaniards, and the men fighting for recognition of independent republics in the New World were commonly referred to as insurgents.[51] Despite the international community's nod toward respecting the nominally internal matters of the Spanish Empire, the "Western question" nevertheless generated profound interest on the international stage.[52] Doubt over the outcome of the Spanish American crisis after Napoleon's downfall in 1815 met existing and calculated competition among Atlantic powers to dominate trade with the region, with Britain attaining the most threatening position. Diplomacy and prorepublican propaganda proved essential to the independence project, given the context of France's Bourbon Restoration, the rise of the Holy Alliance, and attendant European resistance to revolutionary movements.[53]

This second section traces how a group of elite proindependence Spanish American creoles, namely, Colombia's Liberator Party, and their supporters in Paris, turned to printed cartography in ways that went far beyond crafting military invasion strategies like those concocted by Miranda. This second-generation Colombian cartographic independence effort was intended to shape transatlantic geographic imaginings, but it did so in ways that would shift the newly emerging geopolitical order in their favor, allowing them to gain recognition for independence, and, in the process, as much territory for Colombia as they could. The Liberator Party's vision had some staying power. Printed maps of a Colombia that brought Venezuela together with Quito and New Granada began emerging in 1822 and continued circulating well into 1840 throughout the United States and Europe even after Colombia had dissolved by the early 1830s.[54] But a "Colombia Prima" that encompassed the entirety of the South American continent had little to do with these cartographic visions. As it turns out, "Colombia Prima" actually threatened the Colombian Republic's interests. As Restrepo observed in his 1820 letter to General Francisco de Paula Santander:

Although I write on the history of New Granada, this is still just a mere essay that I need to incorporate into a much vaster plan. The map of La Rochette I think needs many reforms. Therefore, please do not omit from collecting as many maps as you are capable of gathering from the different Provinces. At the moment that there may be some rest, perhaps two or three young engineers may work on the map. That way, when we publish the history, we may publish a better map of Colombia, reforming that of La Rochette at all the points necessary.[55]

Restrepo had held "Colombia Prima" in his hands. Although no original copy of the map currently exists in the archives or libraries of Bogotá, an original 1807 "Colombia Prima" map does form part of Restrepo's private archive, suggesting the lengths to which Restrepo may have tried to limit circulation of "Colombia Prima," at least within Colombia.[56] But beyond reforming (and silencing) "Colombia Prima," the quote above also suggests a larger point: Restrepo conceptualized his history of Colombian independence cartographically. The cartographic dimensions of Restrepo's *History of the Revolution,* then, are the focus of this section.

Historians have recently demonstrated the significance of Restrepo's *History of the Revolution* for early republican national politics and for transnational efforts to promote Spanish American independence more generally. Sergio Mejía has shown how Restrepo's *History* became the historical voice of the Liberator Party, one that sanctioned the legitimacy of Bolívar's centralist regime and the political and institutional decisions it made while condemning federalist opposition.[57] Daniel Gutierrez Ardila has argued that Restrepo's historical work should also be understood within a larger Spanish American propaganda effort to obtain and secure the recognition of independence in Europe.[58] But the text of Restrepo's narrative did not slip so easily into the propagandistic needs of diplomats in France, in part because Restrepo's principal, driving purpose was to construct a solid monument to the independence cause destined for posterity.[59] The history that Restrepo wrote in Bogotá suffered little if any identifiable alteration to the original format, argument, or specific historical moments when it went to press in Paris in 1827. As this chapter shows, however, Bolívar's supporters in Paris had a freer hand tweaking the history Restrepo told in his atlas, fomenting a proindependence propaganda effort through cartography.

As suggested by the discussion of "Colombia Prima" above, the making of printed maps required the labor of several geographers, cartographers, printers, engravers, and editors. Who was specifically involved in the making and printing of maps therefore tends to be occluded from the historical record. We do know that Francisco María Restrepo, Restrepo's brother, went to Paris in 1825 on a government commission for the Colombian mint.[60] We also know that when in Paris, Francisco María handed his brother's manuscript to Andrés Bello so that the Venezuelan philologist could write a review of the work.[61] Three Colombian merchants, the brothers Carlos and Rafael Álvares del Pino and Miguel Saturnino Uribe accompanied Francisco María was on his voyage to Paris.[62] Although these men most likely had the manuscript map along with the narrative history printed, there may have been several other people involved in this endeavor. As Gutierrez Ardila has shown, the publishing house that printed Restrepo's oeuvre was located in the "Colombian neighborhood" of Faubourg-Poissonnière, where several prorepublican and proindependence diplomats, merchants, and exiles with ties to the Colombian Republic lived.[63] Carefully mapping out the differences between the two-sheet manuscript 1825 map drawn in Bogotá, titled "Carta Corografica de la República de Colombia con sus divisiones politicas de departamentos y provincias" (hereafter, *Carta*), and the 1827 Parisian-printed "Historia de la Revolución de Colombia, Atlas" (hereafter, *Atlas*) reveals how those who intervened in the making of the printed version of the map in Paris were decidedly supportive of a Bolívar-centric narrative of independence.

The resulting differences between the two versions of the map are not radical. Both defended Colombian territorial claims against British incursions, and both identified battlegrounds in ways that worked to legitimate Bolívar's command over other insurgent generals that threatened his centrality. Nevertheless, subtle differences in format and cartographic details reveal the ways the Liberator Party in Paris altered Restrepo's *Carta* for the purposes of a propaganda campaign that desperately needed historical and scientific legitimacy. While the map face of the *Carta* only offers a summary of the cartographic sources it drew on, the printed *Atlas* provides three detailed pages that established the map's scientific pedigree. The *Atlas* is also bolder when it comes to territorial claims: while the *Carta* ambiguously uses insets to convey boundaries, the *Atlas* provides a contiguous presentation of the Colombian territory, running solid boundaries from Central America, across to Brazil, the Guayanas, and back to Peru. Finally, a close comparison of the two maps in terms of the number of battles identified (and erased), and which towns and cities were highlighted (and forgotten), reveals how Colombian diplomats in Paris downplayed the most difficult, two-pronged challenge to Bolívar's central authority: the Federal Congress at Cariaco on the Caribbean coast of Venezuela and insurgent general Manuel Piar's victories along the Orinoco River basin in

1817. The subtle interventions on the *Atlas* emerged precisely at the time when news about these events had started circulating throughout the Atlantic. The neutral, transparent language of cartography would provide tangible evidence before the court of international opinion regarding Bolívar's impeccable record in leading the moral Colombian independence cause.

DIPLOMATIC DEMAND FOR A MAP OF COLOMBIA IN BRITAIN

In 1823, José Rafael Revenga, Colombia's diplomat to Great Britain, reported to José Manuel Restrepo in no uncertain terms that Colombia's territory needed to be defended. The threat was not Spain, however; it was Britain. "Despite the Treaty of Munster and of that passed the 23 of June 1791, these [British] geographers have extended the limits of Demerary [Dutch Guyana] to Barima Point at the mouths of the Orinoco River, and have marked as British a territory extending the Mosquito Coast from the Cape of Honduras to [a place] between the Rivers Baliza and Hondo."[64] Restrepo, who had held "Colombia Prima" in his hands, understood perfectly the threatening British geographic visions Revenga described.[65] As Revenga was right to worry, and as Restrepo was well aware, with time the kind of cartographic claims Britain made in maps like "Colombia Prima" could be cited as title of property.

> Since it is convenient to cut the problem off at its root, nothing would be more effective than the formation of a map of Colombia that is as exact as it should be. Please permit me to propose that the government consider sending here or to the United States all the materials that would contribute to that effort. The business would not be expensive at all, and would be of the greatest utility as far as the maps more accurately convey the true topography of the country.[66]

Revenga, in short, demanded a clear cartographic display that defended the boundaries of the emerging republic against British imperial designs:

From 1823 to 1825, José Manuel Restrepo proved an excellent choice to oversee the project of mapping out Colombia's independence history and defending its territory. During his youth Restrepo developed close ties with the martyred criollo ilustrado, or enlightened creole patriot, Francisco José de Caldas. Restrepo drew several maps and plans of Antioquia, his natal province, and consulted their accuracy with Caldas. The kinds of scientific relations that Restrepo cultivated through his geographic and cartographic work eventually

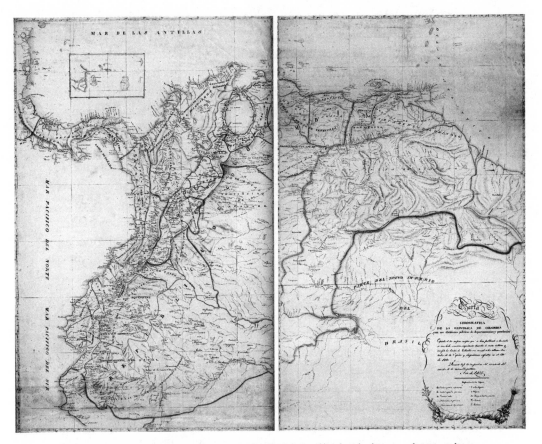

FIGURE 3.7. Jose Manuel Restrepo, *Carta Corografica de la República de Colombia con sus divisiones politicas de departamentos y provincias* (Bogotá, 1825). Manuscript map. 29 × 40 cm. Archivo Histórico Restrepo (AHR), Fondo XII.2, vol. 17. F. 11A–11B. Available at http://www.bibliotecanacional.gov.co/ultimo2 /tools/marco.php?idcategoria=45203.

translated into a position of leadership with the republican regime as secretary of the interior and exterior relations.[67] Restrepo's personal and economic security soon came to depend on the revolution's success. Restrepo wrote his history and oversaw the drawing up of the manuscript form of the "Carta Corográfica" from 1821 to 1825 in Bogotá. This coincided with the critical period when Bolívar reached the peak of his military and political powers. Restrepo believed the best way to "satisfy the curiosity of the enlightened men of Europe" was by introducing them to the grandeur of the Colombian territory, its limits, climate, and resources.[68] The result was impressive. By October 11, 1825, Restrepo's cabinet of cartographers had finished the manuscript map (fig. 3.7). The bounded topographic image makes careful cartographic arguments about Colombia's boundaries, which reflected Restrepo's own diplomatic delineation of what Colombia could rightfully claim.

Restrepo's introduction to his *History* immediately set the territorial record straight. The northwestern boundary was marked off at "Cape Nasau, or rather from the Esequebo River, the former limit of Dutch Guayana."[69] Restrepo's *Carta* erased any trace of Dutch possession at Barima Point, as had been suggested by "Colombia Prima" and other British maps. Colored lines clearly distinguish the border between Colombia and the former Dutch Guyana, which by then was under de facto British control (fig. 3.8). Restrepo, his cabinet of cartographers in Bogotá, and the Liberator Party who printed Restrepo's map in Paris all agreed on what the westernmost boundary of Colombia needed to be: Barima Point appears on the printed version of the *Atlas* much as it does in the manuscript *Carta* (fig. 3.9). Colombia's boundary with Central America was a different story, however.

How far the boundary of Colombia went up into Central America was, at least in 1825, still unclear. Restrepo's text identified the northwestern limit at "Cape Gracias-a-Dios in the province of Honduras, fifteen degrees north, and including the islands of Margarita, San Andres, Vieja-providencia, and other smaller ones." The problem was that "from Cape Gracias-a-Dios the interior limits have not been yet fixed with accuracy, and there is a need for an agreement with the government of Guatemala; but the line that divides falls on the Pacific near the lake of Nicaragua on the Gulf Dulce."[70] These vague limits were difficult to place on a map, at least for Restrepo. The manuscript version of Restrepo's map uses pink watercolor to highlight a dotted boundary line beginning at "Gfo. Dulce" that runs across the isthmus just to the east of Bocas del Toro, not far from the current national boundary between Panama and Costa Rica. Restrepo identifies "Costa Rica" on the western side of the pink border. The map then employs an inset to bring Cape Gracias a Dios, which is much further to the northwest on the isthmus, into the viewer's field of vision in order to stake a claim on it for Colombia (fig. 3.10). Restrepo's cartographic diplomacy is impressive in its subtlety. His manuscript map used the inset convention to both claim and disclaim Colombian rights over disputed territories in Central America.

Those involved in the *Atlas*'s printing in Europe made much more ambitious claims for Colombia in Central America (fig. 3.11). The *Atlas* does not use insets. Instead, the *Atlas* allows for continuity along the isthmus well into Central America to Cape Gracias a Dios. To underscore these extensive claims, the map of the department of the isthmus is the first map the viewer sees upon opening the *Atlas*. All contiguous lands up to Cape Gracias a Dios are claimed

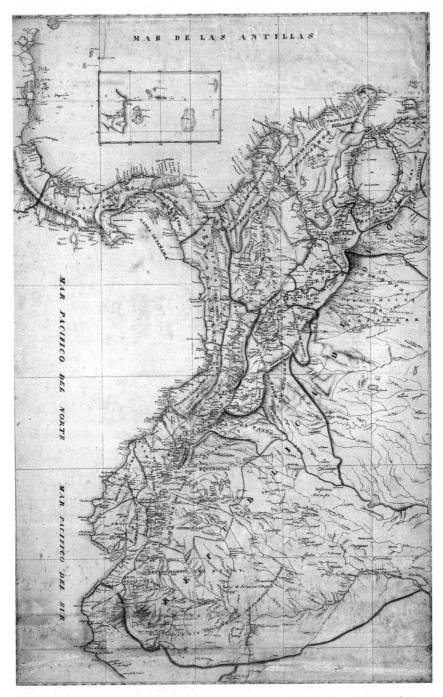

FIGURE 3.8. Restrepo, *Carta Corografica*, 1825 (detail of boundary with Dutch Guyana). Note that Barima is depicted next to the mouth of the Orinoco River. Note the three battles at Maturín, yet none along the Orinoco River. Note the presence of Cariaco city, and the same size lettering for the towns along the Caroní River basin.

FIGURE 3.9. Restrepo, "Atlas," 1827 (detail of Guayana boundary from figure 3.2). David Rumsey Map Collection.

for Colombia's Department of Panama. No trace of Costa Rica is evident. The Liberator Party in Paris did not hedge its territorial bets with Guatemala, as Restrepo had done in Bogotá. They made the grandest territorial claims the printing press allowed them.

To stake these stark, bold claims against British incursions, the Colombia *Atlas* needed to establish its scientific legitimacy. One possible problem for the French-based printers could have been the fact that Restrepo's manuscript map cartouche was short on explanations; it simply states: "Copied from the best maps that have been published, but these have been corrected in important ways by drawing on unpublished maps. The limits of Colombia have been corrected according to the latest treaties made by Spain and their valid dispositions up through 1810." We do not learn what those published "best maps"

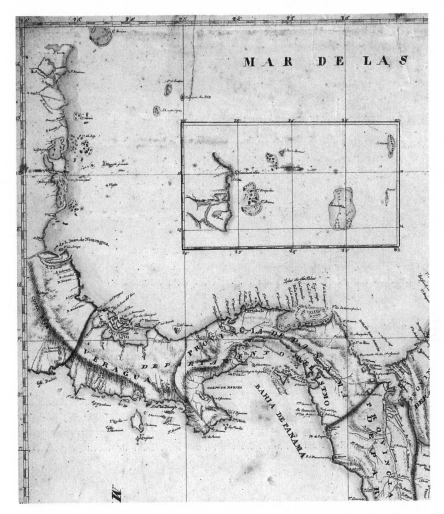

FIGURE 3.10. Detail from Restrepo, "Carta," 1825, on isthmus boundary. Available at http://www
.bibliotecanacional.gov.co/ultimo2/tools/marco.php?idcategoria=45203.

are, or what manuscript information the mapmakers drew on. The printers
of the *Atlas* in Paris had much more information on hand about the making
of the map than that available on the cartouche. The three-page introduction
explains how famed mathematician José Lanz drew the cartographic base. Due
to Lanz's absence from Colombia during the final stages of drafting the map,
"various intelligent people" continued the project, which Restrepo oversaw.
The list of expert maps and measurements included some of the usual suspects, such as Alexander von Humboldt and Francisco José de Caldas. Vicente
Talledo's work on the eastern mountain range and the provinces of Cartagena,

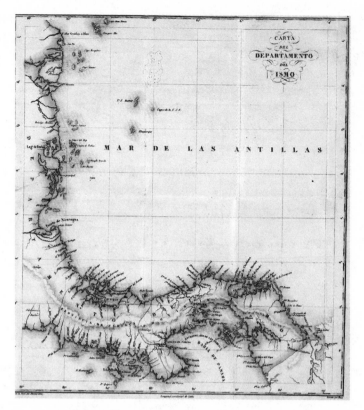

FIGURE 3.11. Jose Manuel Restrepo, "Carta del Departamento del Ismo," in "Historia de la revolucion de la Republica de Colombia, Atlas" (Paris: Libreria Americana, Calle del Temple, no. 69, 1827)
(verso of title page). Imprenta de David, Calle del arrabal Poissonniere, no. 6 en Paris. 35 × 32 cm. Scale
1:2,750,000. David Rumsey Map Collection.

Santa Marta, and Rio Hacha joined that of José Manuel Restrepo's own maps
and measurements of Antioquia and Cauca. Gabriel Ambrosio de la Roche and
Rafael Arboleda's maps of Chocó and Popayán, and Pedro Maldonado's "very
exact" maps of Quito, rounded out the southwestern portion of Colombia.
The detailed list was long and impressive.[71] Not surprisingly, given Restrepo's
personal missive to Santander, and Revenga's alarm at printed British maps,
De la Rochette, "Colombia Prima," Francisco Miranda, and William Faden
appear nowhere in this introduction.

With the *Atlas*'s scientific pedigree established, the members of the Liberator Party in Paris could go beyond defending territorial claims for Colombia
against the British and turn to cartographic narration of historical events. After
all, the *Atlas* illustrated Restrepo's *History*. The *Atlas*'s introduction stated in no
uncertain terms what made it most valuable for international audiences: "The

very important improvement that is marked on the map is [that it indicates] the places where the major battles of the war of independence of Colombia were carried out . . . [making it] . . . preferable to any other published until the present day."[72] Both the *Carta* and the *Atlas* signaled where the major revolutionary battles were won with upraised swords; blades pointing down indicated battles lost; and blades to the side meant a draw. Not surprisingly, both manuscript and print editions pointed toward a clear military insurgent victory over the royalists. As the rest of this chapter demonstrates, however, the differences between the two maps illustrate how the men who had the *Carta* printed in Paris were desperate to confirm a Bolívar-centered historical narrative about independence, and the legitimacy of Bolívar as a leader, by erasing inconvenient truths from Colombia's territory.

ERASING INCONVENIENT HISTORICAL TRUTHS FROM THE MAP

By the mid-1820s Colombian diplomats had made independence synonymous with the moderating elite-led influence of Bolívar. They did so to distance it from the excesses of the French and Haitian revolutions. They also believed Bolívar's wealth, his family's elite status, his personal connections to Europe, and his already existing prestige abroad would be indispensible for a cause that in many ways was still little known in Europe, especially as compared to Greece's contemporary fight for independence from the Ottoman Empire.[73] Making Bolívar synonymous with independence would convey to international audiences that their movement was one of just moderation, distant from the excesses of Jacobinism. Several Colombian diplomats, including Leandro Palacios in New York, wishing to project an elite version of independence, were irked by images that made Bolívar's face look "summarily disfigured, as if it were that of a mulatto."[74] These pro-Bolívar diplomats policed newspapers for stories that worked against Bolívar's image. The problem was that by the mid-1820s, the controversial decisions Bolívar had made in the fields of battle and of politics threatened this dimension of international diplomacy.

If insurgent propaganda tried to make Bolívar the honorable authority at the center of the revolution, then any attempt to dissuade international audiences from believing in Bolívar's honorability and legitimacy was potentially deadly for the Colombian cause. H. L. Ducoudray-Holstein's "secret history" about the Liberator became one disturbing threat.[75] Ducoudray-Holstein was a French general who had served under Bolívar but whose damning, ironic narrative claimed that "Bolívar was not beloved, and his vanity, pride and coldness, rendered him unpopular."[76] Ducoudray-Holstein held up a plethora of examples of Bolívar's imperiousness, lack of belief in democratic institutions, and marked tendency to claim the victories of others as his own: "To any experienced military man, the following reflections will give a convincing proof of Bolívar's weakness and small capacity as a commander-in-chief. Instead of employing every means in his power to compel Mariño and Piar to do their duty, he approved, in an official manner, their defection, which naturally encouraged them to act in an isolated and independent way."[77] In this author's rendition, Bolívar's military operations near Caracas became a "childish predilection . . . extremely injurious to the cause of independence in Venezuela."[78] Manuel Piar, supported by Admiral Louis Brion's naval blockade, decided against continuing campaigns along the Caribbean coast, as Bolívar insisted, and instead coordinated an attack on Guyana, where royalist forces were spread thin. Delivering a striking victory at San Felix, Piar opened St. Tomas de Angostura to insurgent control. "This brilliant and eventful conquest was effected without the knowledge or the order of general Bolívar. It was owing entirely to the courage and exertions of two foreigners, Brion and Piar. It resulted in vast advantages to the republic. And what was their recompense? The former died poor and broken hearted in Curiaco [sic]; the latter was shot by order of the supreme chief."[79]

The fickle Bolívar of Ducoudray-Holstein also lacked prowess in politics. The author detailed how patriot chieftains disliked Bolívar's assumption of supreme power while neglecting to call for a congressional assembly. That is why in May 1817 these men called for a congress at Cariaco in Bolívar's absence, one that was well attended. "Admiral Louis Brion, the intendant [Francisco Antonio] Zea, Jose Cortes Madariaga, better known under the name of the Canonicus of Chili, addressed the assembly, showing the necessity and urgency of establishing a Congress."[80] The congress at Cariaco was "received with unanimous approbation" and represented the reinstallment of democracy. "As soon, however, as [Bolivar] learned of what was done, he fell into a violent passion, and not only annulled the proceedings, but persecuted the members appointed, especially the Canonicus of Chili, against whom his hatred seemed more particularly directed."[81] The canon, Dr. Cortes de Madariaga, an old friend of Miranda's, along with several recognized generals and diplomats had set up this congress in Cariaco, much to Bolívar's dismay. Ducodoray-Holstein's narrative threatened Colombian diplomatic efforts to place Bolívar at the center of a successful Colombian struggle for independence.

If Ducoudray-Holstein's story were to be believed, Colombia's position would grow even more precarious in places like Bourbon Restoration France.

Already in 1822, the Venezuelan native Tiburcio Echevarría had written to Bolívar from Paris concerned that France lacked knowledge of Colombia's territory, population, or the "immortal actions" of the Bolívar-led patriot armies. He urgently inquired about the status "of the history that Mr. Restrepo was writing, or of any other that is ready, since it would be of utmost importance to have it in hand, and immediately I would have it printed."[82] France was one place where diplomats had hung their hopes for French recognition of independence on a Bolívar-centric ideal. By the late 1820s, in a bid of desperation, these same diplomats were flirting with the idea of installing Bolívar as monarch and marrying him off to a French princess, if only this would mean recognition and, perhaps, a greater measure of social stability on the ground.[83]

By the late 1820s, Ducoudray-Holstein's narrative was difficult to dispute in light of significant documentation that gave credibility to his tale. Some tried discrediting Ducoudray-Holstein's authority by questioning his Napoleonic military credentials.[84] Proindependence diplomats and allies abroad ultimately needed to alter the historical record in ways that diminished the significance of the events that Ducoudray-Holstein had narrated. Several books, pamphlets, and histories papered over the messier aspects of independence history and Bolívar's role in that mess. Restrepo's *History*—and its *Atlas*—became a valuable weapon in the Liberator Party's arsenal. The Bolívar-led revolution would finally be legitimized through an objective historical narrative. The primary sources, and, of course, the *Atlas* indicating where battles were fought, won, lost, and tied, were all intended to make this narrative all the more transparent and believable.

FORGETTING THE CARIACO FEDERAL CONGRESS

In order to construct Bolívar as the "Creator and Liberator" of Colombia, the Liberator Party needed to marginalize those who threatened Bolívar's centrality. Generalissimo Francisco de Miranda, and his brand of federalism became one prime target. Rather than identifying Miranda as a key founding father of independence, Restrepo's 1827 history only referred to the general in order to narrate Bolívar's quest and achievements. Miranda's "first, most important decision during the revolution," according to Restrepo, "was to confer the command of the important fort of Puerto Cabello on the Colonel Bolívar."[85] But Bolívar lost control of the fort in July of 1812. The two crossed swords pointing downward near Puerto Cabello represent this defeat, honoring the claim that this map displayed key battles in a historically accurate way. Without Puerto Cabello, Miranda considered he had no choice but to surrender to the Spanish. In a rather unexpected turn of events, Bolívar participated in the arrest of Miranda following this surrender, however. Miranda subsequently was imprisoned in Cádiz, where he died in 1816.

Bolívar's role in handing Miranda over to the Spanish caused much historical debate throughout the nineteenth century. Bolívar, for one, made sure to remind his biographers that he would have executed Miranda for treason himself had it been possible.[86] By vehemently forcing this memory of events onto his contemporaries and on posterity, Bolívar figuratively tried and executed Miranda for treason in the public's historical imagination. The imagined execution, to be sure, helped Bolívar deflect his own responsibility, for it was his loss of the strategic fortified port of Puerto Cabello that left Miranda with no choice but to surrender. But it also meant Miranda would be remembered as a traitor to the independence cause, rather than its originator, at least in patriotic, pro-Bolívar histories.

Inconveniently enough for Bolívar, Canon Madariaga, a significant supporter of Miranda during the First Republic, escaped imprisonment in Cádiz and returned to Venezuela by April 1817 with the help of British support. Restrepo played down the 1817 arrival and instead, in 1827, remembered the Chilean's visit as "the first step toward the union of Venezuela and New Granada, but one that did not produce any favorable effect."[87] Restrepo argued that for the two republics to unite politically, they demanded much more than a mere treaty. Venezuelans and Granadinos could only see the value of unity after suffering through long, drawn-out battles against the Spanish. Perhaps the only other way to unite the two pueblos was through "a well-authorized Congress. . . . But the installation of such an assembly was still remote [in 1811]."[88] Restrepo's observation makes a nod to the "well-authorized" congress at Angostura of 1819, whose proceedings he described at length and in detail. But Restrepo's comment that "only a well-authorized" congress could have achieved unity begs the question, was an "unauthorized congress" possible? The federal congress at Cariaco of 1817, which was called by Cortes de Madariaga and presided over by General Santiago Mariño, one discussed by Doucodaray-Holstein and others—but notably *not* by Restrepo's 1827 history—may have been exactly that.

To understand how much was at stake in Cariaco, we need to address the precarious position of patriot troops, officers, and rank and file. A combination of loyalty and circumstance dictated which patriot generals pledged loyalty to the Cariaco congress, which to Bolívar, and when. Those embroiled in drawn-out and losing battles along the Caribbean welcomed the boost that international recognition of Venezuelan independence would bring to the patriot cause. This is precisely what Madariaga had brought with him to Cariaco in 1817. Still, disastrous losses in Venezuela's major coastal cities made several patriot soldiers nervous. Some, led by General Manuel Piar, had already fled to a different field of battle: the Orinoco River basin. In April of 1817, as Mariño and others held the fort at Cariaco, Bolívar left for Angostura, beckoned by Piar's promising victories. Bolívar soon realized he was at a serious disadvantage after arriving in the Orinoco River basin. He had not personally led any troops to victory in the region, and the generals in Cariaco, encouraged by Madariaga's call to form a federal congress that the British could recognize, seemed to be conspiring against Bolívar's leadership.

The generals present at the Cariaco congress understood how important and yet how precarious their legitimacy was, what with a royalist attack looming and Bolívar far away. Most hoped that by invoking Bolívar's tacit approval, and by translating the proceedings of the congress into English for immediate dissemination in London and the United States, Venezuela's independence would be assured.[89] The problem for Bolívar was his second in command, Santiago Mariño. General Mariño had grown weary of Bolívar's exhaustive losses from 1814 to 1816 and had challenged the Liberator's authority several times. Mariño's leadership of the proceedings at Cariaco on May 8, 1817, meant that Bolívar was losing control of the patriot project, if ever he had it.[90] If recognized, the Cariaco congress, under Mariño's control, would access diplomatic channels that Bolívar so anxiously tried to maintain but from his weakened position in the Orinoco had trouble doing so. Bolívar was becoming marginal by the day. The stakes were high. Rivalries flared. Bolívar denounced the Cariaco congress as illegitimate. But it was too late. Already announcements of the Cariaco congress had left for London.

Restrepo's 1858 version of independence history emphasized how a scheming Mariño grabbed onto every word uttered by Cortés de Madariaga, installing a federal congress detrimental to the needs of independence. Restrepo chastised the participants, arguing that "they did not take into account that by [installing a federal government] they were openly contradicting repeated acts consented to by all of the jefes that exercised authority, which had adopted republican unity and the centralization of power."[91] Miranda's brand of federalism reared its head through this Cariaco congress, and Restrepo, writing in 1858, on the eve of a reinstatement of a federalist national state form, warned his readers about the futility and uselessness of such government schemes. "This

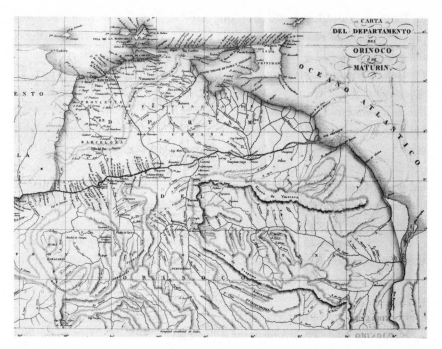

FIGURE 3.12. Jose Manuel Restrepo, detail from "Carta del Departamento del Orinoco o de Maturin. Gravado en Paris por Darmet, 1827. Escrito por Hacq," in "Historia de la revolucion de la Republica de Colombia, Atlas" (Paris: Libreria Americana, Calle del Temple, no. 69, 1827) (verso of title page). Imprenta de David, Calle del arrabal Poissonniere, no. 6 en Paris. 30 × 45 cm. Scale: 1:2,650,000. David Rumsey Map Collection.

farce ultimately bore no fruit, and soon the *congresillo* of Cariaco was forgotten because subsequent military developments and General Morillo attracted all attention."[92] In 1827, however, the "congresillo" of Cariaco had not made it into Restrepo's narrative, not so much because royalist commander Pablo Morillo had dissolved it, but rather because the Cariaco congress proved too destabilizing for Bolívar's ultimate authority. The *Atlas* conveniently added to Restrepo's 1827 silence on the federal congress by forgetting to include Cariaco, the city, on the map of the Department of Maturín. This was despite the fact that Restrepo himself had included this city on the manuscript version of the map (compare figs. 3.8 and 3.12). By refusing to mention the Cariaco congress in the text of Restrepo's 1827 *Historia* and erasing this city from the *Atlas* the Liberator Party conveyed a clear message: Cariaco was tangential to, and insignificant for, Colombia's independence history. These erasures were not innocent, and they were connected to an effort to subdue yet another threat to Bolívar's power: General Manuel Piar.

Modern-day historians not only have adopted Restrepo's derisive term *congresillo* when describing events during the Cariaco federal congress; they have also missed the significant connection between the installation of the congress and Manuel Piar's rebellion against Bolívar in the Orinoco theater of war.[93] Instead, when referring to Piar, historians primarily remember him as the *pardo*, or mixed-race, general who wished to wage a race war to exterminate all whites. Memories of Piar's impressive ability to turn the tide of independence in the Orinoco region are by comparison harder to recall. The tantalizing narrative of Piar's "race war" also elides an alternative explanation for Piar's rebellion against Bolívar. For Piar, the problem with Bolívar was that as soon as the Liberator had arrived in the Orinoco basin, he did everything in his power to undermine Piar's authority so as to bolster his own. One serious problem involved drawing too heavily on supplies from nearby missions to reward patriot troops, threatening the long-term patriot presence in the region. A second, more dramatic action involved a supposed misunderstanding. Bolivar had ordered the removal of several Catalan friars to the Pastora mission on the banks of the Caroni River, but unruly officers, unfamiliar with the area, supposedly believed Bolivar meant the order as a euphemism for sending the friars to their deaths. And yet, this "mistaken" execution nevertheless fulfilled the logic of the *guerra a muerte*, or war to the death, against all Spaniards in ways that satisfied the revenge-hungry, disgruntled war-weary officers and soldiers on the patriot side. After this incident, Piar, who had honored Bolivar's call to end the *guerra a muerte* by protecting the friars, saw his authority in the region start to crumble. Bolívar's solidified. Piar's subsequent (and historiograpically underexamined) decision to flee the Orinoco and pledge loyalty to Mariño's federal congress was met with the full force of Bolívar's wrath. Bolívar, conveniently drawing on Piar's illegitimate birth, accused him of waging a race war that undermined the republican effort and had him executed. The *Carta* and, to an even more emphatic extent, the *Atlas* both sought to elide inconvenient truths about Piar's challenges to Bolívar's moral authority in the contentious Maturín-Orinoco theater of war.

Not all places identified by the *Carta* or the *Atlas* received the same kind of attention. As we saw above, an appearing and disappearing Cariaco, the city where the federal congress occurred, suggested the marginal significance that the map printers wished to give to this city. The *Carta* and the *Atlas* displayed hierarchies of place in other ways. Like other maps, both made manifest the relative significance of towns and cities through typographic letter size. Focusing on how the *Atlas* uses cartographic conventions in the printed "Orinoco or Maturín Department" map as compared to Restrepo's *Carta* illustrates how the more widely disseminated *Atlas* worked to marginalize General Piar's significance as a military leader while bolstering Bolívar's honorable reputation and credit for wise leadership along the Orinoco River basin.

On August 5, 1817, Bolívar made a widely publicized speech that explained how "Maturín buried in its plains three Spanish armies, but Maturín always remained exposed to the same dangers that had threatened her just prior to her triumphs. This is how stupid the jefe that directed her in military operations was."[94] This "jefe" of Maturín was not only stupid in the field of battle; according to Bolívar he also was a rapacious, greedy thief in day-to-day life. "Once he gathered his booty, his valor wanes, and his constancy abandons him. That is what they say at the battlefields of Angostura and San Félix, where his presence was as void as that of a last drum beat."[95] This stupid, incompetent, rapacious, jefe that bungled Maturín three times over, and whose presence created a void along the Orinoco River basin, was none other than General Manuel Piar. With such a sleight of hand, Bolívar turned Piar's three victories at Maturín against him. They reflected the latter's incompetence: there should have been only one Maturín, not three. Bolívar did not stop at issuing this manifesto to defame Piar; his party of followers lent further strength to Bolívar's arguments by displaying them cartographically. The three battles at Maturin were embedded in the landscape Restrepo scientifically drew out in Bogotá, and in the map the Liberator Party printed for national and international audiences in Paris. Similarly, the Orinoco River basin is not only devoid of battles; San Félix itself is nowhere to be found (figs. 3.8, 3.9, and 3.12). Who, then, was this disaster of a general, Manuel Piar?

Historians rightly point out that racial discourse was a critical element in the conflict between Piar and Bolívar. As Alicia Rios has observed, Manuel Piar has been one of the most controversial yet underexamined figures in Venezuelan historiography precisely because of his alleged desire to instigate a race war in 1817 and the speed with which other *pardo* generals allegedly adopted the cause.[96] Calixto Noguera, for instance, had been accused in 1822 of being a seditious enemy of Cartagena's whites for allegedly exalting the memory of *pardo* general Manuel Piar.[97] A decade later, Bolívar lamented the execution of Piar because of the "just clamor with which those of the class of Piar" would complain, especially after the execution of yet another *pardo* general, José Prudencio Padilla, in 1828.[98] Unfortunately, too much emphasis has been given to the supposed "race war" that Bolívar feared that the *pardo* general was conspiring to launch in 1817. As Marixa Lasso has shown, rumors of race war often were rooted in specific political tensions that may have had little to do with race. Spreading such rumors allowed certain public figures to sway public opinion toward one or another political player.[99] Lasso's suggestions allow the Liberator Party's cartography to be examined in light of the military and political tensions that gave rise to Bolívar's accusations of Piar.

Piar, not Bolívar, was the first patriot general to attack royalist positions along the Orinoco River basin in 1817. Realizing the futility of continued attacks on Venezuela's Caribbean cities, Piar rallied proindependence troops from the plains, known as the *llaneros* from Maturín, his regional base of support. Together, they won three consecutive victories, and headed to Angostura (today Ciudad Bolívar). Control over Angostura meant control over the Orinoco River transport system, which went deep into the vast hinterland plains of Guyana Province, drawing out livestock and other natural resources from the interior for trade with merchants in the Caribbean. Piar quickly realized that the Catalan Capuchin missions along the Caroní River, a tributary of the Orinoco, were the most productive in the region. Piar's successful invasion of the missions in February of 1817 secured supply lines for his troops.

Restrepo's *Carta* and the printed *Atlas* refused to register Piar's victories. But despite what the Liberator Party's maps of the Orinoco would have us believe, the taking of the Caroní missions by Piar's troops involved several battles, causing much weariness among his troops. Several men threatened to leave Piar due to unbearable conditions, and in early January a handful deserted in order to rejoin Bolívar along the Caribbean basin. Piar furiously demanded that Bolívar mete out the severest punishment, arguing, "In such quarters clemency is seen as weakness; kindness is mistaken for lack of character and energy."[100] Piar knew the value of ruthlessness on the battlefield. He also recognized the importance of victory to appease his men and keep them loyal. This was why Piar had set his sights on the missions at Caroní. An escaped Capuchin friar later recalled Piar's attack.[101] His report described the missions, their history, and their population. The abundant labor and fertile lands of the twenty-seven towns along the Caroní River provided foodstuffs and supplies to the major port cities of Angostura and Guayana la Vieja. These missions had held out against patriot forces, and the Catalan friars often incited the mission Indians to defend against attacks by patriot armies. Spanish forces at Guayana la Vieja provided the main defense. Piar strategically ordered his troops to blockade that city, and Admiral Brion aided on the naval front. By the middle of February, the plan had succeeded. Piar had the thirty-four Catalan Capuchin friars who ran the mission towns imprisoned and placed under the authority of the Caraqueño vicar of the patriot army, José Félix Blanco.[102] This allowed Piar to lead strikes against royalists in the rest of the region.[103]

Although the Liberator Party refused to locate the Orinoco battles won by Piar, they did locate the mission towns along the Caroní River. Curiously, while Restrepo's manuscript *Carta* identifies individual settlements along the Caroní (fig. 3.8) using similar lettering for all of the mission towns that sprinkled the Caroní River basin, the *Atlas* edited this view. It instead makes subtle, but important typographic changes to the Caroní region, particularly to the small mission town of Pastora. As we will see, it was Pastora's historical significance, not its territorial importance to the region, that garnered such cartographic emphasis.

Just prior to Piar's victories along the Orinoco and Caroní Rivers, both Santiago Mariño and Manuel Piar, along with several other generals, witnessed Bolívar's issuing of a proclamation intended to broaden the appeal of independence. In 1816, after returning to Venezuela from Haiti with renewed supplies and troops, Bolívar promised freedom for all slaves and declared an end to *guerra a muerte*. Bolívar's declaration abolishing slavery was intended to fulfill his obligation to Alexandre Pétion after having obtained soldiers, supplies, and munitions from the Haitian leader. The end to *guerra a muerte* was intended to win moral ground, especially among international audiences that were still unsure which side of the "Western question" they should support. The result on the ground was mixed. Although abolition may have inspired slaves and *pardos* to join the rank and file of the patriot armies, slaveholding whites grew worried. Calling an end to *guerra a muerte*, may have persuaded fence-sitting Spaniards and other Europeans of the righteousness of independence, but it prevented patriot troops from exacting vengeance on the battlefield.

General Piar's decision to take the friars prisoner honored Bolívar's call to end *guerra a muerte*. The problem was that the captured friars had been the same men who had directed the Caroní missions, supplying Spanish troops with the horses, foodstuffs, and supplies that allowed royalists to exact a significant military toll from Piar's troops. Furthermore, the missionaries had themselves fended off patriot attacks. As escaped Catalan friar Nicolás Vich later recalled, "The community had won the indignation of the rebels to such an extreme that they threatened the friars several times with the terrible expression 'From the beards of the Catalan Capuchin missionaries we will make horse harnesses.'" The *llaneros* were less than pleased to have to spare the lives of the captured priests, it seems. Their disdain of the Capuchins was further evidenced by the poor conditions the monks suffered while in the "care" of patriot troops.[104]

Although five escaped, fourteen died in captivity, leaving only fifteen survivors.[105]

Precisely because these events were so controversial at the time, the historical record grows murky after Piar's San Félix victory in April of 1817. From what can be pieced together from various sources, we know that Bolívar left the quagmire in the coastal city of Barcelona to join the victorious General Piar, arriving on May 2, 1817.[106] The resulting partnership was not smooth. As presbyter Blanco later explained, Bolívar "lacked the resonant voice of command and obedience from the troops. . . . Only a few officials recognized him. . . . The Jefe Supremo was isolated."[107] Bolívar's isolation was in part due to Piar's victorious command; Bolívar had yet to prove himself to the troops on the Orinoco field of battle. Piar nevertheless recognized Bolívar, at least initially. He left for Angostura the first week of May to debrief Bolívar about patriot victories along the Orinoco. Blanco also left his post at the missions to join them.[108] In the meantime, Bolívar sent Colonel Jacinto Lara and Captain Juan de Dios Monzón to the mission where the fifteen surviving Capuchin friars were being held.[109] Bolívar's men then ordered the captives executed in a spectacularly gruesome display of *guerra a muerte*. Pro-Bolívar historians have explained the incident as an unfortunate misunderstanding of the Liberator's orders. Bolívar allegedly had ordered his men to send the Capuchin missionaries to another mission called Divina Pastora, or "divine shepardess." Not familiar with any place by that curious name, Bolívar's men understood this as a euphemism for execution.[110]

A small mission town by the name of Pastora did exist. Established in 1737, it was one of the older mission towns, but with its population of 833 Indians in 1816, it constituted only about 4% of the 20,000 who populated the region.[111] This mission was not only small in terms of its population; it was also peripheral to the productive needs of the region.[112] Restrepo's *Carta* reflected the status of the Pastora mission town as equal to all other small missions along the Caroní by using the smallest font (fig. 3.8). The printed version of the *Atlas* told another story; the map reader would come away thinking this town was among the most important in the Orinoco Department (figs. 3.9 and 3.12). The lettering and boldface font is the same used to pinpoint Guyana Vieja, the former capital of the region. By emphasizing Pastora so prominently, Bolívar's (and the Liberator Party's) explanation that the executions were the result of an unfortunate "misunderstanding" contributed cartographic plausibility. Thus, the *Atlas* attempted to shape imagined historical geographies and memories, saving Bolívar's face as leader before audiences that had no direct experiences with the ways war had unleashed a violent, bloody moral economy of its own.

Both Bolívar and Piar understood this logic. But Piar, following orders, did not have the friars executed. Bolívar exploited Piar's "weakness" in the eyes of his troops to his advantage, but did so with plausible deniability in the eyes of the international community. Bolívar never reprimanded Lara or Monzón for their egregious "mistake." On the contrary, Lara was promoted shortly thereafter. Bolívar's promotion of Lara is not an example of an "inexplicable mishandling" of justice by the Liberator, as some historians have argued.[113] The explanation is as simple as it would seem if it had not been conveniently overlooked by a dominant, pro-Bolívar historiography: after the executions of the friars occurred, General Piar's authority among Orinoco patriot troops started to crumble as Bolívar's solidified.

Less than a month after the executions of the friars, the chain of command over the Caroní missions had fallen into complete disarray. Piar wrote frantic letters to Vicar Blanco, begging him to give a clear report on the mission's available resources, and was infuriated by Bolívar's orders that threatened to deplete them.[114] By early June, Piar was so frustrated he petitioned Bolívar to be discharged from military duty so he could assume complete control over the missions from his post in Upatá, a town also highlighted for its significance in the *Atlas* (fig. 3.12).[115] Bolívar eventually assented to Piar's demands, but conveniently, after Piar was discharged, rumors began to circulate that Piar was plotting a race war against the patriot cause. Bolívar did not denounce these rumors; he fueled them.

By buying into Bolívar's vivid language and fearmongering about the potential race war Piar was supposedly conspiring to launch, historians like John Lynch have missed a less exciting but nevertheless important fact.[116] Piar, dismayed at the handling of the Caroní missions and frustrated with Bolívar's clear mishandling of the war effort in the Orinoco basin, left Guyana to join Mariño and Madariaga's federal congress in Cariaco. Piar himself had intimated to Bolívar in June of 1817 that the revolution needed democratic institutions and political authority alongside Bolívar's military authority.[117] Bolívar worried Piar would pledge his allegiance to the federal government, and that Piar would take with him the *llanero* troops loyal to him. Bolívar tried dissuading the powerful general. He wrote to Piar in mid-June, informing him how key generals like Rafael Urdaneta (1788–1845) and Antonio José de Sucre (1795–1830), despite their attendance at Cariaco, had not pledged loyalty to the new "illegitimate government." Without Urdaneta and Sucre's soldiers, Mariño would "have nothing left beyond his personal guard."[118] Unfortunately, Bolívar's letter was captured by royalist forces and never made it to Piar, whose discharge was effective June 30. If Piar had gathered a significant *llanero* military force and reached Mariño, who had fled to Maturín after Spanish attacks on Cariaco, the republic would indeed have been divided, but not necessarily along racial lines. Mariño and Piar together could have dealt a serious blow to Bolívar's claim to leadership over the revolution. That is because Mariño would have had access not only to Piar's support and his *llanero* troops, but also to Madariaga's diplomatic channels back in London. Bolívar ably maneuvered his way around these threats with the aid of circumstance.

Bolívar needed to completely discredit any and all military and patriotic credibility Piar had gained up until 1817. He needed to do it resoundingly. He also needed to be careful. Bolívar could not risk further alienating either elite white classes suspicious of Bolívar's call to end slavery or the *pardo llanero* troops so necessary to the independence cause. Bolívar also needed to make an example of Piar so that other officers would fall into line. Bolívar's "Manifesto" of August 5, 1817, from Guyana is therefore a shining example of how the Liberator fortified a myth of the republic's racial harmony at a particularly difficult political juncture. He did not denounce General Piar for desiring to join a congress that Bolívar had no control over. Instead of attacking the discredited Mariño and the men who participated in the Cariaco congress, Bolívar found a more useful weapon: racial anxiety surrounding the independence wars. Bolívar deflected any possible accusations of racially motivated violence by the insurgent troops by painting General Piar with a racially charged brush. He argued that Piar, born an illegitimate child, wanted to "slander the government, suggesting it was transforming into a tyranny; [and] proclaim the odious principles of a race war to destroy the equality that has been our fundamental base of existence since the glorious day of our insurrection until the present."[119] Piar was not looking for equality among the men of color, according to Bolívar. That was because they already enjoyed equality, as was proven by the fact of the *pardo* Piar's ascendancy. What Piar wanted, according to Bolívar, was to exalt racial differences that could incite a race war that would put *pardos* in control. In late August, Bolívar's men arrested Piar. He was tried for sedition and conspiracy, found guilty, and executed.

Bolívar's plan worked. The execution of Piar convinced other patriot generals that it was in their best interest to declare their loyalty to Bolívar. Mariño quickly returned to Bolívar's fold and accepted the offer to be restored to a position of command. Several other generals also fell in line. José Antonio Paéz (1790–1863), a leading *llanero* caudillo, later explained that he accepted Bolívar's role as supreme chief in deference to Bolívar's military gifts and his international prestige, but above all because of the many advantages accruing from having "a supreme authority and a center that could direct the various leaders."[120]

By remembering the federal congress at Cariaco, and the fact that several patriot generals and civilians swore by its legitimacy, we can better understand why, as Lynch puts it, Bolívar turned from "personalism to professionalism" when he did.[121] Starting in September of 1817, Bolívar called for several military and political reforms and made the workings of the new government transparently known through the *Correo del Orinoco*, the first official paper of the republic at Angostura. These measures uncannily followed the recommendations Madariaga had given to the Liberator in his letter of 1817, which had begged Bolívar to set up a formal government and disseminate news about that government so that international powers, and especially the British, could recognize it.[122] But Restrepo and other pro-Bolívar historians ridiculed the Cariaco congress that Madariaga had inspired. Instead, the Angostura congress of 1819 is identified as the "duly authorized congress" that brought unity to Colombia. Both Restrepo's *Carta* and the printed *Atlas* accordingly locate Angostura and display it prominently (figs. 3.8, 3.9, and 3.12).

CONCLUSION

The maps discussed in this chapter can be considered portable monuments intended to shape the viewer's historical understanding of a "Colombian" independence landscape. Much like immovable monuments implanted in a particular place, these transportable maps silenced some histories while privileging others. While "Colombia Prima" cleansed regions of imperial contests that challenged British designs on South America, the printed *Atlas* helped international readers of maps visualize the kind of Colombian Republic the Liberator Party wished to naturalize at home and abroad.

Despite their claims to permanence, these monuments were nevertheless dependent on too many contingent variables, and beholden to too many rapidly changing circumstances. The 1807 continental vision of "Colombia Prima" never materialized. The crisis of sovereignty generated by the Napoleonic-induced interregnum starting in 1808 opened the way for mushrooming autonomous juntas throughout the Spanish Empire that only with difficulty ceded authority to another power. A united, continental "Colombia Prima" increasingly became unthinkable as junta battled junta for autonomy prior to Fernando VII's return. Restrepo's *History* and his *Atlas* of 1827 testify to how the return of Fernando VII in 1814 further polarized relations between Spain and its New World territories, culminating in independence for large swaths of Spanish America. Ironically, once Fernando VII's royalist armies were defeated in South America by 1826, the Colombian Republic pictured in 1827 had little reason for being. Fragmenting forces on the ground that had managed to build their own transatlantic networks of recognition broke the Colombian Republic apart.

Comparing and contrasting the 1820s cartographic project of the Liberator Party with the earlier, poorly timed proindependence project hatched by Francisco de Miranda, one that also included a cartographic dimension, offers some intriguing results. Both projects undoubtedly sought to captivate foreign audiences to gain support for emancipatory political projects in Spanish America. But many of Miranda's compatriot Caraqueños viewed him with suspicion in 1807 when "Colombia Prima" was published, given his failed strike against Coro in 1806. Furthermore, the ease with which British cartographers imposed their imperial designs upon Miranda's project cartographically reveals the extent to which Miranda was beholden to his British sponsors. Restrepo, on the other hand, was the minister of the interior of a Colombian Republic already recognized by the United States and Great Britain when his *History* and *Atlas* were printed in Paris in 1827. True, France never recognized Colombia before it fragmented into Venezuela, New Granada, and Ecuador in 1830. But the fact that Restrepo's cartographic-historical project, which announced bold territorial claims and drew subtle attention to historical details that put Bolívar at the center of Colombian independence, suggests that Colombian diplomacy in France, however precarious, worked. At least it worked well enough to challenge British imperial territorial claims made by "Colombia Prima."

Restrepo's *Carta* and *Atlas* did more than challenge British imperial designs, however. Carefully examining and contextualizing these two versions of the same map allows us to better see how the Liberator Party conceptualized a Bolívar-centered narrative cartographically. This approach allows us to uncover some of the historiographical pathways buried deep under the mythic pro-Bolívar, antifederalist history that Restrepo championed, and that Colombian diplomats abroad exaggerated.[123] To a large extent, the clear historical markers that the Liberator Party sought to embed within Colombia's territory bore fruit. Contemporaries and subsequent historians have laughed at the *congresillo* at Cariaco, gasped at Piar's desire to wage a war of extermination against all whites, and lamented the unfortunate massacre of the Capuchin friars.[124] Historical memory usually isolates each event from the others. It is by taking a close look at what Restrepo's *Carta* depicts, and what the *Atlas* emphasized and erased, that we may better allow the overlapping shadows cast by the ghosts of Miranda, Piar, and the Catalan friars to emerge.

ACKNOWLEDGMENTS

Thanks to Nancy Appelbaum, Jorge Cañizares-Esguerra, Santiago Muñoz, Sebastian Diaz, and Daniel Gutierrez Ardila for insightful critiques, questions, comments, and suggestions. This article was made possible in part by a Fulbright-Hays Scholarship to Colombia and the Jeannette D. Black Fellowship in the History of Cartography at the John Carter Brown Library in 2010. All translations from Spanish-language sources are my own.

NOTES

1. Michel-Rolph Trouillot, *Silencing the Past: Power and the Production of History* (Boston: Beacon Press, 1995).

2. Robert Proctor and Londa Schiebinger, eds., *Agnotology: The Making and Unmaking of Ignorance* (Stanford: Stanford University Press, 2008); Neil Safier, *Measuring the New World: Enlightenment Science and South America* (Chicago: University of Chicago Press, 2008); and Jorge Cañizares-Esguerra, "Landscapes and Identities: Mexico, 1850–1900," in *Nature, Empire, and Nation: Explorations of the History of Science in the Iberian World* (Stanford: Stanford University Press, 2006), 129–68.

3. For the "impartiality" of cartography illustrating patriotic histories of the Independence period, see German Colmenares, *Las convenciones contra la cultura: Ensayos sobre historiografía hispanoamericana del siglo XIX* (Bogotá: Tercer Mundo Editores, 1997), 33–48.

4. Sergio Mejía, *La revolución en letras: La Historia de la revolución de Colombia de José Manuel Restrepo (1781–1863)* (Bogotá: Uniandes-Ceso, Departamento de Historia Universidad EAFIT, 2007).

5. Jose C. Moya, "Introduction: Latin America—the Limitations and Meaning of a Historical Category," in *The Oxford Handbook of Latin American History*, ed. Jose C. Moya (Oxford: Oxford University Press, 2011), 1–24.

6. Matthew Brown and Gabriel Paquette, *Connections after Colonialism: Europe and Latin America in the 1820s* (Tuscaloosa: University of Alabama Press, 2013).

7. Rebecca Earle, *The Return of the Native: Indians and Myth-Making in Spanish America, 1810–1930* (Durham: Duke University Press, 2007); Mark Thurner, *History's Peru: The Poetics of Colonial and Postcolonial Historiography* (Gainesville: University Press of Florida, 2011); and Michael Gobat, "The Invention of Latin America: A Transnational History of Anti-imperialism, Democracy, and Race," *American Historical Review* 118, no. 5 (December 2013): 1345–75. For New Granada, see Mauricio Nieto Olarte, *Orden natural y orden social: Ciencia y política en el Semanario del Nuevo Reino de Granada* (Bogotá: Ediciones Uniandes, 2009); and Renán Silva, *Los ilustrados de Nueva*

Granada, 1760–1808: Genealogía de una comunidad de interpretación (Medellín: EAFIT, 2002). See also: F. J. Caldas, "Estado de la Geografía del Vireynato de Santafé de Bogotá . . . ," *Semanario del Nuevo Reino de Granada* 1 (Santafé [de Bogotá]), January 3, 1808): 2–3.

8. José Manuel Restrepo, *Historia de la revolución de la República de Colombia: Atlas* (Paris, 1827), 5–7.

9. Ramaswamy, *The Goddess and the Nation: Mapping Mother India* (Durham: Duke University Press, 2010), 283–98.

10. Jeremy Adelman, "Iberian Passages: Continuity and Change in the South Atlantic," in *The Age of Revolutions in Global Context, c. 1760–1840,* ed. David Armitage and Sanjay Subrahmanyam (New York: Palgrave Macmillan, 2010), 59–82.

11. José Manuel Restrepo, *Historia de la revolución de la Republica de Colombia* (Paris: Librería Americana, 1827), 1:iii.

12. Karen Racine, *Francisco de Miranda: A Transatlantic Life in the Age of Revolution* (Wilmington, DE: Scholarly Resources, 2003).

13. Francisco de Miranda, *Diary of Francisco de Miranda: His Tour of the United States, 1783–1784,* ed. William S. Robertson (New York: Hispanic Society of New York, 1928).

14. Matthew Dennis, "The 18th-Century Discovery of Columbus: The Columbian Tercentenary (1792) and the Creation of American National Identity," in *Riot and Revelry in Early America,* ed. William Pencak et al. (University Park: Penn State University Press, 2002); Trouillot, "Good Day, Columbus," in *Silencing the Past,* 108–140; and Elise Bartosik-Vélez, *The Legacy of Christopher Columbus in the Americas: New Nations and a Transatlantic Discourse of Empire* (Nashville: Vanderbilt University Press, 2014).

15. *Gentleman's Magazine* 8 (June 1738): 286.

16. Dennis, "The 18th-Century Discovery of Columbus," 211.

17. Francisco de Miranda, *Segunda Sección, El Viajero Ilustrado, 1781–1788,* vol. 7 of *Colombeia* (Caracas: Ediciones de la Presidencia de la Republica, 1983), 405.

18. José María Rojas, *El General Miranda* (Paris: Libreria de Garnier Hermanos, 1884), 191.

19. Miranda edited the newspaper *El Colombiano* and several other proindependence tracts in London that contributed to this enterprise. For the way print capitalism influenced the rise of nationalism, see Benedict Anderson, *Imagined Communities: Reflections on the Origins and Spread of Nationalism* (London: Verso, 1991).

20. Francisco de Miranda, *Tercera Sección, Revolución Francesa, 1790–1792,* vol. 9 of *Colombeia* (Caracas: Ediciones de la Presidencia de la Republica, 1983), 55.

21. Francisco de Miranda, "Plan militar formado en Londres en Agosto, 1798," in *Cuarta Sección: Negociaciones, 1797–1799,* vol. 18 of *Colombeia,* 102–13.

22. Marshall Smelser, "George Washington Declines the Part of El Libertador," *William and Mary Quarterly,* 3rd ser., 11, no. 1 (January 1954): 42–51.

23. Faden to Miranda, August 23 and October 15, 1804, cited in William Spence Robertson, *The Life of Miranda* (Chapel Hill: The University of North Carolina Press, 1929), 2:274.

24. Thomas Byrne, "British Army, Irish Soldiers: The 1806 Invasion of Buenos Aires," *Irish Migration Studies in Latin America* 7, no. 3 (March 2010): 305–12.

25. "Conferencias con los Ministros de S. Mag. Brit.," October 13–16, 1804, cited in Robertson, *Life of Miranda,* 1:276–77.

26. Robertson, *Life of Miranda,* 1:279.

27. Racine, *Francisco de Miranda,* 156–72.

28. See Lina del Castillo, "Embellishments and Erasures: Placing Basque Merchant Power in Cruz Cano's Map of South America," in *Cartographic Conversations from the John Carter Brown Library*, ed. Jordana Dym, available at http://www.brown.edu/Facilities/John_Carter_Brown _Library/cartographic/pages/castillo.html See also Thomas R. Smith, "Cruz Cano's Map of South America, Madrid 1775: Its Creation, Adversities and Rehabilitation," *Imago Mundi* 20 (1966): 48–78; Walter William Ristow, "The Juan de la Cruz map of South America," in *Festschrift: Clarence F. Jones*, edited by Merle C. Prunty, Jr., Northwestern University Studies in Geography, no. 6 (Evanston, IL, 1962). For Miranda's role in the reproduction of the Cruz Cano map see Alexander Hamilton, *The Papers of Alexander Hamilton*, ed. Harold C. Syrett (New York: Columbia University Press, 1975), 22:208 and 526.

29. For the culture of map commerce at the end of the eighteenth century, see Mary Pedley, "Maps, War, and Commerce: Business Correspondence with the London Map Firm of Thomas Jefferys and William Faden," *Imago Mundi* 48 (1996): 161–73.

30. Racine, *Francisco de Miranda*, 143–47.

31. Merle E. Simmons, *Los escritos de Juan Pablo Viscardo y Guzmán, precursor de la independencia Hispanoamericana* (Caracas: Universidad Católica Andres Bello, 1983).

32. Georges L. Bastin and Elvira R. Castrillón, "La 'Carta dirigida a los españoles americanos,' una carta que recorrió muchos caminos," *Hermeneus* 6 (2004): 276–90.

33. Simmons, *Los escritos de Juan Pablo Viscardo y Guzmán*, 68–70.

34. "Representación de Fr. Manuel Sobreviela destinada a que el Virrey informe al Rey la necesidad de la misión de los religiosos," in *Juicio de límites entre el Perú y Bolívia: Contestación al alegato de Bolívia, Prueba Peruana presentada al gobierno de la república Argentina por Víctor M Maurtua, abogado plenipotenciario especial del Perú, tomo 6 Misiones Centrales Peruanas* (Buenos Aires: Compañia Sud Americana de Billetes de Banco, 1907), 276–79.

35. Kenneth Maxwell, *Conflicts & Conspiracies: Brazil and Portugal, 1750–1808* (New York: Routledge, 2004), 150–72.

36. José Joaquim da Rocha, *Geografia histórica da capitania de Minas Gerais: Estudo crítico*, ed. Maria Efigénia Lage de Resende (Belo Horizonte: Fundaçao Joao Pinheiro, 1995); Junia Ferreira Furtado, "Cartographic Independence," in *Mapping Latin America: A Cartographic Reader*, ed. Jordana Dym and Karl Offen (Chicago: University of Chicago Press, 2011), 114–19.

37. Francosco Marques de Sousa Viterbo, ed., *Dicionário histórico e documental dos arquitetos, engenheiros e construtores Portugueses ou a Serviço de Portugal*, vol. 1 (Lisbon: Imprensa Nacional, 1899), 241.

38. Juan de la Cruz Cano y Olmedilla, "Mapa Geográfico de America Meridional . . ." (reprint, London: Fadden, 1799).

39. For the swapping of territories between the Dutch and the British between 1781 and 1814, see Robert Montgomery Martin, Esq., *History of the Colonies of the British Empire in the West Indies, South America, North America, Asia, Austral-Asia, Africa and Europe; comprising the Area, Agriculture, Commerce* . . . (London, 1843), 118.

40. Francisco de Miranda, *Cuarta Sección: Negociaciones, 1797–1799*, 102–13.

41. Ricardo D. Salvatore, "Imperial Revisionism: US Historians of Latin America and the Spanish Colonial Empire (ca. 1915–1945)," *Journal of Transnational American Studies* 5, no. 1 (2013): 1–54.

42. The literature on this tumultuous period is vast. For a brilliant overview of past and

recent historical literature, see Gabriel Paquette, "The Dissolution of the Spanish Atlantic Monarchy," *Historical Journal* 52, no. 1 (March 2009): 175–212.

43. Racine, *Francisco de Miranda* (n. 12 above), 196–206; John Lynch, *Simón Bolívar: A Life* (New Haven: Yale University Press, 2006), 49–63.

44. Alfredo Ávila, Jordana Dym, and Erika Pani, eds., *Las declaraciones de Independencia: Los textos fundamentales de las independencias americanas* (Mexico: El Colegio de México; Universidad Autónoma de México, 2013).

45. Juan Germán Roscio, "Vicios legales de la Regencia de España e Indias deducidos del Acta de su instalación el 29 de enero en la Isla de León," *Gaceta de Caracas* 105 (June 29, 1810).

46. Arístides Rojas, *Los Hombres de la Revolución, 1810–1826: Cuadros históricos; El canonigo José Cortés Madariaga, el General Emparán* (Caracas: Imprenta de Vapor de La Opinion Nacional, 1878), 11.

47. Daniel Gutiérrez Ardila, *Un Nuevo Reino: Geografía política, pactismo y diplomacia durante el Interregno en Nueva Granada (1808–1816)* (Bogotá: Universidad Externado de Colombia, 2010), 486–96.

48. José Acevedo Gómez, "Relación de lo ocurrido con motivo de la llegada del Enviado de Caracas" (Santa Fe, March 18, 1811), Archivo Restrepo, Fondo 1, Rollo 8, folio 25, John Carter Brown Library.

49. Suplemento al No. 6 del Semanario Ministerial del Gobierno de la Capital de Santafé de Bogotá, Nuevo Reino de Granada, Relación de lo ocurrido con motivo de la llegada del Enviado de Caracas. Santafe Marzo 22 de 1811. Archivo Restrepo, Fondo 1, Rollo 8, folio 42–43, John Carter Brown Library.

50. Clément Thibaud, *Repúblicas en Armas: Los ejércitos Bolívarianos en la Guerra de Independencia en Colombia y Venezuela* (Colombia: Editorial Planeta, 2003).

51. Rafe Blaufarb, "The Western Question: The Geopolitics of Latin American Independence," *American Historical Review* 12, no. 3 (June 2007): 742–63.

52. Ibid.; and Karen Racine, "'This England and This Now': British Cultural and Intellectual Influence in the Spanish American Independence Era," *Hispanic American Historical Review* 90, no. 3 (August 2010): 423–54.

53. Daniel Gutiérrez Ardila, *El reconocimiento de Colombia: diplomacia y propaganda en la coyuntura de las Restauraciones (1819–1831)* (Bogotá: Universidad del Externado, 2012).

54. The first printed map of the Colombian Republic formed part of H. C. Carey and I. Lea, *A Complete Historical, Chronological, and Geographical American Atlas . . .* (Philadelphia, 1822). Restrepo's atlas of 1827 is the first to depict the territorial divisions of the republic effective in 1824. Printed maps of Colombia thereafter depicted these divisions, even after 1830. See, for instance, Jeremiah Greenleaf, "Colombia," in *A New Universal Atlas; Comprising Separate Maps of all the Principal Empires . . .* (Battelboro, VT, 1840); A. R. Fremin, C. V. Monin, and A. Montemont, "Colombia et Guyanes," in *L'Univers. Atlas Classique Et Universel De Géographie Ancienne Et Moderne* (Paris: Bernard; Mangeon; Laguillermie, 1837); Joseph Thomas, "Colombia," in *Thomas's library atlas, embodying a complete set of maps, illustrative of modern & ancient geography . . .* (London: Joseph Thomas, 1835). All available at http://www.davidrumsey.com.

55. J. M. Restrepo to Santander, Rionegro, Julio 5, 1820, *Boletín de Historia y Antigüedades: Órgano de la Academia de Historia Nacional* 3, no. 25 (January 1905): 181; emphasis added.

56. Restrepo's descendants have made his archive, including the Colombia Prima map, avail-

able to the public by allowing microfilmed copies to form part of the Biblioteca Luis Angel Arango and the AGN in Bogotá, and the John Carter Brown Library in Providence, RI.

57. Mejía, *La revolución en letras* (n. 4 above), 7–9.

58. Gutiérrez Ardila, *El reconocimiento de Colombia*, 145–54.

59. Ibid., 154.

60. *Instrucción que se ha de observer el Sr. Francisco Restrepo en el encargo que se le hace para prover las casas de moneda de la república de las máquinas y utensilios que se necesitan* (Bogotá, 11 de octubre de 1825), AGN, MRE, DT8, box 507, folder 7, fol. 55.

61. Andrés Bello, "Historia de la revolución de Colombia por el Sr. José Manuel Restrepo," in *Obras Completas, Temas de Historia y Geografía,* vol. 23 (Caracas: Fundación La Casa de Bello, 1982), 395.

62. Gutiérrez Ardila, *El reconocimiento de Colombia*, 152.

63. Gutiérrez Ardila, "Los primeros Colombianos en Paris," in *El reconocimiento de Colombia*, 201–36.

64. Revenga to the Minister of External Relations, Pedro Gual (London, October 8, 1823), AGN Ministry of Exterior Relations, Delegations, Transferences, 2, t. 299, fol. 219.

65. Examples of other contemporary British maps like "Colombia Prima" that expanded British claims in this way include John Cary, "A new map of South America (north sheet), from the latest authorities," in *Cary's new universal atlas containing distinct maps of all the principal states and kingdoms throughout the world: From the latest and best authorities extant* (London: John Cary, 1808).

66. Ibid.

67. Mejía, *La revolución en letras*, 50–62.

68. Restrepo, *Historia de la revolución* (1827; n. 11 above), 1:12–13.

69. Ibid., 1:13–14.

70. Ibid., 1:13–14.

71. José Manuel Restrepo, "(Text Page in) *Historia de la revolucion de la republica de Colombia, Atlas*" (Paris: Librería Americana, Calle del Temple, no. 69, 1827), i–iii; Jean-Baptiste Boussingault, *Memorias*, vols. 1–4 (Bogotá: Banco de la República, 1985).

72. Restrepo, *Historia de la revolución . . . Atlas* (1827; n. 8 above), 7.

73. Gutiérrez Ardila, *El reconocimiento de Colombia* (n. 53 above), 337–71.

74. Leandro Palacios to Santander (New York, October 6, 1825), in *Correspondencia dirigida al general Francisco de Paula Santander: Compilación de Roberto Cortázar de la Academia Colombiana de Historia* (Bogotá: Talleres Editoriales de Librería Voluntad, 1964–70), 6:111–13.

75. Henri La Fayette Villaume Ducoudray-Holstein, *Memoirs of Bolívar, President Liberator of the Republic of Colombia; and of his Principal Generals; secret history of the revolution, and the events which preceded it, from 1807 to the Present Time* (Boston: S. G. Goodrich & Co., 1829).

76. Ibid., 39.

77. Ibid., 178.

78. Ibid., 193.

79. Ibid., 193–216.

80. Ibid., 196–99.

81. Ibid., 198.

82. Oficio de Tiburcio Echevarría a Simón Bolívar (Paris, February 1, 1822) in AGN, Ministerio de Relaciones Exteriores, Delegaciones, Transferencias 2, t. 115, fols. 13–16, cited in Gutiérrez Ardila, *El reconocimiento de Colombia*, 146.

83. By 1829 French agent Charles Bresson and British chargé d'affaires Patrick Campbell had suggested outright to Bolívar's Council of Ministers that Bolívar be made king of Colombia, with a European prince succeeding him. Richard Slatta and Jane Lucas de Grummond, *Simón Bolívar's Quest for Glory*, Texas A&M University Military History Series, no. 86 (College Station: Texas A&M University Press, 2003), 286.

84. Leandro Palacios to Fernández Madrid (Paris, December 16 and 17, 1829, and January 14, 1830), AGN Ministerio de Relaciones Exteriores, Delegaciones Transferencias 8, box 508, folder 16, fol. 162, 164–65; box 509, fol. 17, fol. 22, cited in Gutiérrez Ardila, *El Reconocimiento de Colombia*, 133–45.

85. Restrepo, *Historia de la revolución* (1827), 3:129.

86. Salvador de Madariaga, *Bolívar*, 2 vols. (Buenos Aires: Editorial Sudamericana, 1959); David Bushnell, *Simón Bolívar: Hombre de Caracas, proyecto de América; Una biografía* (Buenos Aires: Editorial Biblos, 2002), 38; Lynch, *Simón Bolívar* (n. 43 above), 62–63. See also Daniel Florencio O'Leary, *Memorias de General Daniel Florencio O'Leary: Narración*, 3 vols. (Caracas, 1952), 1:113–14.

87. Restrepo, *Historia de la revolución* (1827), 2:249–53.

88. Ibid.

89. For notices of the federal congress at Cariaco British newspapers of the period including Sylvanus Urban, ed., "Sept. 1817," *Gentleman's Magazine and Historical Chronicle* 87 (London: Nichols & Son, 1817): 270; and *Monthly Review or Literary Journal* 89 (London: Pall Mall, 1819): 171–72.

90. Those assembled included Admiral Luis Brion, commander of the naval forces; the intendant general Francisco Antonio Zea, from New Granada and recently arrived from Europe; citizen José Joaquin de Madariaga, canon of the Cathedral Church of Caracas; citizen Francisco Javier Mayz, in charge of the Executive Department of Caracas; and citizen Francisco Manuel Alcalá. Other attendees included Diego Ballenilla, Diego Antonio Alcalá, Manuel Isaba, Francisco de Paula Navas, Diego Bautista Urbaneja, and Manuel Maneiro. "Congreso de Cariaco: Acta de la Congregación Convocada para el 8 de Mayo de 1817," in Daniel Florencio O'Leary, *Memorias del General O'Leary: Documentos* (Caracas: Imprenta de la Gaceta Oficial, 1881), 15: 250–52.

91. José Manuel Restrepo, *Historia de la revolución de la República de Colombia en la América Meridional* (Paris: Besanzon, J. Jacquin, 1858), 2:396.

92. Ibid., 397.

93. The following historians describe the Cariaco Congress and Piar's rebellion but do so in terms of how "preposterous" and "little" the congress was while highlighting the racial motive of Piar's rebellion: Daniel O'Leary, *Memorias* (1883), 1:429; Felipe Larrazabal, *The Life of Simón Bolívar, liberator of Colombia and Peru, father and founder of Bolivia, carefully written from authentic and unpublished documents* (New York: Edward O Jenkins, 1866), 1:298–325. Subsequent historians have tended to repeat this view, especially Lynch, *Simón Bolívar*, 104–7; Bushnell, *Simón Bolívar*, 83–85. Jaime Rodríguez O. offers a more nuanced interpretation in *The Independence of Spanish America* (Cambridge: Cambridge University Press, 1998), citing José Gil Fortoul, *Historia Constitucional de Venezuela*, vol. 1 (Berlin, 1907).

94. Simón Bolívar, "Manifesto del Jefe Supremo a Los Pueblos de Venezuela," in Augusto Mijares, ed., and Manuel Pérez Vila, *Doctrina del Libertador Simón Bolívar*. Available at Biblioteca Virtual Miguel de Cervantes: http://www.cervantesvirtual.com/servlet/SirveObras /79150596101682496754491/p0000002.htm#I_23_.

95. Ibid.

96. Alicia Rios, "La época de la independencia en la narrativa venezolana de los ochenta," *Hispamérica* 22, no. 64/65 (April–August 1993): 49–54.

97. AHNC, República, Guerra y Marina, 14, fol. 115, cited in Marixa Lasso, *Myths of Harmony: Race and Republicanism during the Age of Revolution, Colombia 1795–1831* (Pittsburgh: University of Pittsburgh Press, 2007), 137 and 147.

98. Aline Helg, *Liberty and Equality in Caribbean Colombia, 1770–1835* (Chapel Hill: University of North Carolina Press, 2004), 209, citing Bolívar to Briceño, November 16, 1822, and Bolívar to Páez, November 16, 1828, in Simón Bolívar, *Obras Completas*, ed. Vicente Lecuna (1947), 2:505–8.

99. Lasso, *Myths of Harmony*, 137.

100. Piar to Bolívar, San Felipe, January 31, 1817, in O'Leary, *Memorias*, 15:150–55.

101. R. P. Fr. Nicolás de Vich, fray, "Victimas de la anárquica ferocidad . . ." (Vich, Spain: En la Imprenta de Felipe Tolosa, 1818). John Carter Brown Library.

102. José Féliz Blanco and Ramón Azpurúa, *Documentos para la historia de la vida pública del libertador de Colombia, Peru y Bolívia* (Caracas: Imprenta de "La Opinion Nacional," 1876), 5:617; and Vich, "Victimas," 11.

103. O'Leary, *Memorias*, 15:195–228.

104. Vich, "Victimas," 16–22.

105. Lynch, *Simón Bolívar* (n. 43 above), 103; and Vich, "Victimas," 11.

106. Slatta and Lucas de Grummond, *Simón Bolívar's Quest for Glory* (n. 83 above), 155.

107. "Declaración del venerable José Félix Blanco . . . 1817," in Blanco and Azpurúa, *Documentos*, 5:647.

108. O'Leary, *Memorias*, 15:190–228; and Blanco and Azpurua, *Documentos*, 5:646–48.

109. "Declaración . . . ," in Blanco and Azpurúa, *Documentos*, 5:647.

110. Restrepo, *Historia de la revolución* (1858), 2:402. See also Bushnell, *Simón Bolívar* (n. 86 above), 83; Lynch, *Simón Bolívar*, 103–4, 282.

111. Vich, "Victimas," 10.

112. Blanco and Azpurúa, *Documentos*, 1:462.

113. Lynch, *Simón Bolívar*, 103–4.

114. Blanco and Azpurúa, *Documentos*, 5:663–66.

115. Ibid., 6:109.

116. Lynch, *Bolívar*, 104–7.

117. "Carta de Bolívar a Piar, fechada en San Félix el 14 de Junio de 1817," *Gaceta de Carácas*, July 2, 1817.

118. Ibid.

119. Simón Bolívar, "Manifiesto del Jefe Supremo a los Pueblos de Venezuela," Cuartel de Guayana, August 5, 1817. Biblioteca Virtual Miguel de Cervantes.

120. José Antonio Paéz, *Autobiografía* (New York, 1870), 1:136, 141.

121. Lynch, *Simón Bolívar*, 110–12.

122. "Carta de José Cortes de Madariaga a Simón Bolívar, 25 April 1817, Margarita Island," in Blanco and Azpurúa, *Documentos*, 5:625.

123. German Colmenares, "La Historia de la revolución por José Manuel Restrepo: Una prisión historiográfica," in *La Independencia: Ensayos de Historia Social*, by German Colmenares et al. (Bogotá: Instituto Colombiano de Cultura, 1986), 10–11; Alexander Betancourt Mendieta, *Historia y Nación: Tentativas de la Escritura de la Historia en Colombia* (Medellín: La Carreta, 2007),

27–38; and Jorge Orlando Melo, *Historiografía Colombiana: Realidades y Perspectivas* (Medellín, 1996), 107:46–56.

124. David Brading, *The First America: The Spanish Monarchy, Creole Patriots, and the Liberal State, 1492–1867* (Cambridge: Cambridge University Press, 1991), 607; Lynch, *Simón Bolívar*, 102–10; Bushnell, *Simón Bolívar*, 82–85; Slatta and Lucas de Grummond, *Simón Bolívar's Quest for Glory*, 158–60; Lasso, *Myths of Harmony* (n. 97 above), 137–47.

CHAPTER FOUR

DEMOCRATIZING THE MAP

THE GEO-BODY AND NATIONAL CARTOGRAPHY IN GUATEMALA,
1821–2010

Jordana Dym

Les limites de la république guatémalienne ne sont pas aussi faciles à déter-
miner qu'on le croirait, en jetant les yeux sur la carte de ces régions.

PHILIPPE FRANÇOIS DE LA RENAUDIÈRE, *Mexique et Guatemala* (Paris:
Firmin Didot Frères, 1843), 254–55

Can a caricature of a map arouse nationalism, royalism, or other serious sentimental responses?

THONGCHAI WINICHAKUL, *Siam Mapped*, 138

In July 2010, one bookseller at Guatemala's annual book fair covered its stalls
for the night with a banner by a group called "Convergence for Human
Rights" (fig. 4.1). The banner called for the military to follow President Alvaro
Colom's order to open its archives and to end government impunity and cover-
ups of genocide. The "Massacres" map of Guatemala accompanying the text
on a blood-red background locates sites and tallies a dozen killings perpetrated

FIGURE 4.1. Banner on display at FILGUA, July 2010. Photograph by the author.

throughout national territory during over thirty years of civil war. Clearly, a Guatemalan at the book fair would see that the map shows the state acting against its people in communities located throughout the country, with both internal, departmental boundaries and external, international frontiers clearly marking the Guatemalan spaces affected by the conflict.

This map, then, implies not only a Guatemalan territory but also a Guatemalan people. However, territory and people are less unified than the map seems to suggest. Guatemala officially claims Belize. Its Instituto de Geografía Nacional (IGN) legally holds a monopoly on making or approving maps of the national territory. Yet Belize is not part of this map's territory, which lacks the IGN's seal and authorization. Further, by placing the victims of armed conflict in both indigenous and ladino (Westernized) departments, the map blurs real ethnic and even class distinctions among victim communities, and also hides the country's urban-rural divide.

Guatemala's Maya population is generally considered the most grievously wounded by the conflict; hundreds of thousands were displaced or disappeared, particularly in the country's northwest highland districts. Instead of ethnic and regional distinctions, however, the fine print tallies the number of boys and girls caught up in the military's murderous clutches, perhaps to make the point that the victims were innocent, no matter where or in what period of the conflict the murders took place. By emphasizing child victims and eliding problems of race and class, the map draws on the past and a common understanding of what it means to be Guatemalan—having lived through and suffered in the civil war—while erasing some of the most salient features behind the conflict. So while seemingly representing a "done deal" and a common violation of human rights, this map also reveals the contingent in the fixed.

As French traveler and mining engineer Philippe de la Renaudière (1781–1845) observed in 1843, "The limits of the Guatemalan republic are not as easy to determine as one might think, when casting an eye on a map of these regions." Then, Guatemala lacked agreed boundaries with its neighbors and had already divided internally from seven (1825) into twelve departments (now twenty-two). Today, the comment still rings true. Guatemala has two geo-bodies, or national territories, the official and unofficial.[1] Since the 1860s, the most visible toggle has been with land: is Belize part of Guatemalan territory or not? Government and civil society have created both maps. Yet, as the 2010 book fair map shows, inclusion and exclusion of different Guatemalan populations, particularly the indigenous majority, has been equally complicated. Language spoken is one of several strategies used to show and hide distinct national and subnational peoples.

This essay explores the long and complex process of establishing Guatemala's national geo-body and map, and the equally long experience of presenting and teaching them. It begins with the establishment of the state in 1825, follows a long nineteenth century of establishing and marking international boundaries and internal political and administrative divisions, and concludes in the twenty-first century as different groups adopt the competing geo-body territories and peoples, seemingly unconsciously, to establish their own claims to belong to and participate in Guatemalan debate and development.

I argue that over these two hundred years, a two-step process of "decolonization" and "democratization" shaped the national territorial map. For Guatemala, decolonization occurred in two phases as the government identified land and people over which it claimed sovereignty, representing that claim cartographically, and then successfully administering it. For the map to be fully decolonized, I suggest, the government must not only create national maps, but also employ national agents and institutions to create maps considered "accurate" or correct both internally and externally. In Guatemala's case, the first national map based on work of a Guatemalan mapmaker and printed in Guatemala dates to 1832. However, full decolonization of the map happened in the early twentieth century, when the government developed a permanent and recognizable geo-body—"decolonizing" the land from an undefined shape whose internal districting slowly abandoned colonial divisions to the wounded land seen in figure 4.1 with clear national and district boundaries—and assigned mapping to state-run cartography institutions. The second process, which I call "democratization" and Sarah Radcliffe considers "re-mapping the nation," occurs when the tools of map reading and mapmaking become sufficiently widespread such that the government, regardless of any claim to hold an absolute monopoly on public mapmaking and consumption, is only one of many cartographic producers, and the nation also participates in map production as well as consumption.[2]

The evidence is the map of the full national "geo-body" of Guatemala as it evolved from the nineteenth century to the early twenty-first century, including representation of Guatemalan citizens on that national space. For "decolonized" maps, the study relies on private and national archival sources. I supplement such sources for "democratized" maps with maps found through eBay and Internet browsing, and cameras of itinerant colleagues Karl Offen and Matthew Taylor, to offer a sense, however incomplete, of how private citizens have adapted the national geo-body for their own purposes. The evidence shows that the adoption of cartographic techniques and the internally divided geo-body by a range of actors in civil society, from authors of textbooks to professional associations to telephone companies to indigenous labor movements to newspapers, reflects an active and effective appropriation of the geo-body for national as well as state purposes that had begun by the mid-nineteenth century and became increasingly widespread.

This argument owes much to Benedict Anderson's understanding of mapmaking as a tool deployed by modern states—along with censuses and museums—not just to claim but also to demonstrate sovereignty over specific territories and peoples as part of a single imagined community.[3] Beyond that, I adopt Michel-Rolph Trouillot's approach to underline the importance of distinguishing between state and nation when talking about a truly "indigenous" map. Although Trouillot frames an argument about government acting against society in twentieth-century Haiti, the parallels he finds in a small country whose ethnic, racial, and economic divisions are arguably an extreme are a relevant model for Guatemala.[4]

Put in cartographic terms, for a map to be of the nation and not just of the state, I argue that the "imagined communities" themselves must accept the geo-body and use it in their own representations. This analysis concurs with Thongchai Winichakul's insight that a map caricature or logo can "arouse nationalism, royalism, or other serious sentimental responses" even when it becomes a floating map (an outline on a blank background divorced from territoriality).[5] It resonates with recent scholarship about the development of nineteenth-century North and Spanish American national mapping, rise of national cartographic missions and later institutes, and development of schoolbook geographies with national maps.[6] It also resonates with work of geographers like Sarah Radcliffe and Karl Offen on adoption and adaptation of national maps by "nonstate" actors, generally meaning disadvantaged populations, through formal organizations like CONAIE in Ecuador and the Process of Black Communities in Colombia.[7]

There is no scholarly study of Guatemalan cartography from independence in 1821 to the present.[8] However, there is increasing interest in both the historical territory and its representation. Two informative twentieth-century government compilations, one prepared by the 1929 Limits Commission and the other engineer Florencio Santiso's 1944 report to the second Pan-American meeting on geography and cartography, offer a chronological overview of cartographic development from codices to aerial surveys.[9] More recently, Arturo Taracena's 2002 analysis of Guatemala's territorial formation traces the development of the national territory (and indirectly the geo-body) from 1821 and 1935.[10] Following cartographic and territorial development, this paper evaluates Guatemala's cartographic decolonization and democratization as taking place within three broad periods: forming the national body (1821–1900), establishing national cartographic institutions and the public presentation of the geo-body (1900–1935), and proliferating and diversifying cartographic representations of the geo-body by both citizens and national agencies (1935–2010).

MAPPING THE STATE OF GUATEMALA: HYBRID MAPS AND CARTOGRAPHIC INDEPENDENCE (1821–32)

Guatemala's current geo-body has a distinctive shape with borders defined both by the jagged lines of natural features including oceans, rivers, and mountain ranges and carefully plotted straight lines and angles reflecting the measurements of boundary commissions (see fig. 4.1). That shape is neither natural nor predestined. The fixing of national borders seems, today, to be an inevitable consequence of establishing sovereign societies, but as Alexander Diener argues, there is really nothing natural about a border or its features.[11] Guatemala's complex establishment of both internal and external borders is a case in point. Guatemalan governments spent much of the country's first hundred years working with and against foreign powers to claim and mark international boundaries, largely through the work of border commissions and treaties. Concurrently, governments legislated to divide the country internally to favor central control, rewarding regions that supported the national government while dividing and weakening districts, such as the western highlands, with breakaway movements or popular uprisings that tried to overthrow the central leadership.[12] Cartographically, we can see the transition starkly by comparing a map showing the colonial divisions of Central America right before independence, which offers no preexisting shapes recognizable as the contemporary national territory, with a twentieth-century silver charm whose outline is fully formed and instantly identifiable (figs. 4.2 and 4.3).

Within a decade Guatemala achieved first political independence from Spain and then initial cartographic independence, a two-map and one-geography process when it was a state in the short-lived Central American Federal Republic (1825–39). The first map, by British envoy George A. Thompson, showed the overall shape of the country within the federation. A few years later, the geographic essays by Honduran-born statesman José Cecilio del Valle inspired Guatemala's president to commission the first Guatemalan-made and Guatemalan-printed map from naturalized citizen Miguel Rivera Maestre. Together, these maps displayed Guatemala's initial territorial shape and content, both far from the iconic form that emerges at the end of the century.

By 1825, Guatemala was one of five federal states in the fledging republic, forged from a half-dozen districts surrounding Guatemala City, Central America's colonial capital.[13] Guatemala's initial seven "departments" are named on the map produced in 1829 for diplomat Thompson's travel narrative; it was the first map of both the state and the federation. Thompson credits Valle, a federal official, with helping trace the original map onto British cartographer Aaron Arrowsmith's colonial map (probably fig. 4.2).[14] Thompson's map was a hybrid of an imperial and a "decolonized" map: produced by foreigners who used local knowledge. Intended for an English-speaking public, Thompson's map helped Europeans and Anglo-Americans visualize the outlines of Central America's fledgling states, with brightly colored lines drawing attention to the new nation's representation of international and state boundaries. Although the map's subject is independent Central America, this map takes the pulse of Guatemalan geographic, commercial, and ethnographic understanding of national territory and limits in 1826, as well as the already-extant focus on internal divisions and the representation of Guatemala not just as an independent polity but also as situated within and as part of a Central American region.

On this map, Guatemala's territory bears little resemblance to its twentieth-century geo-body. In the Petén area, not a single straight, surveyed line separates Guatemalan from Mexican districts (resolved by treaty and survey in 1882) or from the British "establishment" of Belize (not resolved as of 2016).[15]

FIGURE 4.2. A. Arrowsmith, Map of Guatemala, London, 1826 (detail). Courtesy of the Library of Congress, Geography and Maps Division.

Guatemala's internal divisions are named although not drawn by area; the missing departmental lines suggest the limits to local knowledge or Thompson's interest at that time; his local informant, José Cecilio del Valle, later claimed to have offered "negligible" input.[16] Externally, two disputed districts, Sonsonate and Chiapas, are drawn as separate from Guatemala but not yet part of Mexico. Perhaps most important for both Thompson's intended British audience and Central American cartography, Belize's territory is limited to the area authorized for logging in the 1786 treaty between Spain and the United Kingdom; later nineteenth-century maps by Guatemalan, Mexican, and British cartographers all show Belize extending further south. In addition, the map emphasized the outsider's signal preoccupation with the Central American isthmus as a transit point for world commerce and a potential trading hub, a cartographic theme that Central American governments would soon pick up in their own cartography. Perhaps for that reason, the only indigenous peoples to appear on the map are the Caribbean Miskitu peoples who controlled territory not yet claimed by Honduras and Nicaragua.[17] The Maya absence, whether by choice

FIGURE 4.3. Guatemala, 1951. Guatemala's national bird, the Quetzal, affirms the nationalist message in this map, as does inclusion of Belize, albeit with a line indicating division from the rest of the country. Private Collection.

or oversight, tends to corroborate Valle's disclaimer, since their presence was an integral element in his description of Guatemala's cartographic past, present, and future.

In 1830, Valle's detailed "geographic description" was serialized in the monthly bulletin of the country's Sociedad Económica, an institution of modernizing elites pushing development of Guatemala's peoples and commerce. Valle's chapters itemized the geographic position and elevation of each of Guatemala's seven departments alongside information on topography, languages, climate, population, agricultural production, and principal institutions and buildings.[18] Valle highlighted Guatemala's indigenous populations through linguistics and history. Demographically, he discussed ethnic diversity through the thirteen indigenous languages spoken by people making up two-thirds of the country's population. However, much like later reformers, he viewed native languages as "an obstacle opposed to the *indios'* civilization, a wall of separation" distancing them from "cultured and enlightened men."[19] Not incidentally, Valle insisted the country needed not one, but three maps, one for each stage of its history—pre-Columbian, colonial, and independent—each showing internal divisions of (respectively) kingdoms, provinces, and departments. It seems likely that he wanted his data and view of indigenous history as Guatemalan history used for a map of "our actual state [and] the real location of the pueblos," which, he argued, "should be brought to view to legislate and govern."[20] Valle anticipated historian Raymond Craib's argument for mid-nineteenth-century Mexico that without "a reliable national map," a new government "could hardly begin to conceive of, let alone carry out, any political reorganization of the territory."[21]

Between them, Thompson and Valle created an agenda that Guatemalan governments would promote for official and popular maps. Although no nineteenth-century Guatemalan government created all three of Valle's proposed maps, interest in geographic, historical, and ethnographic Guatemala became integral to subsequent official geographies and cartographies, as did the commercial and economic agenda seen in Thompson's map and embedded in the extensive lists of departmental production in Valle's geography text.

Shortly after Thompson's visit and Valle's geography, Guatemalan president Mariano Gálvez (1831–38) essentially adopted Valle's cartographic program. He commissioned histories of colonial and independent Central America from Francisco de Paula García Peláez and Alejandro Marure, respectively, and a Guatemalan atlas by surveyor Miguel Rivera Maestre (1783–1856), who apparently was already preparing a map for the Sociedad Económica.[22] For most scholars, this 1832 *Atlas guatemalteco*, not Thompson's collaboration, initiates Guatemala's postcolonial cartography. It certainly achieves its cartographic independence.

The atlas was an ambitious first project, with one map of the state (fig. 4.4) and one of each of Guatemala's seven departments filling in mountainous terrain. Although the initial atlas did not include maps depicting pre-Columbian and colonial territories, some copies add Rivera Maestre's later state-commissioned maps of Maya archaeological sites.[23] Lacking formal cartographic training, Rivera Maestre probably relied on existing maps of Central America and Valle's "Descripción Geográfica" as well as other government-compiled statistics. Long-time residents José Casildo España and Francisco Cabrera engraved the atlas maps, which were printed in Guatemala.[24]

Focused on Guatemala's internal divisions, Rivera Maestre's department maps rendered mountains, cities, roads, and rivers visible; his state map showed the government its sovereign territory and reflected Guatemala's tacit acceptance of the loss of Sonsonate to El Salvador in 1824. The atlas was less successful in other areas, including technical proficiency and fixing external boundaries. Historian Roger Claxton points to distortions from drawing equidistant parallels and meridians, misalignment of mountains in the department maps, and a host of other smaller errors.[25]

FIGURE 4.4. Miguel Rivera Maestre, *Carta del Estado de Guatemala en Centro-América, Año de 1832*. In *Atlas guatemalteco en ocho cartas formadas y grabadas en Guatemala*. Engraved by José Casildo España (1778–1848). Bibliothèque Nationale de France, Manuscrits Orientaux, MS Angrand-12, p. 3.

Guatemala's lack of settled international boundaries is both visible and invisible. The line denoting the border with Honduras and El Salvador to the south seems firm, but there was no discussion of, let alone agreement on, the exact limits during the conflict-ridden federal period (1824–39). Indeed, Honduras and Guatemala disputed the exact path of their borders for almost a hundred years.

The map, with a tentative and incomplete northern border, better captured ongoing uncertainty about Guatemala's limits with Mexico, and the role the Maya played there, and with Belize. Chiapas effectively joined Mexico in 1823, rejecting Central American overtures to be the federation's sixth state. Yet the map names it a "state" alongside El Salvador and Honduras. It is a subtle dig, since that identifier alone distinguishes Chiapas from the Mexican districts of Tabasco and Yucatán. Further, the map presents an "undefined boundary" with Yucatán, home of the "Lacandones" (a Maya people), and land south of the Tabasco boundary in the hands of "Los Mayas." In other words, Rivera

Maestre consigns substantial territory between Totonicapán and Verapaz to peoples and areas seemingly beyond either country's control.[26] Valle's preoccupation with indigenous people as not civilized, and speaking different languages, is presented by Rivera Maestre as a political rather than cultural or linguistic problem.

The map does suggest the beginning of a northern boundary with Belize in an incomplete line that shows Belize topographically united with Guatemala by river systems and roads. The name "Belize" hovers only to the north of the Sibún River, rendering invisible or ambiguous both British and Guatemalan claims to the area between the Sibún (northern) and Sarstún (southern) rivers with their British settlements.[27] Overall, the map presents an amorphous national shape, which helps explain its importance both in 1832 and for governments and scholars since; Rivera mapped what the government claimed, and left unsettled or ambiguous what was not clear.

Rivera Maestre's atlas and map offered the first official cartography of independent Guatemala and fulfilled Valle's agenda of showing the political divisions of the "actual state," although without addressing the pre-Columbian or colonial cartographic content laid out in Valle's treatise. The country map was the first national map published in Guatemala. Prepared by a (naturalized) Guatemalan, it reflected a governmental claim to a national territory that was adamant about internal authority, as demonstrated in the lines and words that identified and placed the country's departments on the national territory, and by the decision to leave out the mountain elevations that figured prominently on the atlas's departmental maps. Yet the map also showed awareness of limited control over certain Maya peoples and lands and honestly depicted external, "undefined" limits with neighbors. Depicting political authority in a map designed to "be brought to view to legislate and govern" was the map's principal claim, and it gave graphic form to state assertions of the right not only to defend a certain territory from external claims but also to divide the new state based on national interests. That is, Rivera Maestre's atlas declared Guatemala's cartographic independence.[28]

MAPPING THE REPUBLIC OF GUATEMALA (1850–1900): FOREIGN EXPERTISE AND NATIONAL TEXTBOOKS

Within a decade after publication, the Rivera Maestre map, made for a federal state and a compilation of others' data, was out of date and outmoded, reduced to a point of departure. The federation dissolved in 1839, and Guatemala proclaimed itself a sovereign republic in 1847. The national government commissioned a new territorial map based on the mapmaker's own on-the-ground travels to verify and measure with modern instruments the latitude and longitude of cities, rivers, and mountains. At first, this meant partial recolonization and reliance on foreign expertise to collect, analyze, and print data in cartographic form. The late nineteenth century, in addition to recolonizing the Guatemalan map, also initiated its democratization, as semiofficial school geographies began to put the increasingly well-defined and delimited geo-body in the hands of citizens and schoolchildren. These official and semiofficial maps reflected also political developments, including decrees increasing the number of departments and treaties with Great Britain (1859) and Mexico (1882) fixing international boundaries.

Notable among these partially "decolonized," nineteenth-century official maps are maps by two German surveyors: the 1859 map produced for conservative president Rafael Carrera by Maximiliano von Sonnenstern, who also drew El Salvador and Nicaragua's first official national maps, and the 1876 map by Herman Aú for liberal president Justo Rufino Barrios.[29] The New York–engraved Sonnenstern map used extensive measurements made by Belgian would-be colonist and engineer Agustin van de Gehüchte and received the encomium of "well detailed, and one of the best from the time" eighty years later.[30] That praise was probably well earned. As Van de Gehüchte wrote in a letter to London's Royal Geographic Society, he had spent eight years making measurements, checking others' facts, and recalculating locations after his first explorations in the "interior" showed "that all the maps of the country were bad" because "they were not the results of measurement as they ought to be, but compilations made in offices, from a mass of false data, and from those, each more incorrect than the other."[31] Sonnenstern's map married his drafting capabilities to the Belgian's impressive data to depict a clear, contiguous shape and greatly improved information on departmental areas, mountain ranges and rivers. The topographical features highlighted Belize as a natural extension to Guatemalan territory, fixing the government's claims to that territory on paper.[32] This map, based on extensive fieldwork, brought respected methods to Guatemalan official cartography.

Almost twenty years later, Herman Aú's 1876 map "drawn and published by order of the government" was printed in Hamburg and reflected the focus of a liberal state on its political claims and increasingly economic aims. As Stefania Gallini has shown, this map contributed to Guatemala's cartography by naming the "Costa Cuca," highlighting features that promoted investment in the country's nascent coffee industry, and minimizing problematic features, including slopes and rivers, that suggested a difficult trip from farm to port.[33] Brightly colored border lines seem to reduce the increasingly accurate topographic representation to backdrop of a country whose man-made divisions are more eye-catching than natural features. Aú uses blue for international borders, bright red to show departmental limits, and yellow to mark economically important telegraph lines. The Maya—so prominent as territorial claimants within Guatemala in Rivera Maestre's map—disappear, and the state seems increasingly in control and in communication across its territory.

The northern international border problem also seems headed for resolution. By 1876, the Guatemala-Belize border seemed settled, although the issue continued to surface, most notably in the 1930s, as Guatemala argued that Britain had failed to fulfill obligations in the 1859 Aycinena-Wykes Treaty to build a road.[34] In contrast with the very visible Belize border that reflected the (temporary) resolution, both maps seemed to indicate Guatemala's northern lands extending to infinity—or at least off the map—abdicating responsibility for showing a northern boundary. The "Lacandon tribe" placed by Sonnenstern in some areas is shown by Aú as "independent Lacandon Indians," hinting perhaps that the problem in resolving Guatemala's northern border seen in Rivera's 1832 atlas might have less to do with Mexico than with the forested area's residents. Aú's map does suggest the growing certainty of the western Guatemala-Mexico Border, dropping claims to Soconusco and connecting the Usumacinta River to a straight line of a frontier "traced" in 1811; this general outline, not yet mapped by Sonnenstern in 1859, laid the foundations for the final border agreed in 1882, although Mexico succeeded in incorporating territory substantially east of the river.

The Sonnenstern and Aú maps offer the best official hybrid "decolonized" cartography of the third quarter of the nineteenth century; foreigners drew and printed the maps but used significant local information and advanced government aims. Interior divisions received as much attention as international boundaries, and the areas identified as belonging to the Maya were slowly being reduced to a particular group, the Lacandon, who had never been under Spanish control and whose problematic status was increasingly located on the Mexican side of the border. If the geo-body was still amorphous in the north, interior space was increasingly under state control, and problematic peoples moved off the map.

While the Guatemalan government sought to fix the state's territorial shape on the map and display that geo-body in officially sponsored cartography, decolonizing it, a parallel process by private individuals pushed maps of Guatemala into schoolrooms and the popular imagination through geography textbooks and foreign commercial and travel accounts, making the map the work of the nation as well as the state. This was the first step in democratizing the Guatemalan geo-body, or getting it into citizen and children's hands and minds.[35] The Guatemalan government first promoted and mandated geographic instruction, "principally that of the republic," in 1836.[36] However, rather than commission textbooks, it purchased them after publication for use in classrooms.[37] By the 1860s, early school geographies' maps were a first element of the democratization of Guatemalan mapmaking, for Guatemalan educators drew and printed their maps in the country. By century's end, however, the same demands for accuracy and better printing technology sent school geography maps, like official maps, to Europe for engraving and printing.[38]

––––––––

The most notable case of cartographic development is that of Francisco Gavarrete, whose *Geografía de Guatemala,* with accompanying foldout map, went through three editions (1860, 1868, and 1874).[39] The maps followed the same arc in terms of shape and production seen in the transition from Rivera Maestre to Sonnenstern and Aú. Gavarrete—who headed the Guatemalan archives in the 1840s, was a member of the Sociedad Económica in the 1860s and served later as archivist and librarian of Guatemala's archbishopric—prepared his own rudimentary maps, and, for the first two editions, had them engraved and printed in Guatemala as black-and-white foldouts in the textbook; the third edition's more sophisticated foldout map was printed in Paris and published by Emile Goubaud, a French-born Guatemalan coffee grower and bookseller. Juan (Justo) Gavarrete took the project to its final stage, with a colored wall map, from approximately 1880 and also engraved in Paris, that brought the map out of the textbook (fig. 4.5).

In many ways, the schoolbook cartography shared the same goals and trajectory as the official maps. Each national map demonstrated the official geo-body with national borders and demarcations revealing internal political, economic, and cultural divisions. Still, perhaps because they were aimed at children and citizens, there were differences. Nineteenth-century "official" maps prioritized accurate topography and placement of roads, rivers, cities, and international boundaries. Department capitals and borderlines were present but, excepting Aú's 1876 map, difficult to see. Schoolbook geographies, however, inked thicker lines, colored departments, or subtracted topographical information to highlight internal political divisions. Further, while many reflected ambivalence about national borders by following Rivera Maestre and Aú's example and leaving out or minimizing international boundary lines, Gavarrete's

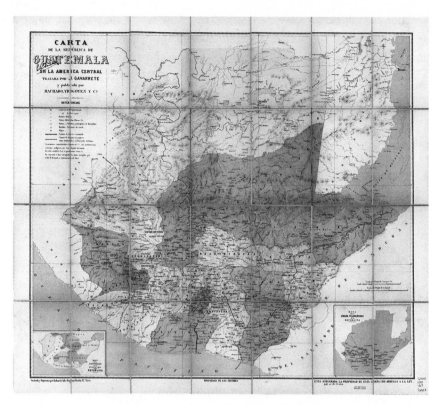

FIGURE 4.5. Juan Justo Gavarrete, *Carta de la República de Guatemala en la América Central* (Paris: Ernard, n.d.), G4810.1880.G3. Note Soconusco's incorporation into Mexican national territory and inset maps showing Guatemala's language and telegraph lines. The Bancroft Library, University of California, Berkeley.

maps of Guatemala gave the Petén a clear triangular (1874) and then horizontal (ca. 1880) northern border with Mexico.[40]

Gavarrete's textbook sequence also brings into focus the three options used by later Guatemalan mapmakers to address the evolving but still unresolved status of the boundary with Belize: showing Belize fully merged with Guatemala, merging Belize but showing a fixed or tentative border line, or marking Belize as a separate entity. Gavarrete's first map showed no boundary line, but the 1868 and 1874 maps included a tentative line separating Belize from Guatemala that reflected 1859 treaty terms. By 1874 Belize was topographically Guatemalan as well. In terms of economics and indigenous peoples, Gavarrete's map treated ports like other towns and identified no specific commercial ventures. The "Lacandones" appeared in the 1874 schoolbook edition in northwest Petén; the only indigenous group on the map seemed situated outside Guatemalan territory. This map did introduce one new element to the public:

elevation. For the first time, the map named Guatemala's principal volcanoes, topographic features that interested scientists, travelers, and residents alike and made their way onto most subsequent topographical maps. Although these maps showed the geo-body as it was understood at the time, the detail emphasizing internal rather than external structures hinted that the decolonized Guatemalan citizen and Guatemalan map should be less concerned with where the country stopped and more engaged with its internal makeup.

Juan Justo Gavarrete completed the sequence and incorporated substantially more information. His wall map became a foundation for twentieth-century maps.[41] Geographically, it followed Valle's approach of more is more; more cities, rivers, and very clear internal boundaries connected Guatemalan cartography with the 1830 textual description. Valle's 1830 sensitivity to internal administrative divisions appeared graphically in a map that distinguished parish and departmental capitals, "dependent villages," and even rural districts, including plantations. The map and its legend highlighted not only sites of ruins and battles, or Guatemala's history, but also economic and commercial projects, including existing and projected roads, mines, and ports. The map touted past, present, and future. In addition, this wall map also addressed Valle's demographic and commercial agendas in inset thematic maps: telegraph routes and the "distribution of aboriginal languages in the Republic" notably characterizing the southern half of the country, a shape that more commonly was used to show railroads and other land-based transportation, which were concentrated in this region.[42] Not only were the Petén's independent Lacandon peoples, of such concern to earlier governments and cartographers, not on the map, but no indigenous peoples seemed to populate the area.

The idea of representing indigenous *language* groups rather than indigenous *peoples* continued Valle's initial categorization but here seems to have served a double purpose. Guatemala's indigenous population could be recognized not as historic or politically autonomous, but as linguistically and spatially distinct from Spanish speakers. The latter's location needed no map, presumably because they lived throughout the country. Where Valle criticized indigenous languages, Gavarrete's map seems to celebrate them. Conceptually, Gavarrete's approach included the majority population within the national territory and community while it marked its members as culturally separate. For this, as the legend noted, Gavarrete could draw on scientific (foreign) expertise (the information came from "Dr. H. Berendt"), suggesting international approval. In marrying geographic knowledge with Valle's original agenda of incorporating the country's past, people, resources, and commercial opportunities on a single sheet, this map set the stage for Guatemala's twentieth-century offi-

FIGURE 4.6. "Guatemala," tobacco card, ca. 1900–1920 (El Perú: Fábrica de Cigarrillos de Roldán y Ca., Lima). Private Collection.

cial and popular cartography. The late nineteenth-century Guatemalan geo-body based on successive legislation and the 1859 and 1882 treaties that would become the national form was already being adopted by popular cartography, transformed into a map-as-logo in a Peruvian cigar card during the presidency of Manuel Estrada Cabrera (1898–1920) (fig. 4.6).

Guatemala completed its nineteenth century with strengthened cartographic expertise, yet the century's final map remained a joint effort. Interestingly, the mapping of Guatemala's Mexican border by foreign experts working along-side graduates of new engineering and military schools revealed the country's political weakness. Despite the boundary agreement of 1882 and "scientific commissions" appointed by both countries "to properly trace the boundary line on trustworthy maps and to erect upon the territory monuments which will show the limits of both Republics,"[43] the map both countries accepted in 1899, which largely resembles today's geo-body, yielded to Mexican pressure rather than treaty imperatives. Implementing the treaty as written would have substantially reduced Guatemala's northern territory, but the country lost a little more territory around the waist to post-treaty Mexican interventions (areas 1 and 2 in fig. 4.7).

Miles Rock, the North American civil engineer who headed Guatemala's team, opposed the decision. His sketch map documenting multiple violations curved the word "Guatemala" determinedly across the areas claimed by Mexico, with the final "LA" firmly in the triangular northern territory reaching into the Yucatán. This map, and the 1895 published version,[44] pointedly sup-

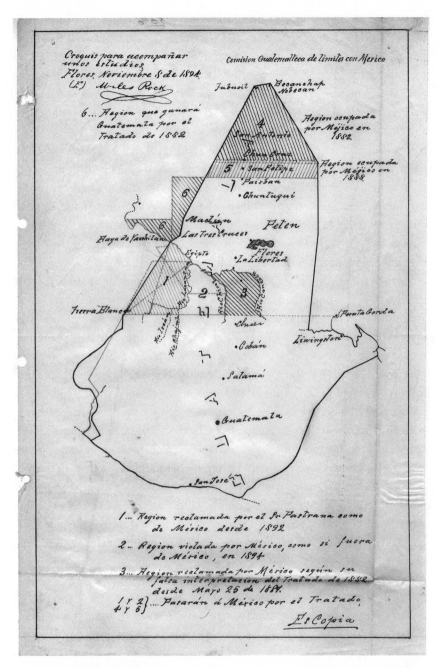

FIGURE 4.7. [Miles Rock], manuscript copy of a sketch map of Guatemala, showing areas (1–5) that he considered should by law be part of Guatemala; of these, Mexico only gave up the area in 3. Private collection.

ported the nation's claims, showing that Guatemala lost territory principally in the Lacandon jungle shown as "los Mayas" in Rivera Maestre and "Lacandones" in Gavarrete and Sonnenstern.[45] This area to the south and east of the Usumacinta River had been visible cartographically as Guatemalan even in renowned Mexican cartographer Antonio García Cubas's atlases and national maps in the 1850s and 1860s.[46] In one way, however, Rock's map did reflect the Guatemalan state's growing authority; the Petén showed "milpas," subsistence agriculture plots, where the Maya and Lacandon had previously been named. By 1895, if Guatemala was losing territory and finally assuming a permanent geo-body, the state could at least show clear administrative authority and subsume indigenous groups into productive agriculture, even if not cash crop production. A 1902 catalog of the country's settlements by the statistics bureau charged with the national census included a plan to demarcate interior divisions, suggesting the state's inwardly turning attention.[47]

As the nineteenth century closed, the decolonization of Guatemala's territory and cartography and the formation of its geo-body were thus largely accomplished. The country was no longer a collection of amorphous provinces whose national territory and shape were hard to define or recognize, whose maps and histories had yet to be written. From the work of the Mexico-Guatemala boundary commission, the current shape—perhaps fancifully imaginable as the silhouette of a round-bottomed ship—had largely emerged. The state by 1895 had a specific territory and could represent it, with Guatemalan-trained engineers like Claudio Urrutia, who replaced Rock, working alongside foreigners who made Guatemala their home and applying modern techniques and team-based mapping of complex projects. In the next century, the only significant variation in the shape would come as the government's interest in publicizing its claim to Belize waxed and waned. The final stage of a decolonized and democratic map—one produced by a Guatemalan engineer for a public audience and made in Guatemala—arrived with the new century.

MAPPING GUATEMALA FOR AND WITH THE PUBLIC: THE TWENTIETH-CENTURY GEO-BODY IN CONCRETE AND ON PAPER

The public cartography of engineer Claudio Urrutia (1857–1934) at the turn of the twentieth century offers insight into the public acceptance of Guatemala's (almost finalized) geo-body and the government's engagement with the public on representations of national territory. One of Guatemala's most prolific and published cartographers, Urrutia participated in and later headed the Guatemalan boundary commission team; later still, he headed similar commissions to finalize borders with El Salvador and Honduras. Urrutia's contributions to the 1905 relief map constructed in Guatemala City and to 1916 and 1923 printed maps of Guatemala show the evolution of the Belize issue under the Estrada Cabrera dictatorship, and the transformation of the geo-body from the property of state mapmakers and textbook geographies to documents and monuments for public consumption and debate. Urrutia's cartography targeted both state and nation and paved the way for decolonization of the map and democratization as, for the rest of the twentieth century, multiple groups adopted and adapted the national map for their own purposes.

Urrutia was born in Costa Rica to Spanish parents but spent most of his life in Guatemala. He worked as an engineer and later chief of the border commissions for Mexico (1889–), Honduras (1912–), and Belize (1927–), while also serving as dean of the University of San Carlos's engineering faculty from 1891. Yet while he was an official surveyor for the state, Urrutia's public cartography generated sufficient controversy to demonstrate that, even in the early twentieth century, Guatemalan citizens were coming to feel as proprietary about the national geo-body, and its portrayal in influential maps, as the government.

Urrutia won a larger audience for his cartography than Rivera Maestre or even Gavarrete. Greater literacy, particularly in the country's growing cities, and improved printing technology brought venues and platforms for map display and distribution out of government offices and schoolrooms. Urrutia participated in Guatemala's most notable and unique public cartography project, the relief map, a three-dimensional concrete map built on the outskirts of Guatemala City at a 1:10,000 (horizontal) × 1:2,000 (vertical) scale inaugurated by President Estrada Cabrera in October 1905 (fig. 4.8). Estrada Cabrera commissioned the work from engineer Francisco Vela, who in turn contracted Urrutia in 1904 to undertake the relief map's design, projection, and building in Zone 2, just north of the city's original center.[48]

In essence, the relief map was Guatemala's first "logo map," consciously constructed to conflate geo-body with *patria*, or homeland, and the map succeeded admirably. Urrutia himself considered it "an artwork constructed with great care and precision, that attracts the eye and engraves the country's configuration easily in memory . . . in short, a synoptic painting of Guatemala."[49] He understood the map's purpose as both didactic and celebratory, teaching not just accurate information about specific mountains or rivers but "engraving" the geo-body in visitors' memory, instructing them to identify the country with its shape. Although state sponsored, this construct quickly became, as

FIGURE 4.8. M.M.S., Race Track and Relief Map, Guatemala City. ca. 1910. Postcard. Private collection.

intended, a national icon. The 1908 *Pan American Magazine* complimented a "work that merits a visit from every traveler arriving in Guatemala, whatever may be the object of his expedition."[50] Still an obligatory visit for today's schoolchildren and one of the few sites to attract foreign visitors to downtown Guatemala City, the relief map laid out the nation's topography, departments, and international frontiers in concrete. Standing above in wooden viewing stations, the visitor or citizen would instantly see the territory that accepted the loss of Chiapas and Soconusco but still claimed Belize. Just as the nineteenth century saw the transition from Valle's "geographic description" of 1830 aimed at literate adults to the promulgation of school geographies for children, twentieth-century citizens used the Relief Map for their own purposes. To name just one later example, in 1983, Guatemala's Club Andino (a mountaineering club) published a guide to the country's volcanoes, inviting readers to visit the Relief Map to see "in one view the entirety of Guatemalan soil in its different features . . . [and] the geography of our country."[51]

Although few seem aware of it, this nationalist map was not, in its creator's view, complete. Urrutia, like Valle in 1830, connected Guatemala's physical geography with its economic and commercial development. He unsuccessfully proposed building a twin "agricultural and commercial" map to highlight the country's products and road systems, even as he emphasized that the map was more than just "an Orographic and Hydrographic Map," or geographic representation.[52] And while Guatemala's indigenous people did not appear on the map, nor did Urrutia suggest a demographic map to complete the project, Urrutia identified the manual workers on the project as "of indigenous race," perhaps to find a way to include the majority population as creators as well as receivers of the map. At a time when the nation's body was selling the Positivist "order and progress" notion of the age, this national and nationalistic map had no room for separate groups or agenda and signaled a unity of purpose important in an age of nationalism.[53]

After the 1905 public project, Urrutia's government-sponsored cartography served traditional map consumers. He produced a basic black-and-white geographic map for government use (1916) and a colored wall map for schoolroom and perhaps office use (1923, reissued 1934). Urrutia's 1916 national map showed basic geography, topography, and road networks and a clear international border with Belize as a separate territory.[54] This map seems to have served as a base map for both military and civilian government cartography for the next twenty years, an outline national and departmental map with only the major mountains and rivers as topographical features. Seven years later, Urrutia published a commercial color wall map, printed in Germany and based on cartographic data prepared for the government, with gradient tint for relief (fig. 4.9).[55] While "compiled by disposition of the superior government," this was not an official map. Like the school geography maps, it meant to bring the latest information to national and international public audiences, and was a successor to Juan Justo Gavarrete's 1878 wall map.

While Urrutia's 1916 map is arguably a fully decolonized Guatemalan map, it is the public discussion of the 1923 map that demonstrates the democratization of cartography of the ongoing formation of the geo-body, which had become a sensitive national issue not just to the government but also to citizens. In territorial terms, the map remains ambiguous about Guatemala's claims to Belize; the font used to show the name is the same size and shape as that of Mexico, El Salvador, and Honduras, but the topographical features and color schemes used for Guatemala continue into Belize alone among the country's neighbors. The public reaction to the map was strong, but not because of Belize. Instead, a May 1928 article in the *Diario de Guatemala* heatedly rejected the map as "not appropriate for the country" for its "misrepresentation" of the Guatemalan-Honduran border. Urrutia promptly penned a letter to the editor to defend his work, noting that the Guatemalan representative in the ongoing border discussions had used and praised the map. Experts had asked to consider a smaller version for school use and told him that the map "seems perfect to us," asking only that the frontier with Honduras not be drawn as a finalized

FIGURE 4.9. Claudio Urrutia, *Mapa del Estado de Guatemala, República de Centroamérica* (Hamburg: L. Friedrichsen & Co., 1923). Scale: 1:400,000. Courtesy of the Mapoteca Manuel Orozco y Berra, Servicio de Información Agroalimentaria y Pesquera, SAGARPA, Mexico City, CGCAV2–10-CGE-O-A.

line to avoid giving Honduras an argument to use in negotiations.[56] Urrutia seemed to have followed the reviewers' suggestion to note "frontier not fixed" over the relevant mountain ranges and used color through the disputed area and to label disputed territory.[57]

The popular reaction to Urrutia's map—even comments in a national newspaper that might have been shaped by official political interests—and his own defense of it suggest that by the 1920s, Guatemala's geo-body really had been adopted by some citizens as well as the state. By this time, academics were including the national map without Belize in their published work.[58] That is, the map had been democratized to the point that adults could argue publicly about its merits rather than have official engineers or bureaucrats make executive decisions about how to show territory both permanent and still "undefined" to meet only state interests. The reaction also reveals that in the first quarter of the twentieth century, Belize—included or excluded, ambiguous or not—was not naturally a flash point for either government or people. After one hundred years of living with an uncertain border and fluctuating Guatemalan and British interest in resolving its location, Belize could appear both on and off the map; Guatemalans recognized and accepted either geo-body depicted.

The multiple insets on Urrutia's map make it a protoatlas. They show the national territory as sufficient for the geographic, scientific, and political understanding of Guatemala, but only a partial base for a full demographic and historical story. Urrutia inserted maps of the country's spoken languages in the upper left, the historical territory and peoples of what became the Kingdom of Guatemala (1521–1821) at the time of "discovery and conquest" in the lower right, and mapped mountain and volcano elevations in the lower left-hand corner. In addition, he filled oceans and neighboring countries with tables identifying the geographic position of "principal limit points," the height at which plants such as wheat and coffee grew, statistics about the area, capital, height, and population of each department, and monthly averages of national rainfall and temperature in Guatemala, Quetzaltenango, Salamá, and Puerto Barrios. These geographic, political, and environmental features address only the national territory. The conquest-era map and the map's name, however, maintain the country's isthmian context. The language map shows contemporary Guatemala, with the languages occupying not just different departments, but also the spaces lost to Mexico and Belize, as if the nation (if not the territory) were not only rooted in the contemporary political structure but also more ample than the geographic map can accommodate.

Juan Justo Gavarrete's 1878 map initiated this multimap or map-as-atlas presentation of Guatemala; Urrutia updated and expanded the topics. This map essentially achieves Valle's 1830 agenda of three maps to represent Guatemala. Although it does not show territorial divisions of Guatemala's pre-Columbian or colonial kingdoms, it situates contemporary and pre-Columbian indigenous peoples on separate maps and presents the colonial area "at the time of conquest" as a space with native peoples hovering over undemarcated territory and colored lines tracking the routes of the conquerors which the modern country, with interior divisions, has occupied and defined.[59] The irrelevance of the colonial provinces (which are not mapped) to the contemporary state is the one way in which this map takes issue with Valle; modern Guatemala needs its indigenous past and present, but its Spanish colonial past fades except as a regional area in which several countries were founded. From his work on the national commission, to authorship of "official" maps for government and public use, Urrutia was the Guatemalan cartographer who set the stage for subsequent Guatemalan national mapping and Guatemala's twentieth-century achievement of lasting cartographic independence.

FROM BASE MAPS TO FLOATING MAPS: DISSEMINATING AND ADOPTING THE GEO-BODY, 1930–2010

Guatemala's nineteenth-century cartographers decolonized the national map, mastering the shape and contours of the consolidating geo-body. Their early twentieth-century successors, Guatemalan engineers and cartographers, engaged with Guatemalan citizens feeling confident and (responsible) enough to challenge or discuss cartographic decisions in the press. The 1930s transformed the national map and geo-body into a standard base map and subsequent logo-map, "floating" unanchored to its neighbors. Three important changes took the map into its final stage: lithographic printers, institutionalization of a cartographic bureaucracy, and increased literacy combined with the spread of print media. These changes made the map of Guatemala's geo-body, with and without coordinates or other technical details, accessible to nonspecialists as well as professional and official cartographers. Although it is not within the scope of this essay to offer a comprehensive treatment of this increasingly decentralized process, this section shows how democratization of the map has led, perhaps paradoxically, perhaps obviously, to a ubiquitous emptied-out logo-form that is recognizably Guatemala and employed broadly, with either a blank interior or one showing departmental markers or national symbols.

Before considering these elements it is worth noting that Guatemala's case is unusual. Most countries have a single geo-body. Guatemala's "recognizable" geo-body may include and exclude Belize in official and popular representation without provoking or confusing Guatemala's people. They use logos with and without that part of the silhouette in everything from professional associations to advertising illustrations. Perhaps the most striking example of the inconsistency of the government position is the country's five-cent stamp, which in 1935–36 separated Belize only to redraw it in 1948 as part of the national territory (figs. 4.10 and 4.11). The stamps also show the government using the geo-body to project its message into any household posting or receiving letters, at home and abroad. In the era when government news probably reached most homes by radio rather than print media, letting the people know which

FIGURES 4.10 AND 4.11. These two five-cent Guatemalan postage stamps are from 1935–36 and 1948. Private Collection.

geo-body the government supported on postage stamps was clever. Ironically, the first stamp was issued during the dictatorship of Jorge Ubico, who insisted Belize belonged to Guatemala and incited nationalist passions.[60] The second appeared during the progressive presidency of Juan José Arévalo (1945–51), whose own 1936 atlas did not include Belize as national territory. Did the director of the post office feel free to ignore presidential policy?

Also starting in the 1930s, government offices drew on Guatemala-educated engineers and experts to produce their own thematic maps, several of which were printed by Byron Zadik's Guatemala City lithographic printing shop (opened in 1930). Zadik made quality, color map printing available in Guatemala for both public and private projects, marking the return to cartographic independence experienced briefly in 1832. Zadik published (among others) maps of transportation networks from the Ministry of Public Works and Roads and the Ministry of Agriculture's 1933 map promoting "the best coffee in the world."[61] Using Urrutia's 1916 black-and-white outline map (or similar), the miltiary imposed multiple themes on the national territory, especially transportation and communication (from road and air routes to telegraph lines).

The final stages in decolonizing the map began when Guatemala founded its first official mapmaking institution, the Dirección General de Cartografía (DGC), in 1934 and when, in 1964, its successor, the Instituto de Geografía

Nacional (IGN), began lithographic printing.[62] Even though the US provided the IGN with substantial support as part of a hemispheric initiatve, the Guatemalans saw the base map as an important national project and achievement, and also as part of a regional initiative; starting in 1956, Central America's sister institutes began meeting for "cartographic weeks" to coordinate policy.[63] The IGN produced Guatemala's first map with data from modern geodesy and triangulation, a 1-sheet and later 12-sheet "preliminary" map at 1:200,000 (1945, 1958), and later a 197-sheet set of 1:50,000-scale photogrammetry maps. On top of these "basic" maps, the IGN published Guatemala's second atlas (1964 [preliminary], 1972) over a hundred years after the first, and several thematic maps. One of the first, in 1962, mapped Guatemala's indigenous languages using the familiar style of shading language families onto the backdrop of departmental outlines. But instead of using foreign scientists' data, the IGN drew information from the country's Instituto Indigenista Nacional, run by Antonio Goubaud Carrera, a US-trained anthropologist and grandson of the bookseller Emile Goubaud, who had printed Gavarrete's maps.[64] The stand-alone theme maps that followed included geologic, hypsometric, climate, forest, and land use maps (1964–66), folk crafts (1966), a 4-sheet map for school use (1971/6), and a national road map for tourists (1980). As a military office from 1983 to 1996, the Instituto Geográfico Militar (IGM) updated the road maps (1983–97) and added a map to archaeological sites (1991). Once merely insets on a larger map, theme maps became separate and powerful displays of state interest and knowledge.

Those maps could be and were put to multiple uses by residents, reaching and influencing a broader audience than nineteenth-century official maps, which seem targeted at foreign investors as much as for national use. Whatever the focus, by the late 1960s, the Instituto ledgers document substantial popular interest in its output. There were bulk sales of maps to national agencies, international petroleum companies working in the Petén, and US AID, plus smaller quantities purchased by Guatemalan engineers, architects, companies, schools, individuals, and clubs, including the Club Andino, which drew on DGC and IGN materials for its 1983 guide to the country's volcanos.[65]

———————

Zadik's lithography also supported a 1930s revolution in schoolbook geographies, when Juan José Arévalo's *Geografía elemental de Guatemala* (1936) echoed the government's message of national pride. Arévalo, a schoolteacher who as Guatemalan president in the 1940s promoted progressive policies favoring workers, land reform, and industrialization, takes the Guatemalan geo-body as a container limited by neighboring countries and filled by its own departments and uses color to layer themes on floating outline maps, including population centers, communications, department capitals, forests, climates, and mountain ranges. Unlike its nineteenth-century predecessors, this anti-rote-learning geography rarely used words in the images—schoolchildren received what Arévalo called "mute" maps next to brief explanations, requiring engagement and application of geographic knowledge. To help teachers adopt interactive learning, one exercise invites students to plan trips within the country from their home department. Another asks them to identify the departments on the border with Guatemala's neighbors (including Belize, not shown as part of the geo-body) and then to describe their size and shape and other characteristics.

Further democratizing the map, Arévalo emphasized not just map reading and geographic knowledge, but also mapmaking as something children should do. Arévalo's instructions for a teacher's use of the *Geografía Elemental* included instructions for the child to draw the plan of the classroom, the town, and then the department, and only then a map of the republic, at which point the student would be ready for the geography lessons. For lessons with maps, Arévalo recommended that teachers draw the maps on the blackboard during the lesson for maximum "educative and solid" impact and urged teachers away from rote learning and overemphasis on details and toward having students understand and interpret what they see. Finally, Arévalo recommended that if materials were available, the child should make his own map for each lession, collecting them in his own atlas, as the most rewarding manual and intellectual experience "to know the general map of Guatemala."[66]

The only themes Arévalo did not map were Guatemala's cultures and languages, relegating Guatemala's heterogeneity to a few paragraphs and a photograph at the end of a 120-page book. This approach, which incidentally dismissed Afro-Guatemalan presence and contributions and considered native languages primitive, might have reflected the author's focus on class and commerce over race as the way to develop the country.[67]

However, schoolchildren, scholars, and government agencies soon had access to maps of Guatemala's indigenous languages prepared by Goubaud Carrera, who not only worked with the IGN, as discussed above, but also authored a 1946 map of "present-day indigenous languages" that has been reproduced in school geographies,[68] numerous academic studies, and in 1964 in an IGN edition published for the Seminario de Integración Social. The Seminar's secretary general indicated in a statement printed on the map that part of the print run was intended as a teaching instrument for schoolchildren to "better know the human makeup of their country." Achieving this result, or "positive reality," would require "the active cooperation of Guatemalan schoolteachers."[69]

A few years later, Goubaud Carrera adapted this map into a pamphlet published by Guatemala's Instituto de Antropología e Historia (IAH), *Idiomas Indígenas de Guatemala* (1st ed., 1949; 2nd ed., 1984), to help young readers know and thus love the homeland's "historical and cultural treasure," put "intelligence" and "sentiments" at its service, and "with pride call yourself a GUATEMALAN (sic)."[70] The text is a conversation between a student surprised to hear indigenous languages on the street and his teacher, who explains that the twenty-three Maya languages in Guatemala are part of a rich heritage and should inspire pride. When the student asks where each language is spoken, the teacher consults maps of the "general regions" for each, shown as shaded areas in sections of the country's departments—essentially Goubaud Carrera's national map broken into regions. Like Arévalo, the IAH and Goubaud Carrera wanted readers to learn by doing, although here the emphasis is on learning to respect Guatemala's living peoples, not how to make maps. Exercises include visiting local markets to ask the indigenous there where they are from; when visiting indigenous areas, to take notes about language, economic activity, buildings, and so forth; and to use color to paint the areas shown for each group on a national map.[71]

Arévalo's geography is now hard to find. It had only one print edition, perhaps because its glossy, colored, lithographed pages made it prohibitively expensive. Even Arévalo's descendants today apparently don't own a copy.[72] Fortunately, Arévalo's geography was soon joined on the shelves by more accessible maps.

Julio Piedra Santa, from Quezaltenango, founded the press Editorial Piedra Santa with his wife Oralia Díaz in 1947. Both were teachers imbued with the revolutionary ideals of the era, and they shared the goal of providing useful and affordable cartographic and other didactic materials to Guatemalan and other Central American schoolchildren. Unlike Arévalo's hardcover, big-format, heavy-paper geography, Piedra Santa's small, paperback, newsprint textbooks have gone through numerous editions and can be found throughout Guatemala. The first edition covered all of the Americas (1976), the second narrowed to Central America (1980), and, finally, the third concentrated on Guatemala (2001–present; see fig. 4.12).[73] In many ways, Piedra Santa shared the motivations of Arévalo and their nineteenth-century predecessors. His principal audience was "professors and students," whom he expected to find the work "of much utility as much for its clarity and exactness as for its rigorous updating."[74] Starting in 1976, the *Geografía Visualizada* series of geography textbooks was

FIGURE 4.12. Cover of the 2007 *Geografía de Guatemala: Cuaderno de mapas . . .* Note the map's emphasis on physical features and political divisions, the use of Piedra Santa's logo where one might expect to see the word Belize, and the mix of historical and contemporary architecture and peoples on the map, with a reflection on urban and rural life. Courtesy of Irene Piedra Santa.

published to complement the *hojitas,* or blank maps for studying and filling in, that Piedra Santa had sold for pennies and that are now sold in *cuadernos* (notebooks) for "annotation and exercises," costing about $3.00 (15–18 quetzales) in 2010.[75] Piedra Santa, like Arévalo, was not just decolonizing the map, making it the product and property of Guatemala, but also democratizing it, by putting an affordable "mute" map in hands of schoolchildren who could make it "talk" with their additions.[76]

Over the years, the Piedra Santa geographies have adapted to Guatemala's changing political context, with the end of the civil war in 1996 suggesting a transition point on both territory (Belize) and population (indigenous and other). Consciously or not, the *Geografía Visualizada* included Belize on the geo-body on the cover in 1976 and 1991 and made it ambiguous on the cover in 2001. In terms of citizens, the 1991 Central American geography mentioned the indigenous as a majority population in passing, but then only listed the number of inhabitants per country, mapping no peoples in the region. By 2001, the *Geografía Visualizada* accompanied a discussion of the connection of Mayan languages to those of Guatemala's neighbors with an updated Goubaud Carrera linguistic map. In the 2007 edition, the text cites a 2003 law of national languages and the institutionalization of bilingual education, bringing the modern Maya onto the map and into the political community. This latest edition also maps the United Nations Development Programme's indices for human development and population density by region, allowing those with interest to connect linguistic or ethnic status with socioeconomic achievements.[77]

MAPPING GUATEMALA BY AND FOR GUATEMALANS

The Guatemalan government was at the forefront of defining and mapping the national territory in the nineteenth century, commissioning maps from foreigners for official depictions and from Guatemalans for geography textbooks. By the 1930s, however, more Guatemalans—scholars, teachers, and businesspeople—were publishing increasing numbers of unofficial maps of the country, and, like the government, filled the outline or logo with themes that trumpeted their pride in their nation and its territory as a home. The geo-body became a logo used by organizations ranging from professional associations to advertisement agencies, since government and people knew this appeal to a common territory (however detached from physical neighbors it became) would resonate.

Guatemalans seem unperturbed by their double geo-body, and logos both

FIGURE 4.13. Guatemala's geo-body as a logo, with and without Belize, and other national symbols, such as the quetzal, the national bird.

include and exclude Belize, amid other national symbols (see fig. 4.13). Among those logo geo-bodies including Belize, the Geological Society of Guatemala (1974) lays a pickaxe across the geo-body and the Instituto Geográfico Nacional lays a light longitude line where the border should be. Of the logos that don't show Belize in the geo-body, the Asociación Bibliotecológica de Guatemala (1948) opens a book across and the territory and the federation of coffee cooperatives (FEDECOCAGUA, 1969) shows a coffee bean in central Guatemala flanked by pine trees. Government agencies and affiliated associations, such as those of Guatemala's municipalities and municipal firemen, have adopted the geo-body, generally including Belize in their logo. The Instituto Geográfico Nacional showed its mission by making the aerocartography project its logo; both the IGM in the 1980s and the IGN today retained that logo, although they now map from satellite images. In 1987, members of the military-organized rural Comité Voluntario de Autodefensa Civil (voluntary civil self-defense committees) were issued an identity card with a ten-point conduct code, whose cover reflects both how the state used the map as a national symbol and the expectation the map had meaning for those who

FIGURE 4.14. Identity Card from a Comité Voluntario de Defensa Civil (ca. 1987). Courtesy of Matthew Taylor.

would see it. The front cover shows the geo-body of Guatemala in outline. In front, a soldier (in camouflage) and patrol member (in peasant work clothes) face the viewer standing shoulder to shoulder, each holding a rifle at the ready. The national flag waves, planted firmly in the ground behind them, with its stripes covering where the Guatemala-Belize border should be, conveniently masking where such a line might go, and explicitly claiming the territory for Guatemala. On the back cover, underneath another unfurled flag, readers are exhorted to "remember always what we swore under our national flag: to defend it up to the loss of our lives and to not abandon that which is ordered in military action."[78] The word "Guatemala" does not appear on the front or back cover, signaling the military's expectation that anyone holding the booklet would recognize the national shape (fig. 4.14).

The logo map is also visible on the landscape. The 1905 Relief Map is still a must-see tourist, school, and family destination; most Guatemalans I have met in the capital recall visits in their youth and taking their own children in the present. But there are many other spaces outside the classroom and Zone 2 that a Guatemalan is expected to see and identify with the national territory. To take just one example, a billboard on the road to Antigua proclaimed from the 1990s to around 2010 that "Guatemala speaks with Comcel," an important telephone service provider, next to a road map floating on a blue background. Although right outside the national capital, this sign tucked its text right where Belize should have appeared.[79]

An important step in democratization was achieved by the end of the civil war: maps of Guatemala for Guatemalans were not just *about* the indigenous but, to an increasing extent, *by* them. Indigenous communities themselves rather than foreign or elite scientists became privileged sources for information in nonofficial maps intended to reach the country's general population, and civil society rather than state institutions increased map production. Piedra Santa was part of this transition; in 1988, its wall map in English and Spanish showing the "Languages of Guatemala and Belize" (Spanish and English included) distributed select images of Maya men, women, and children on the edges of the map, in what could have been blank spaces, suggesting on the surface an exclusionary map of the kind studied by Sarah Radcliffe in Ecuador.[80] However the publication information on the back of the map (along with a text describing each linguistic group) reveals that the cartographer and text author are, respectively, Narciso Cojti and Margarita López Raquec of the Linguistic Project of Francisco Marroquin University, whose surnames suggest indigenous origins. So while the map is under the artistic direction of Raul Piedra Santa and the watercolors are by Carmen de Petterson of the Ixchel Museum of indigenous dress in Guatemala City, here is a map that reflects an expanding mapmaking community.

Six years later, *Prensa Libre*, a leading Guatemalan newspaper, which now publishes an annual national map and departmental maps in weekend supplements, produced a similar poster map, "Maya Languages of Guatemala," with the support of the Marinela Bakery, the fourth in a series of educational maps. Like the Piedra Santa and earlier maps, this one shows Maya languages mapped onto the national and departmental geo-body (including Belize) and also takes a more integrated approach to Guatemala's multicultural heritage. If the map shape and features are Western, the compass rose shows not only NSEW but also four Maya glyphs. Less visible, but more important, this map, like its predecessors, claims to derive from the most recent research on Maya names and territory and to reproduce a map by the Centro Educativo y Cultural Maya CHOLSAMAJ, diversifying authority. It also recalls the human element of Guatemala's Central American context, although the "floating map" does not show neighboring countries. Instead, the text underlines that Guatemala Maya speak twenty-one of thirty Maya languages spoken in Mesoamerica, and that their languages are related to those of Mexico, Belize, and Honduras.[81] It also tracks changes in the government, for the bibliography not only includes Maya rights documents produced by Maya organizations, but also cites a national agency on bilingual education (PRONEBI).

This more inclusive way of understanding Guatemala's multilingual and multiethnic heritage has spread beyond the indigenous working on or with academic projects. After the combatants in the internal conflict laid down their arms and signed a peace accord in 1996, the Comisión para el Esclarecimiento Histórico (CEH) completed extensive investigations and published a report. The cover, designed by Servigráficos, a Guatemalan graphics firm, offers an alternate way to demonstrate the value of Guatemala's many languages. It shows a green silhoutte of Guatemala with a Maya glyph in the center. The background is not cartographic; Guatemala's neighbors are not shown as empty but contiguous spaces. Rather, the text "Guatemala, Memory of Silence" repeats in Spanish, Kekchi, and Caqchikel languages, over and over, forming a wall of words around the map. What is striking about this use of the map is the integration of indigenous and ladino cultures through words and space; unlike traditional maps meant to spatialize "native language groups," this map takes as its starting point that the country is both a physical space with a Maya past (as evidenced in the glyph) and national present. Both language and geo-body are inclusive—all the peoples and languages are Guatemalan. Belize and English are, notably, absent.[82]

Today, however, language is not the only way to represent Guatemala as a country by or about indigenous residents. Take an example of adoption of the geo-body to tell a counternarrative of organizing for political and social change, rather than depicting the state territory as external or problematic. In 2007, Editorial Rukemik Na'ojlil in Guatemala City produced the pamphlet *Lucha, resistencia e historia* (Struggle, resistance, and history) for the CUC (Comité de Unidad Campesina), a leading Guatemalan peasant organization.[83] Here the national map, with its internal divisions and including Belize, illustrates a discussion of CUC education workshops of the 1970s. A young woman, who seems to wear traditional garb, points to three districts—the central highlands, southern coast, and Huehuetenango—as sites of the group's earliest organizing. The map shades the departments to identify the districts, thus recognizing both the official political administration and the peasant organizers' regional operational divisions (fig. 4.15a). Putting the peasant struggle in national, regional, and departmental context showed how official territories had become shared spaces for activists, adopting the technique in the 1949 IAH pamphlet to opposite effect. The same editorial team also produced the telling *Historia de Guatemala: Desde un punto de vista crítico* (History of Guatemala:

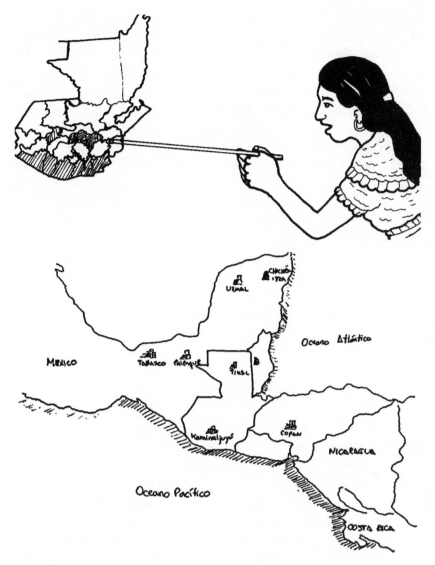

FIGURE 4.15A–B. *Lucha, Resistencia e Historia*, p. 17, and *Historia de Guatemala*, p. 10.

From a critical perspective), with a map to illustrate the country's indigenous and colonial origins. Strikingly, the map placed the Guatemalan geo-body with its fixed contemporary limits into a map depicting Mayan territoriality— anachronistic but telling (fig. 4.15b).[84] Maya and national Guatemala were thus, cartographically, united.

More recently, the government and private groups have presented multi-racial Guatemala on the map, using the map-as-logo to expand beyond the multilingual map. The Guatemala Joven campaign of the Ministry of Education shows three figures throwing up their arms energetically—white largest and centered, black and grey on either side, standing on the geo-body. Other national campaigns for national reconstruction, scholarships, and family aid show four hands—black, white, red, yellow—on the geo-body, consciously or unconsciously suggesting that Guatemala is a "cosmic race," strong because of its African, European, Amerindian, and Asian origins.[85] The use of Maya glyphs or stylized figures on the Guatemalan national map to signal pride in and acceptance of or connection to Guatemala's indigenous past and present can be seen in the emblems of professional organizations; the Colegio de Médicos y Cirurjanos de Guatemala, for example, shows a kneeling Mayan figure conveniently covering Belize, and INDEGUAT (Guatemala's nutritionists' association) shows a jeweled female figure in profile looking at corn and other foods. Both use glyphs on the map to represent links between modern science and ancient traditions or native products. In different ways, each of these logos seeks to be inclusive and to incorporate Guatemala's multiethnic population, putting the indigenous as well as mestizo and arguably Afro-Guatemalan (Garifuna) on the map.

CONCLUSION

Although José Cecilio del Valle penned the first geography text and imagined the national map that government would use to administer Guatemala in 1830, and Miguel Rivera Maestre and his countrymen published the first national map in 1832, Guatemala's cartographic independence took longer to consolidate than to declare. Creating the Guatemalan nation's geo-body was a century-long process, beginning with independence in 1825 and largely concluding by 1934 when permanent borders were agreed with Honduras and responsibility for the basic national map went to the first national mapping office. In this period, within Guatemala, the government not only took over the task of creating national maps suited to national interests, but also increased state capacity to create scientifically accurate documents, drawing on local knowledge and local cartography. The government sought to identify and map national and international limits, peoples, and historical territories in "official maps," at the same time it actively supported the creation of maps for public consumption, primarily in geography textbooks and wall maps by nonstate authors.

In terms of democratizing the map, President Manuel Estrada Cabrera, inaugurating a relief map in Guatemala City in 1905, ushered in a century of increasing map creation for and by a general public, as focus on knowing and depicting the physical attributes of the geo-body continued with increasing thematic representations of Guatemala's people, transportation networks, resources, archaeology, and more. The rise of private printing houses like Byron Zadik & Co. and Editorial Piedra Santa (both still in business today) and the growth of the periodical press, especially *Prensa Libre*, contributed to putting the national map into more peoples' hands.

After a nineteenth century dominated by state mapping and largely private, official, and educational consumption, the twentieth-century map and nation came together publicly. The Guatemalan map and geo-body have become identified with both state and popular initiatives. The map is democratized, as more public and private organizations and institutions use the geo-body for professional association logos, national advertising campaigns, and even military recruitment, and mapmakers put "Guatemalans" and not just ethnic or linguistic groups on the map. Even if finalizing Guatemala's national shape still requires official resolution of the Belize question, the appeal of the geo-body to the executive branch, government agencies, big and small business, and organizations of civil society as an inclusive space, despite practical exclusion of many groups, seems clear. If successors to President Colom (2008–12) succeed in resolving Guatemala's relationship with Belize, the country will finally have a single geo-body, although the evidence shows that the existence of two maps does not trouble the government or private groups on a day-to-day basis. The map-as-logo works with and without Belize.

In this context, the 2010 massacre map discussed at the beginning of this essay makes more sense: despite ethnic divisions, many kinds of Guatemalans see themselves on the map and have done so at least since 1905. If indigenous peoples are often reflected on the map via their languages, their own agencies teach community organizing using the Guatemalan national map, identifying specific departments as sites for organizing or imposing the geo-body on a colonial Mesoamerican past. Similarly, peasant as well as professional groups and big industries take the geo-body for a logo or use it in an ad campaign. If the northeastern border remained a matter of concern, 2010 was not 1933, and the president seemed to want to settle rather than exacerbate this issue. So even though state ministries and institutes still uniformly use a logo that shows a geo-body with Belize, and some Guatemalan citizens still take up the case, in 2009, Mrs. Colom smilingly accepted a plaque from the forestry, timber, and environmental resources worker's union that covered a Belize-free geo-body in trees.[86] So while La Renaudière's 1848 comment that Guatemala's limits cannot be determined while looking at a map holds true, the answer to Thongchai's question about whether a "caricature of a map" can "arouse nationalism, royalism or other serious sentimental responses" is undoubtedly yes. When the national geo-body is decolonized and democratized and belongs to both the state and its people—enabling the delineation of borders and the population of interior spaces with information of the people's choosing—then maps of this geobody can truly be tools of the nation-state as well as sources of opposition to it.

ACKNOWLEDGMENTS

The author warmly thanks the participants in the 2010 Nebenzahl Lectures, especially organizer Jim Akerman, for comments and suggestions. This research relied on materials in individual and institutional collections. Particularly helpful were materials and librarians at the Library of Congress, Tulane University Latin American Library, University of California Bancroft Library, and New York Public Library in the United States; Bibliothèque National in France; and Instituto de Geografía Nacional, Instituto Guatemala de Turismo, CIRMA, and Archivo General de Centroamérica in Guatemala (AGCA). Thanks to Ana Carla Ericastillo (director, AGCA), Guisela Asensio Lueg (director, CIRMA), Thelma Porres (CIRMA) for going above and beyond, and to descendants of Guatemalan mapmakers, particularly Irene Piedra Santa and Rolando Urrutia, who generously opened family archives, networks, and stories.

NOTES

1. See Thongchai Winichakul's *Siam Mapped: A History of the Geo-body of a Nation* (Honolulu: University of Hawai'i Press, 1994), 16–17.

2. Sarah Radcliffe, "Re-mapping the Nation: Cartography, Geographical Knowledge and Ecuadorean Multiculturalism," *Journal of Latin American Studies* 42 (2010): 293–323. Radcliffe shows what happens in a nation-state when nonstate actors appropriate cartographic "spatial practices."

3. Benedict Anderson, *Imagined Communities: Reflections on the Origins and Spread of Nationalism,* rev. ed. (New York: Verso, 1991); James Scott, *Seeing like a State: How Certain Schemes to Improve the Human Condition Have Failed* (New Haven: Yale University Press, 1998); and Jens Andermann, *The Optic of the State: Visuality and Power in Argentina and Brazil* (Pittsburgh: University of Pittsburgh Press, 2007).

4. Michel-Rolph Trouillot, *Haiti: State against Nation: The Origins and Legacies of Duvalierism* (New York: Monthly Review Press, 1990).

5. Thongchai, *Siam Mapped,* 138.

6. For North America, see Martin Brückner, *The Geographic Revolution in Early America:*

Maps, Literacy and National Identity (Chapel Hill: University of North Carolina Press, 2006); and Susan Schulten, *The Geographical Imagination in America, 1880–1950* (Chicago: University of Chicago Press, 2001). For Spanish America, see especially the work of Raymond Craib, Hector Mendoza, Carla Lois, Lina del Castillo, and Magali Carrera. For an overview of national cartography in nineteenth-century Latin America, see Jordana Dym, "Presentación, Mapeando Patrias Chicas y Patrias Grande: Cartografía e Historia Iberoamericana, siglos XVIII–XX," *Araucaria* 24 (2010): 99–109, http://www-en.us.es/araucaria/monograficos.htm (consulted February 1, 2015).

7. Karl Offen, "Making Black Territories," in *Mapping Latin America: A Cartographic Reader*, ed. Jordana Dym and Karl Offen (Chicago: University of Chicago Press, 2011), 288–92; Sarah Radcliffe, "National Maps, Digitalization and Neoliberal Cartographies: Transforming Nation-State Practices and Symbols in Postcolonial Ecuador," *Transactions of the Royal Geographic Society*, n.s., 34 (2009): 426–44; and Radcliffe, "Representing the Nation," in Dym and Offen, *Mapping Latin America*, 207–10.

8. See Sylvia Sellers-García's book *Distance and Documents at the Spanish Empire's Periphery* (Stanford: Stanford University Press, 2013); and Sophie Brockmann's "Surveying Nature: The Creation and Communication of Natural-Historical Knowledge in Enlightenment Central America" (Ph.D. dissertation, University of Cambridge, 2013). Early national mapping is important as well to Roxanne Dávila, "Los primeros pasos de la arqueología Maya: Exploradores y viajeros en el siglo XIX," in *XX Simposio de Investigaciones Arqueológicas en Guatemala, 2006*, ed. J. P. Laporte, B. Arroyo and H. Mejía (Guatemala City: Museo Nacional, 2007), 179–86.

9. For a useful bibliography of nineteenth-century Guatemala maps, see P. Lee Phillips, *A List of Books, Magazine Articles, and Maps Relating to Central America . . . 1800–1900* (Washington, DC: Government Printing Office, 1902). For early Central American cartography, see Jens P. Bornholt, *Cuatro siglos de expressiones geográficas del istmo centroamericano, 1500–1900* (Guatemala City: Universidad Francisco Marroquín, 2007). For a general introduction to geographic institutions, see Noé Pineda Portillo, *Desarrollo de los institutos geográficos de Centroamérica . . .* (Tegucigalpa: Instituto de Geografía Nacional, 1997) . For over 150 historical maps, many from the Library of Congress, see Comisión de Límites, *Cartografía de la América Central* (Guatemala City: Tipografía Nacional, 1929). For a nationalist overview, see Florencio Santiso, *Informe acerca de la cartografía de Guatemala presentado . . . ante la II Reunión Panamericana de Consulta sobre Geografía y Cartografía* (Guatemala City: Tipografía Nacional, 1944), esp. 1–12.

10. Arturo Taracena Arriola, Juan Pablo Pira, and Celia Marcos, "La construcción nacional del territorio de Guatemala, 1821–1934," *Revista Historia* 45 (2002): 9–33, esp. 11; and *Los departamentos y la construcción del territorio nacional en Guatemala: 1825–2002* (Guatemala City: ASIES, 2003). See also Carolyn Hall and Hector Perez-Brignoli, John V. Cutter, cartographer, *Historical Atlas of Central America* (Norman: University Press of Oklahoma, 2003).

11. Alexander C. Diener, ed., *Borderlines and Borderlands: Political Oddities at the Edge of the Nation-State* (Lanham, MD: Rowman & Littlefield, 2010); Peter Sahlins, *Boundaries: The Making of France and Spain in the Pyrenees* (Berkeley: University of California Press, 1991).

12. Central America achieved independence from Spain in 1821 and from Mexico in 1823. Guatemala was initially a state (1825–38) in the Central American Republic. After the federation disbanded, Guatemala became a sovereign country and declared itself a republic in 1847.

13. See Biblioteca Nacional de España, "Mapa que comprehende la mayor parte del reyno de Goatemala," ca. 1783, MR/43/195.

14. For additional analysis of this map, see Jordana Dym, "Initial Boundaries," in Dym and

Offen, *Mapping Latin* America, 144–47; and Dym, "De Reino de Guatemala a República de Centro América: Un Periplo Cartográfico," *Boletín de la AFEHC*, vol. 48 (January 2011), http://afehc-historiacentroamericana.org/index.php?action=fi_aff&id=2590.

15. For Chiapas, see Mario Vázquez, "La disputa por Chiapas y el Soconusco: Formación del Estado y gestación de la frontera entre Centroamérica y México, 1821–1842" (paper presented at Conference on Latin American History, January 2008, Washington, DC).

16. José Cecilio del Valle, "Carta Geográfica," *Mensual de la Sociedad Económica de Amigos del Estado de Guatemala*, no. 3 (June 1830): 59.

17. Karl Offen, "Creating Mosquitia: Mapping Amerindian Spatial Practices in Eastern Central America, 1629–1779," *Journal of Historical Geography* 33 (2007): 254–82.

18. José Cecilio del Valle, "Descripción Geografica," *Mensual de la Sociedad Económica de Amigos del Estado de Guatemala*, no. 1 (April 1380): 9–24, presents the country; "Continua la Descripción Geográfica del Estado de Guatemala," *Mensual de la Sociedad Económica de Amigos del Estado de Guatemala*, no. 2 (May 1830): 27–52, offers departmental information.

19. Valle, "Descripción Geográfica," 22.

20. Valle, "Carta Geográfica," 55–63.

21. Raymond Craib, "A National Metaphysics: State Fixations, National Maps, and the Geo-historical Imagination in Nineteenth-Century Mexico," *Hispanic American Historical Review* 82, no. 1 (2002): 36.

22. See Robert Claxton, "Miguel Rivera Maestre: Guatemalan Scientist-Engineer," *Technology and Culture* 14, no. 3 (1973): 384–403; and Oswaldo Chinchilla Mazariegos, "Archaeology and Nationalism in Guatemala at the Time of Independence," *Antiquity* 72 (1998): 372–86. Although born in Spain, Rivera Maestre moved to Guatemala as a four-year-old and spent the rest of his life there.

23. Rivera Maestre traveled, mapped, and sketched the Maya ruins at Copán, Iximche, and Utatlán as part of Gálvez's efforts to incorporate the pre-Columbian past into Guatemala's history and cartography, meeting US diplomat and archaeologist John Lloyd Stephens. See Claxton, "Miguel Rivera Maestre," 394: and Chinchilla Mazariegos, "Archaeology and Nationalism."

24. Claxton, "Miguel Rivera Maestre," 391–93.

25. Ibid., 393.

26. Depicting indigenous peoples' presence by floating their names over otherwise unmarked space was common in colonial-era mapmaking, backhandedly acknowledging indigenous autonomy. See Offen, "Creating Mosquitia," 254–82.

27. For discussion of the claims, see Wayne M. Clegern, "New Light on the Belize Dispute," *American Journal of International Law* 52, no. 2 (1958): 280–97.

28. Foreign companies investing in Verapaz areas also sponsored mapping. See Francis Herbert, "Guatemala Maps," *Journal of the International Map Collectors' Society* 107 (2006): 6–12.

29. Orient Bolívar Juárez, *Maximiliano von Sonnenstern y el primer mapa oficial de la República de Nicaragua* (Managua: Editorial Vanguardia, 1995). See Sonnenstern's *Mapa General de la República de Guatemala*, at http://www.loc.gov/item/2012586638/ (consulted February 1, 2015).

30. Santiso, *Informe* (n. 9 above), 17.

31. Agustin van de Gehüchte, "On the Latitude and Longitude of Some of the Principal Places in the Republic of Guatemala," *Journal of the Royal Geographical Society of London* 28 (1858):

359–62. See also Ernesto van de Gehüchte, *Carta de los estados de Centro-América con todos los proyectos de las diversas vias de comunicacion inter-oceanica . . .* (manuscript, 1862), at www.loc.gov/item /99466745/.

32. For Van de Gehüchte's start as a Belgian colonist and civil engineer in Guatemala City, see Ralph Lee Woodward, Jr., *Rafael Carrera and the Emergence of the Republic of Guatemala* (Athens: University of Georgia Press, 1995), 415.

33. Stefania Gallini, "Coffee Grounds," in Dym and Offen, *Mapping Latin America* (n. 7 above), 168–71.

34. In the 1859 Aycinena-Wykes Treaty, Guatemala recognized Belize as a British territory reaching the Sarstún River in exchange for a British-built road to the Petén. With the road unbuilt, Guatemalan governments repudiated treaty provisions.

35. For the geographies, see Emilie Mendonça, "Espejos y reflejos de Guatemala: Manuales de geografía a finales del siglo XIX," *Boletín de la AFEHC*, vol. 41 (June 2009), http://afehc -historia-centroamericana.org/index.php?action=fi_aff&id=2201. For travelers, see Jordana Dym, "More Calculated to Mislead Than Inform': Travel Writers and the Mapping of Central America, 1821–1945," *Journal of Historical Geography* 30, no. 2 (2004): 340–63.

36. Guatemala, "Estatuo de la Instrucción Primaria," August 31, 1835, title 1, article 3.

37. Gavarrete dedicated the first edition to president Rafael Carrera, citing his support for public education. The third, 1874, edition noted that it had been adopted "almost generally" in Guatemalan primary and high schools. Only with the 1874 edition of Roderico Toledo's *Geografía de Centroamérica* (Guatemala City: Imprenta dela Paz) did the text claim official government adoption for school use.

38. Interestingly, most geographies nominally presented Central America but focused on Guatemala, reflecting a continuing economic and emotional, if not political, regionalism. See Mendonça, "Espejos y reflejos."

39. F.G. [Francisco Gavarrete], *Catecismo de geografía de Guatemala: Para el uso de las escuelas de primeras letras de la República* (Guatemala City: Imprenta de la Paz, 1860); *Geografía de la República de Guatemala*, 2nd ed. (Guatemala City: Imprenta C. de Guadalupe, 1868); *Geografía de Guatemala* (Guatemala City: Imprenta Goubaud, 1874). The first edition had maps of a two-hemisphere world and Guatemala.

40. European-made maps appeared in Central American geography texts including José María Cáceres, *Geografía de Centroamérica* (Paris: Garnier Hermanos, 1880); and F.L., *Lecciones de Geografía de Centroamérica* (Guatemala City: Librería de Antonio Partegas, 1896).

41. Comisión de Límites, *Cartografía de la América Central* (n. 9 above), map 105, "Justo Gavarrete's Map."

42. See, for example, William Tufts Brigham, *Guatemala, the Land of the Quetzal: A Sketch* (New York: Charles Scribner's Sons, 1887); Alejandro Prieto and R. Piatowski, "Ferrocarriles de Guatemala," in *Ideas Generales sobre el ferrocarril interoceánico de Guatemala . . .* (Guatemala City: Imp. de Taracena é hijos, 1880); and "Mapa Parcial de Guatemala con las principales carreteras excluidos el Petén y Belice," in *Patria: Prontuario para 326 municipios de Guatemala*, by Ramón Barrera Blanco and Brígido Cabrera Meza (Guatemala City: Tipografía Nacional, 1979), 32.

43. Guatemala-Mexico Boundary Treaty of 1882, as translated in *The Questions Between Mexico and Guatemala* ([Guatemala?]: El Mensajero de Centro-America, 1895), 10.

44. Miles Rock, *Mapa de la República de Guatemala* (Philadelphia: AH Mueller, 1895), with

cartographic representations of the Guatemala-Mexican border shown by Maestre, Gavarrete, Sonnenstern, Aú, García Cubas, and others. Scale, 1:866,666, http://www.loc.gov/item/2007627461/.

45. Miles Rock, "Mapa de la República de Guatemala . . . Guatemala, 12 de enero de 1895," in *Cuestiones entre Guatemala y Mexico* (Guatemala City: Tipografía Moderna, 1895), 58. For diplomatic consequences, see Manuel Angel Castillo, Mónica Toussaint Ribot, and Mario Vázquez Olivera, *Espacios Diversos, Historia en Común: México, Guatemala y Belice; La construcción de una frontera* (Mexico City: Secretaría de Relaciones Exteriores, 2006), 146–53. For Rock's career and defense of Guatemala's territory, see William Eimbeck, "The Late Miles Rock," *Science*, n.s., 13, no. 338 (June 21, 1901): 978–80. Rock completed his commission work in 1898, remaining in Guatemala until his death in 1901.

46. For Mexico's mapping, see Antonio García Cubas's 1858, 1885, and 1886 atlases. Mexican maps noted the undefined border, but, before the 1886 atlas, seemed to share the Guatemalan government view that its territory reached to the Usumacinta.

47. Casimiro Rubio, Dirección General de Estadística, *Demarcación política de la República de Guatemala* (Guatemala City: Tipografía Nacional, 1902), preface.

48. The Vela and Urrutia families today dispute the division of labor and authorship of the project and which one provided the map's geographic information.

49. Claudio Urrutia, Autobiografía #1, Urrutia Family Papers, CIRMA.

50. "Special Issue on Guatemala," *Pan-American Magazine* 7, no 2 (December 1908): 74.

51. Carlos E. Prahl Redondo and Migual Suárez Flores, *Guia de los Volcanes de Guatemala* (Guatemala: Club Andino Guatemalteco, 1983), 2–3.

52. Urrutia, Autobiografía #1, "El Mapa Agricola y de Comunicaciones." Urrutia also credited contributions by engineer Ernesto Aparicio, draftsmen Eduardo Castellanos, Salvador Rosa, and Eugenio Rosa, artist Domingo Penedo, and master builder Cruz Zaldaño.

53. Miguel Angel Asturias's *Guatemalan Sociology: The Social Problem of the Indian . . .* (Tempe: Arizona State University, Center for Latin American Studies, 1977), his undergraduate thesis, exemplifies eugenics' influence on early twentieth-century "solutions" to the "problem" of unassimilated Maya.

54. Claudio Urrutia, "Mapa de Guatemala, 1916," scale 1:705,000. Archivo General de Centroamérica, Mapas y Planos. This map excludes Belize.

55. *Mapa del Estado de Guatemala, república de Centroamerica: Compilado por disposición del supremo gobierno/por Claudio Urrutia* (Hamburg: L. Friedrichsen y Cia, 1923).

56. Claudio Urrutia, response to "El mapa del Ing. Claudio Urrutia no conviene" (typescript, 1926), Urrutia Family Papers, CIRMA. For a summary of the border dispute, see Clifford J. Mugnier, "Grids & Datums: Guatemala," *Photogrammetric Engineering and Remote Sensing*, July 2008, 815, 822.

57. This map was reissued in 1934, with the border with Honduras, concluded in the 1932 treaty, unchanged. Copy consulted from the Urrutia Family Papers.

58. See J. Antonio Villacorta and Carlos A. Villacorta, *Arqueología Guatemalteca I: Quirigua* (Guatemala City, 1927), whose map of the "actual" Guatemalan republic showed Belize as a separate country, and Guatemala's border with El Salvador past the southern slope of the Merendon mountain range.

59. It is possible that Urrutia copied another Gavarrete map; see the 1880 map from the pub-

lisher of Gavarrete's wall map, *Carta del antiguo reino de Guatemala: En la época de su descubrimiento y conquista (1500—1540)*, Universidad Francisco Marroquín (Guatemala), AV MP 917.28 C322.

60. Clegern, "New Light" (n. 27 above), 280n1.

61. Jorge Ubico, Dirección General de Caminos, "Mapa de la Vialidad de la República de Guatemala" (Guatemala City: Lith. B. Zadik, 1936), scale: 1:600,000; República de Guatemala, Secretaría de Agricultura, Oficina Central del Café, "Mapa Cafetalera de la República de Guatemala" (Guatemala City: B. Zadik & Co., Lithog.; Campins, Del.), scale: 1:800,000.

62. Within the government, the DGC, transferred to the Ministry of Communications and Public Works (1945), became the Instituto de Geografía Nacional (1964–83), Instituto Geográfico Militar (1983–96), and again IGN (1997–). The institution worked with the US Interamerican Geodesic Service and later Defense Mapping Agency, which provided resources for aerophotogrammetry and promoted hemisphere-wide state and military mapping via the Instituto Panamericano de Geografía e Historia (IPGH).

63. The seventh meeting report, for example, touches on efforts to coordinate nomenclature in Central America, geography institutes' role in teaching, remote sensors, Pan-American development of private land registries (*cadastros*), and IPGH and OAS coordination. *VII Semana Carrográifica de América Central*, Guatemala, November 5–9, 1973.

64. Ministerio de Comunicaciones y Obras Públicas, Instituto Geográfico Nacional, "Distribución de los idiomas indígenas actuales de Guatemala elaborado por el Instituto Indigenista Nacional, Ministerio de Educación Pública, Ano de 1962," in *Contribución a los nombres geográficos de Guatemala* (Guatemala City: IGN, 1965). For biographical information, see John Gillin, "Antonio Goubaud Carrera, 1902–1951," *American Anthropologist* 54, no. 1 (1952): 71–73; Marta Casáus Arzú, "De la incognita del indio al indio como sombra: El debate de la antropología guatemalteca en torno al indio y la nación, 1921–1938," *Revista de Indias* 65, no. 234 (2005): 375–401; and Patricia Alvarenga, "El otro en la Mirada etnográfica, Guatemala (1920–1950)," *Cuadernos de antropología* 24, no. 2 (2014): 3–24, http://revistas.ucr.ac.cr/index.php/antropologia/article/view /17704/17359.

65. Ledger, 1968–1969 map sales, IGN Library, consulted summer 2010; the club used the *Diccionario Geográfico de Guatemala*, 1st ed., 2 vols. (DGC, 1961–62); and 2nd ed., 4 vols. (Guatemala: IGN, 1976–83); a 1957 "catalog of the active volcanoes . . . of Central America"; and the IGN's 1:50,000-scale maps.

66. Juan José Arévalo, *Geografía Elemental de Guatemala* (Guatemala City: B. Zadik 7 Cía, 1936), 121.

67. Ibid., 115–17. Arévalo describes whites as living throughout the country; the indigenous as making up two-thirds of the country—still speaking their "primitive languages," no longer having "primitive customs," and using Spanish in business—and the mixed race as making up a "good portion of the population."

68. Most geographies don't cite the original. See Lucila Morales de Gramajo, *Geografía de Guatemala . . .* (Guatemala City: Ed. Ministerio Pública, 1951), 144; and Daniel Contreras, Hugo Cerezo Dardón, and Oscar Gonzáles Goyri, *Geografía de Guatemala para la Educación Primaria* (Guatemala City: Cultural Centroamericana, 1952), 67.

69. A. Goubaud Carrera and A. Arriaga, *Seminario de Integración Social: Mapa de las Lenguas indígenas actuales de Guatemala* (Guatemala City: Dirección General de Cartografía, June 1964), http://ufdc.ufl.edu/AA00004973/00001/1x.

70. Instituto de Antropología e Historia (IAH), *Idiomas Indígenas de Guatemala,* 2d ed. (Guatemala City: IAH, 1984), 2.

71. Ibid., 15–16.

72. E-mail from Irene Piedra-Santa, May 5, 2011.

73. *Julio y Oralia: Homenaje a su talento, visión y dedicación* (DVD) (Guatemala City: Piedra Santa Editorial, Fundación Mario Monteforte Toledo, 2010). Piedra Santa now publishes a series covering Guatemala, El Salvador, Central America, the Americas, and the world.

74. Julio Piedra Santa, *Geografía Visualizada: Centroamérica* (Guatemala City: Editorial Piedra Santa, 1991), back cover. The first edition was reprinted nine times between 1976 and 1989; the second edition dates to 1990.

75. For example, *Geografía de Guatemala: Cuaderno de mapas para anotaciones y ejercicio*s (Guatemala City: Piedra Santa, n.d.). Piedra Santa online catalog, consulted May 19, 2011, http://www.piedrasanta.com/. The *cuadernos* cost fifteen quetzales and the *Geografías* eighteen quetzales (less than 2 US dollars in 2015). Esperanza de Castañeda updates current printings of the *cuaderno*; Patricia J. Peralta Sánchez coauthors the *Geografías.*

76. In the last six years, I've turned this research on Piedra Santa and Arévalo into an article that is now in print: Jordana Dym, "'Mapitas,' *Geografías Visualizadas* and the Editorial Piedra Santa: A Mission to Democratize Cartographic Literacy in Guatemala," *Journal of Latin American Geography* 14, no. 3 (October 2015): 245–72. Piedra Santa is publishing a Spanish-language version.

77. Piedra Santa, *Geografía Visualizada: Centroamérica*, 2nd ed. (1991), 4; *Geografía Visualizada: Guatemala, Edición Actualizada* (1991), 42; *Geografía Visualizada: Guatemala* (2007), 13, 16–18. Interviews with Irene Piedra Santa, 2010–2012.

78. See Matthew J. Taylor, "Viviendo en 'aquellos tiempos' en Ixcán, Guatemala: La violencia y la vida en las PAC," *Mesoamérica* 54 (2011): 157–88.

79. Thanks to Karl Offen for a photograph of this sign.

80. See Radcliffe, "Re-mapping the Nation" (n. 2 above).

81. INGUAT, Guatemala's tourism agency, preserves this map. Librarian Beatriz Pineda made a rich range of maps available to me for consultation in July 2010.

82. CEH, *Guatemala Memoria del Silencio*, vol. 1, *Mandato y Procedimiento de Trabajo: Causas y orígenes del conflicto armado interno* (Guatemala City: UNPOS, n.d.), cover.

83. Comité de Unidad Campesina (CUC), *Lucha, resistencia e historia* (Guatemala City: Editorial Rukemik Na'ojlil, 2007), 16–17.

84. CUC, *Historia de Guatemala: Desde un punto de vista crítico* (Guatemala City: Editorial Rukemik Na'ojlil, 2007).

85. José Vaconcelos, Mexico's minister of education, theorized Mexico's "cosmic race" in the wake of the Mexican Revolution (1910).

86. "La Primera Dama, recibe un reconocimiento de la Federación Sindical de Trabajadores de la Silvicultura, Madera, Medio Ambiente y Recursos Naturales de Guatemala (FESITRASMMAR)" (9 July 2009), http://www.guatemala.gob.gt/fotogaleria2.php?tipo=&pagina=26; http://www.guatemala.gob.gt/fotos/a/030709%20Reconocimiento.jpg (consulted May 8, 2011).

CHAPTER FIVE

UNCOVERING THE ROLES OF AFRICAN SURVEYORS AND DRAFTSMEN IN MAPPING THE GOLD COAST, 1874–1957

Jamie McGowan

On March 6, 1957, Ghana became the first African country south of the Sahara to win its independence. In the lead-up to this landmark date, British colonial institutions were forced to yield to Ghanaian political agendas and interests. Specifically, demands for self-government led to changes in the constitution and town and national legislative structures, and increased educational and professional opportunities. However, the new nation was also marked by many colonial inheritances. Colonial-era maps, surveying agendas, institutions, and practices were among these legacies. Moreover, many postcolonial surveyors and cartographers maintained a sensibility about their work similar to that of their colonial predecessors. They viewed mapmaking as essentially apolitical in nature. To understand these cartographic inheritances, this chapter pursues the ways that local Africans became surveyors and draftsmen and contributed to the mapping practices that supported the emergence and development of the colony. It also examines local surveyors' and draftsmen's training, opportunities, and perspectives on colonial Survey Department practices, illuminating the continuities and subtle changes as the colony moved toward independence.

Pursuing these themes, this chapter reveals the fundamental importance of

Africans as key actors in colonial mapping and surveying. Second, I argue that the persistence of colonial-era mapping practices was possible in part because of the engagement of African surveyors in these scientific techniques. Finally, this study exposes the workings of a colonial governmentality, in which surveyors carry forward their technical practices but distance themselves from their work's political nature.

Ghana's cartographic construction unfolded to a large extent under colonialism, which was formalized in 1874 with the founding of the Gold Coast as a British colony. The mapping of the colony and the institutionalization of surveying emerged over the subsequent decades. The mapping and the emergence of the Gold Coast took place over three temporal phases: (1) colonial expansion (1874–1901); (2) administration and development (1901–30); and (3) consolidation and decolonization (1930–57). This chapter examines each period, focusing specifically on the engagement of local African surveyors and draftsmen and the ways that surveying initiatives and maps fitted into the broader administrative agendas and colonial needs. I draw upon archival texts, maps, and interviews with fourteen Ghanaian surveyors and a draftsman who worked in the Gold Coast Survey Department during the period of decolonization as evidence and to delineate these three periods.

As a prelude to this study on the role of Africans in what is typically viewed as a European scientific project, I situate Ghana's cartographic history within three broad literatures: cartography, colonialism, and local participation; African intermediaries in European colonialism; and colonial governmentality. This section is further enriched by a discussion of the emergence of surveying agendas and institutions, including the role of local surveyors and cartography during the colonial period. These literatures are reinforcing and at times overlapping, but I delineate salient threads of this scholarship, considering how this social history of surveying in colonial Ghana contributes to them. This study documents the ways that Africans were essential to colonialism's technical projects in Africa and gives voice to the silences surrounding their participation in colonial mapmaking. It looks specifically at the ways that colonialism drew in and trained its African staff, inculcating in them the value of its scientific practices and techniques and making them willing participants in perpetuating its mapping practices.

CARTOGRAPHY, COLONIALISM, AND LOCAL PARTICIPATION

The literature on mapping and colonialism typically focuses on the ways in which cartography fostered the founding, development, and legitimizing of

European colonies.[1] In many studies, the role of local experts in relation to colonial cartography is largely neglected. Jeffrey Stone's research on African colonial mapmaking gives scarce attention to the local population's involvement in the process.[2] D. Graham Burnett's study of empire building in Guyana focuses on British roles in mapmaking.[3] He notes that local informants provided toponyms for British surveyors' work, but that much of their information was not seriously considered. Matthew Edney's work on India provides hints of local participation, but he only briefly engages the involvement of Indian surveyors in British processes of mapping the subcontinent. He does mention, however, that the archive contains many instances of Indian resistance to surveying.[4] Despite the evidence of local involvement in colonial cartography, research largely omits their involvement.

There are a number of exceptions to this silencing of local participation in the colonial and imperial cartographic process. J. B. Harley examined the influence of Native Americans on seventeenth-century American maps.[5] In deciphering the "shadows" of Native American influence on the maps, he suggests the subtleties of their contributions. Their active role in mapmaking exercises and as informants is not substantiated. Karl Offen also studied the influence of Native Americans' spatial practices on colonial maps.[6] He contends that the political power and independence of Mosquito Indians enacted authority over their space, resources, and populations—such that their creation of Mosquitia was also represented in eighteenth-century British and Spanish maps. Two studies by Thomas J. Bassett of the indigenous influence on European mapping of Africa demonstrate that African knowledge was important to the making of European maps and also note how Africans helped to make European maps.[7] Bassett describes the influence of travel reports, place names, "oral maps," and drawings in the sand as indications of the sharing of geographic knowledge. He also explores the ways that European mapping practices influenced indigenous mapping traditions. What these studies neglect and what is pursued in this chapter is the systematic involvement of Africans in colonial mapping. That is, beyond the travel accounts and exchange of knowledge, colonialists trained a cadre of workers to facilitate the mapping of their overseas colonies.

Olayinka Balogun offers an account of the training of Nigerian surveyors at the turn of the twentieth century.[8] Balogun notes that surveying was the first professional career introduced to the Nigerian educational system. His study offers a glimpse into the evolving opportunities for African surveyors, which parallels many of the developments in colonial Ghana. He provides some analysis of the links between changing economics and politics of colonialism to the training opportunities available to Nigerian students, yet his assessment does not relate the training opportunities to broader impacts in the mapping of Nigeria.

This chapter builds on these works to examine the engagement, training, and contributions of Ghanaian surveyors and draftsmen in the mapping of the Gold Coast. I argue that the history of mapping the Gold Coast cannot be understood without considering the roles played by Africans mapping the colony. Specifically, by documenting Africans' work and roles as mapmakers, I interrupt the metanarrative of colonial cartography being a practice of foreign agents conquering, partitioning, and mapping the African colonial terrain. This chapter begins with an examination of the influence on the mapping of the Gold Coast by an early Gold Coast surveyor, George Ekem Ferguson. I then discuss the institutional structures of surveying, the training and employment of African surveyors, and the contributions of these surveyors and draftsmen to the work of the Survey Department until Ghana's independence. Documenting the untold story of African surveyors' work over eighty-three years, this study establishes their interests in and concerns with colonial cartography, including its technical and political rationalities, agendas, and organization. This history demonstrates that African surveyors' participation was essential to the colonial project and the mapping of Ghana. Further, this study also causes one to rethink aspects of Matthew Edney's work, in the sense that local participation and knowledge was essential to the construction of a colonial cartographic panopticon. The imposition of British rationality and British rule, enabled in part through mapping, was a far more complex process—one that directly engaged local knowledge and power.

COLONIAL GOVERNMENTALITY

The second body of scholarship this study engages centers on the Foucauldian concept of governmentality and the related concept of colonial governmentality. Governmentality encompasses modern states' power and rationality, undergirded by the techniques and sciences of the state.[9] Michel Foucault writes of governmentality that it is "the ensemble formed by the institutions, procedures, analyses and reflections, the calculations and tactics that allow the exercise of this very specific albeit complex form of power, which has as its target population, as its principal form of knowledge political economy, and as its essential technical means apparatuses of security."[10] With this lens, such calculations and tactics as mapping are means for deploying power in seemingly subtle ways. Couched as technological or scientific interventions or approaches, these techniques appear more benign or even beneficial in the context of governance.

Though Foucault's conception of governmentality emerges from Western Europe case studies, recent scholarship has reframed it for the colonial context.[11] Gyan Prakash describes colonial governmentality as the configuring and administering of the colonized territory and people—"under the authority of science," and particularly according to the knowledge and tactics of the colonial state.[12] He distinguishes the colonial governmentalized state where

> administration became regularized and extended its reach farther down into the colonized society in its effort to generate new forms of knowledge about the territory and population. As the British produced detailed and encyclopedic histories, surveys, studies, and censuses, and classified the conquered land and people, they furnished a body of empirical knowledge with which they could represent and rule India as a distinct and unified space. Constituting India through empirical sciences went hand in hand with the establishment of a grid of modern infrastructures and economic linkages that drew the unified territory into the global capitalist economy.[13]

Prakash notes that beyond the purely administrative agendas, there is commonly a "developmentalist impulse" of such colonial governmentalist tactics, also seen within this case study. In view of that, administration and development may both be desired outcomes of colonial governmental tactics.

Prakash further characterizes colonial governmentality as limited by the weaknesses of a colonial state, and because of these weaknesses, local intermediaries, whom he calls "subordinate functionaries," are needed to facilitate the techniques of governance. That is, in implementing the colonial states' governmental projects, which included censuses, engineering initiatives, and public health projects, the colonial state drew in local agents as intermediaries, who served as translators and assistants, or provided essential labor to such projects. In the case of mapping the Gold Coast, the training of these functionaries and involvement of these colonized peoples provide avenues for understanding African engagement in surveying and mapping.

James Scott's study of the simplification of statecraft sees technologies of the state as "narrowing the vision" from the complex realities that exist to more simplified, legible forms.[14] Scott uses the example of the simplification of complex land tenure systems by the state through land privatization and regulation. He argues that these static measures and the reductive knowledge encoded in a map serve administrative agendas—for instance, in government planning and taxation. Further, a bureaucratizing and modernizing state seeks to record and control its resources in a more systematically consistent way, such as through cadastral mapping.

In a colonial context, Arun Agrawal's study of forest regulations in India demonstrates that colonial forestry policies and management strategies were initially resisted by the Kumaon community. Yet, over time and into the postcolonial era, the same community embraced such regulatory policies, forming environmental groups and policing its own use of forest resources. Agrawal shows that "modern forms of power and regulation achieve their full effects not by forcing people toward state-mandated goals but by turning them into accomplices."[15] He demonstrates the ways in which governmentality influences people's conduct and questions the value of drawing distinctions between the state and society. He explains, "Instead of examining the boundaries and definitions of the state and society, an analysis of governmentality orients attention toward the concrete strategies to shape conduct that are adopted by a wide range of social actors and how these different actors collaborate or are in conflict in the pursuit of particular goals."[16]

With reference to colonial mapping, this analysis draws attention to the ways that the British state in colonial Ghana used mapmaking to affect the conduct of its subjects in ways that legitimated colonialism. These examinations of governmentality reveal some of the ways that Gold Coasters became involved in these techniques and practices, embracing them and implementing them over time. Drawing on these works, I introduce the notion of *cartographic governmentality* to delineate the scientific practices and processes of cartography that informed governance and the ways in which African surveyors became willing accomplices in the practices of the colonial Survey Department.

AFRICAN INTERMEDIARIES AND COLONIALISM

A third body of literature explores the roles Africans played as intermediaries in European colonialism.[17] Seeking to go beyond research that dichotomizes African responses to colonialism into camps of resistance and collaboration, this study understands people's roles in relation to the context and avenues open to them as employees of colonial enterprises. It draws on the colonial governmentality literature, specifically Agarwal's work, to suggest that the colonial state turned to local agents—not as collaborators, but as accomplices, in implementing the techniques and tactics of governance. As accomplices, local people became vested in both the regulation and outcomes that similarly interested the state. Further, it seeks to understand some of the temporal distinctions that unfolded over the transition from colonial expansion to the beginning of self-rule. Thus, this study draws upon literature that highlights some of the temporal variation in the engagement of African workers.[18]

Research on African responses to colonialism tends to identify two primary positions—collaboration and resistance.[19] This tendency to reduce the complexities of intermediaries' multiple worlds either removes them from colonial histories or makes them instruments of foreign rule. For example, David Turnbull's study of Australian "go-betweens" seeks to uncover their hidden role in colonial histories, but in doing so, he puts forward "the figure of the go-between [who] is always two-sided, always both enabler and betrayer."[20] Turnbull continues to describe the man who could move between "two worlds" but was ostracized and "unable to find a home on either side" of the boundaries that he crossed.[21] Kwame Arhin's study of colonial civil servants in the nineteenth century takes a more positive view of the mediating role that George E. Ferguson played in the Gold Coast's colonization as one that built a bridge between modernity and tradition.[22]

Some recent essays on African intermediaries see them more complexly as "straddling multiple worlds."[23] *Intermediaries, Interpreters, and Clerks: African Employees in the Making of Colonial Africa*, edited by Benjamin Lawrance, Emily Osborn, and Richard Roberts, demonstrates that African colonial intermediaries negotiated multiple contexts. They moved in between the changing interests of local colonial administrations, broader colonial networks, African polities on the ground, the educated African elite, rural communities, and their own families. The case studies in Lawrance, Osborn, and Roberts's volume reveal that the influence of intermediaries waned over the colonial period. During the early periods of colonialism, colonialists depended quite heavily on the Africans employed in their service, and these intermediaries held considerable power to interpret, cultivate, and exploit a particular relationship.[24] They argue that "in the flux of conquest and its aftermath, African intermediaries working closely with European colonial officials (or appearing to) could develop or carve out positions of considerable authority. The "rule" of colonialism had not yet been set or developed." Explaining their evolving role, the authors write: "As the bureaucracy of the colonial state solidified, however, the possibilities for Africans to rise to positions of authority declined. The positions held by Africans became more strictly codified: their duties, ranks, and salaries were regulated by the state."[25] Instead of relating to a particular person, at this stage, African colonial employees rather relied on "their understanding and manipulations of the bureaucracy" as their main point of engagement.[26] The scope of this study will demonstrate the changing influence and relationships of African workers to the colonial state, as their individual influence waned and their positions became more normalized within the Survey Department.

The history of the mapping and surveying of colonial Ghana speaks to multiple audiences. It engages with literatures that explore the relationship between colonialism and cartography, which has largely neglected the role of local peoples in mapmaking. It speaks to the postmodernist literature on governmentality by showing how colonial subjects participated in mapping the confines of colonial rule. And it speaks to the relatively recent literature on the intermediaries of colonization. It is this third theme to which this study most directly contributes. The social history of surveying and mapping documents the ways that colonial processes were not purely endeavors of foreign agents but involved local people. By focusing on Ghanaian surveyors, this study shows their role over the course of decolonization and the stability that they provided in ongoing agendas and trajectories in mapping. Lastly, local practitioners who worked during the decolonization and independence eras distanced themselves from the politics of their practice, but still fit into a broader context in which maps and surveying functioned as tools of rule.

The maps and mapping of Ghana are among the inheritances of its colonial era and continue to influence postcolonial mapping practices and views. To substantiate this claim, I provide a social history in the unfolding of Gold Coast maps demonstrating that Africans played significant roles surveying and mapping the territory throughout the colonial period. I present this history of colonialism and cartography in three broad phases: (1) conquest and expansion, (2) administration and development, and (3) consolidation and dissolution. At the beginning of each section, I provide a brief introduction to the period before examining the emergence and roles of African surveyors and draftsmen and the institutional structures that surrounded surveying. At the close of this chapter, I bring these threads together to substantiate the larger claims to Africans' roles in the mapping of the Gold Coast and in the continuities in scientific mapping practices in the postcolonial period.

It should be noted that the three phases of colonialism charted in this study draw on some rather disparate data sources, and thus the narrative across these periods can seem at times rather disjointed. The data, like the periodization, are fragmented by colonial and global change—political mobilizations, changing leadership, wars, and the global depression, to name just a few of these changes. Also, the data are represented by the ruptures of colonial sources, written by a changing array of British officers. Wherever possible, I draw out African surveyors' or draftsmen's voices from the written documents, maps, and documents informed by Gold Coasters, but those voices are far from continuous or wholly represented within the archive.

Given the fragmentary evidence, this chapter bridges the three phases of colonialism with multiple data sources. These sources include reports by and about George Ekem Ferguson during the first phase of colonial expansion (1874–1901). Ferguson produced a large number of maps, and these sources are also examined. Evidence for the second phase (1901–30) of colonial administration and development emerges mostly from the reports of the colonial government surveying units and related archival records. These records are far less individually focused, and instead feature the bureaucracies and structures of administration. To balance this bureaucratic perspective, I draw from personnel files dating from this period to illustrate African participation in colonial-era mapmaking. The third phase (1930–57) centered on consolidation and decolonization. It lacks the depth of archival documentation, as the colonial record keeping was shallow for this period. Partly to make up for this deficit, I draw upon secondary sources and, most important, interviews with surveyors who first started their training and work during the colonial and early postcolonial era. While the data and narratives can seem disparate and disjointed at times, the role of African surveyors is still evident throughout all three periods. The scope of their contributions can be seen in the cartographic construction of the colony, the unfolding governmental practices of mapping, and the continuity of mapping practices during decolonization—a point that I will return to in the chapter's conclusion.

COLONIAL CONQUEST AND EXPANSION, 1874–1901

British colonialism on the Gold Coast began in the nineteenth century. It took on a more defined political arrangement and geographic coherence after 1874. At this time, the British located administrative offices in Accra and assumed administrative control over a continuous territory along the coast and inland to about 6° 50″ N.[27] The British continued to expand their authority along the coast and in small steps northward. The Berlin Conference of 1884–85 clarified the terms of colonial expansion among the European countries, as a whole, and further catalyzed British strategies to expand northward beyond the Asante Empire.[28] The British previously led several incursions into Ashanti and faced considerable resistance. However, as the race to extend colonial territory developed, the British sought to weaken Asante allegiances and sidestep

FIGURE 5.1. Map of the Gold Coast adapted from a 1907 map. Map made by the author.

Ashanti to execute trade, friendship, and protection agreements in the so-called "hinterlands," north of Ashanti, before the Germans or French established any colonial claims there. The British trained and relied on a key African intermediary, George Ekem Ferguson, to explore and document this region in maps and reports, as well as to execute treaties of trade and protection on behalf of the British. His peaceful negotiations with communities in the Gold Coast hinterlands laid claim to an expansive region on behalf of the British Empire. He built a network of spies who facilitated the reconnaissance of Asante and ultimately supported Britain's war against the empire.[29] Ferguson's work and leadership ultimately helped Britain to capture the Asantehene and other key leaders, establishing British rule over Ashanti.

Over a twenty-seven-year period, this expansion led to the formation of three political entities under British rule—the Gold Coast Colony, Ashanti, and the Northern Territories—which collectively formed the Gold Coast (fig. 5.1).[30] Ordinances codified the formation of the Northern Territories and Ashanti in 1901, in which chief commissioners administered these protectorates. The governor of Gold Coast ruled the littoral colony and oversaw the commissioners based in Ashanti and the Northern Territories. The period 1874–1901 marks the phase of colonial conquest and expansion in the Gold Coast. In the following section I provide general contextual and biographical information on Ferguson before describing his specific role in colonial map-making during this period.

GEORGE EKEM FERGUSON, A GOLD COAST SURVEYOR DURING EARLY COLONIALISM

The number of Gold Coast surveyors working during this period of colonial expansion was limited, as educational opportunities within the colony and West Africa were relatively few. However, George Ekem Ferguson, who ultimately learned surveying skills, rose within the British colonial network and wielded considerable influence in the expansion of the Gold Coast.[31] Ferguson was born around 1865 in Anomabu, near Cape Coast, of African and European heritage. His parents were both Gold Coasters. His paternal grandfather was a Scottish doctor who served in the Gold Coast colonial establishment, and on his maternal side he had a Dutch ancestor. These family connections to imperial and colonial networks probably opened up opportunities for Ferguson to both earn an education and make connections within the colonial administration. He attended school at Cape Coast Wesleyan School as well as the Wesleyan Boys' High School in Freetown, where he excelled in his studies.[32] Shortly after his return to Cape Coast, he began working for the colonial administration at the age of seventeen. He copied maps and received on-the-job training from British colonial officers. Thus, his mapmaking career began. Over his career, Ferguson worked in several contexts, producing a number of large-scale maps for the Public Works Office, assisting the survey of the Anglo-German boundary, where he learned how to compile political reconnaissance maps.

One of his earliest signed maps, "A Sketch Map of the Divisions in the Gold Coast Protectorate" (fig. 5.2) dates from August 1884 and was compiled under the direction of the governor, William Young. The map reflects the governor's interests in a preliminary internal partitioning of the colony, following a scheme for district administration the governor submitted to the Colonial Office in September 1884.[33] Based on that communication, it appears likely that Governor Young commissioned the map for colonial administrative purposes.[34] The map helped establish the distribution of colonial officials posted in the districts and creates a hierarchy among the districts. This was the first map that Ferguson compiled of an enlarged Gold Coast protectorate. He had previously copied a number of other larger-scale maps that had a narrower geographic focus than the entirety of British possessions at the time. The "sketch

FIGURE 5.2. *Sketch Map of the Divisions in the Gold Coast Protectorate*, August 1884. The National Archives (UK). Reproduced with permission.

map" demonstrated Ferguson's skills and contributions and garnered considerable interest at higher colonial administrative levels and in London. It was a turning point that advanced Ferguson's mapmaking career.[35]

In compiling the "sketch map," Ferguson drew on multiple sources, integrating cartographic symbols and geographic knowledge gleaned from other maps. The framing of the region is much like existing British maps of the region. Its east-west and southern extent are similar to maps produced by the Intelligence Division of the War Office (IDWO) maps of 1873 and 1881,[36] although Ferguson's map does not extend as far north. The map's scale was also apparently modeled on these earlier maps. Ferguson listed his as 8.5 statute miles to the inch; the IDWO lists theirs as a fraction, but it is slightly smaller, at 1:633,600. Ferguson also added elements to the map—elements that while new to the Gold Coast map were conventions and standards used elsewhere. He included a compass star, not seen on the IDWO's Gold Coast maps or other maps of the region in recent years. It closely resembled a compass star seen in another set of Ferguson's large-scale maps, drafted several years earlier. Similarly, his informed and critical approach to map compilations was reflected by his selective deletion of topographical information and communities that lay beyond the protectorate borders and his addition of topographical data not previously represented on British maps. Ferguson included some data on river currents, depths, and altitudes of selected points. Ferguson's compilation of his "sketch map" demonstrated a competence that extended beyond that of someone who was merely copying maps. His exposure to maps, surveyors' work, and knowledge of colonial records enabled him to produce a map that was truly his.

Ferguson was particularly attentive to geopolitical hierarchies, using weighted or hatched lines, colors, and different lettering styles and sizes was part of the effort used to differentiate the importance of communities and regions. As the first colonial map to mark internal political divisions within the Gold Coast Colony, Ferguson's "sketch map" suggests both boundaries and alignments among regions and ethnicities. However, his use of these design elements is not consistent throughout the map. Colored lines along the coast inexplicably do not match the colors of regions. Blocks of color often align with district boundaries, but not consistently so. The map's named districts do not match other colonial sources, and ethnicity and town names are interchangeably used for regional names. This possibly reflects the confusion of a newly established colony as well as inconsistencies among the sources consulted. The map thus demonstrates the obstacles faced by mapmakers, who provided documents meant to simplify the administrative plans and hierarchies of the government and yet were unable to represent the complexity they knew to exist. The internal contradictions on the map may also reflect Ferguson's lack of training in mapmaking. The map is nevertheless an important example of an African-made colonial work, engaging a local mapmaker with considerable technical skills in administrative and colonial affairs.

Producing an administrative map under the direction of the most senior British colonial official in the Gold Coast, the governor, underscores Ferguson's emerging value to the administration. Although there were British surveyors working in the Gold Coast at this time, it is significant that Governor Young sought out a Gold Coaster to lead this project. Ferguson reflected on his map and his contributions in a letter to the subsequent governor of the Gold Coast: "My first endeavor on entering the public service was to study the geography of the Country and eventually compiled from information which had been collected in the Governor's office Map of the Gold Coast Protectorate under the supervision of the late Governor Young; and I believe I was the first to make out on it the approximate boundary of the several districts in the Protectorate whence the map took its name."[37] This statement verifies Ferguson's commitment to mapping as well as to the concerns and interests of British colonial governance. Ferguson's dedication was actively supported, as he was promoted, trained, and brought into the fold of British colonial expansion through mapping. Ferguson's maps and reports from the years that followed document his perspectives on and contributions to British colonialism and governance in greater detail.

COLONIAL EXPANSION THROUGH RECONNAISSANCE AND MAP DESIGN

By the 1890s there was increasing pressure among the European colonizing governments to secure spheres of influence on the ground in Africa.[38] In their race to claim territory, the colonizing states needed people to execute treaties with local leaders and to document these arrangements in reports and on maps. Toward this end, the Colonial Office and Gold Coast governor sent Ferguson to the Royal School of Mines and the Royal Geographic Society in London in 1889 for formal training in geology, ethnology, and surveying. After his return to the Gold Coast, the governor recruited Ferguson in 1892 to spy on Asante and to attempt to fracture Asante political alliances. In addition, Ferguson's secretive mission, documented by his reports, letters, treaties, and maps, was to travel beyond Ashanti into present-day northern Ghana to negotiate treaties of protection and trade with other communities. He compiled two maps of his journey, one of which was reissued with some modifications by the Intelligence Division of the War Office in three separate versions the following year. Ferguson's letters, treaties, and reports accompanying the maps were part of exchange of communications between the Colonial, War, and Foreign Offices in the United Kingdom and the governor about his accomplishments.

Before Ferguson's departure for the 1892 mission, Governor W. Brandford Griffith met with Ferguson to discuss possible routes and select the key communities with which to secure treaties. The governor wanted him to target four ethnic groups: Dagomba, Gonja, Gurunsi, and Mossi. The two men relied on French-, German-, and British-made maps to plan the journey, and Griffith cautioned Ferguson against making treaties with communities that lay solely within the Neutral Zone between British and German interests created by an 1888 pact. According to the pact this area would remain open to both European nations; neither could claim exclusive rights to control the trade there.[39] (This region north of Ashanti appears on Ferguson's map "Country between Say and Bontuku" [fig. 5.3], described below, as a blue shaded square.)

Over the course of his five-month journey, Ferguson sent a number of reports to Griffith, promising a full report and map on his return to Christiansborg. The London-based Colonial Offices received copies of these reports but eagerly awaited Ferguson's final report and map. The first of his reports described various communities and "native authorities"; the treaties that he

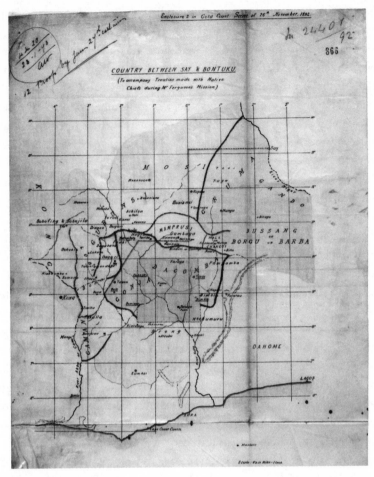

FIGURE 5.3. "Country between Say & Bontuku." The National Archives (UK). Reproduced with permission.

secured with five communities in the north (Boniape, Bole, Daboya, Yendi, and Bimbla); and a map he compiled, "Country between Say and Bontuku" (fig. 5.3). An 1887 map by the German publisher Justus Perthes was the base for Ferguson's new map, which he supplemented with information gleaned from his mission. This map, as well as Ferguson's various reports, maps, letters, and treaties demonstrate both his ability to connect with local communities with whom he negotiated and his awareness of the nuances of the colonial competition in Africa among Britain, France, and Germany. The written record notes that Governor Griffith cautioned him against making treaties with communities in the zone. Ferguson nevertheless crossed into this zone and secured four treaties of political and economic alliance with communities within it. It may be that the governor and Ferguson had a verbal agreement that countermanded the written record and disregarded the treaty with Germany.

His maps support these claims as well. First, he underlined all the communities with whom he made alliances. With the exception of one community, Bole, all of these communities lie within the Neutral Zone.

Second, he labeled ethnic groups under British influence, such as "Gonja" and "Dagomba," in such a way as to extend across their territories and the boundary of the Neutral Zone, apparently legitimizing British incursions into the zone.

Third, colonial agendas in Ferguson's maps are also evident from his use of colors. Situating his maps within the broader context of imperial mapping, in which Great Britain's colonies were colored red, Ferguson used hues of red to signal interest, if not intent, to colonize. He grouped the communities with whom he secured treaties within a reddish-orange territory outlined by a darker red boundary. By doing so, he encased the five communities within a large region asserted to have common interests, and common British interest in them.

Ferguson's use of a lighter reddish orange further asserted that these regions with this shade were under British protection, justified by the political relationships the communities within had with other communities who themselves had treaties with the British. For example, Bole, with whom Ferguson secured a treaty, previously protected the people in Wa from attack. According to Ferguson's report, Wa was by extension under the chiefs of Bole, and having a treaty with Bole entitled British authority over Wa, too. Other areas included Pampamba, Sansanné-Mango, and Gambaga, which were "feudations" of the Dagomba chieftancy based in Yendi. He explained that Walembele and Yariba were dependencies of Daboya. By documenting such relationships, Ferguson made the case for extending British rights of trade and friendship to these areas based on the influence that the five signatories could claim. A similar but lighter shade of orange and a hatched red boundary line visually includes these territories within the scope of British colonial authority. Thus, Ferguson's documentation of the regional political alliances in his reports informed colonial claim making on his maps.

Fourth and finally, Ferguson used colored boundary lines to assert British colonial agendas. Ferguson wrote that the so-called Akba or Como River, today the Black Volta, would be the best natural boundary between French and British interests in the region. He marked this river on the map with a green line, which he contrasted to the yellow hatched line representing a recent French proposed border. France had previously used green to depict its proposed boundary on a map, and Ferguson adopted the use of the same color for marking his countermapping of an Anglo-French boundary. His map, likewise, shows the French boundary cutting across the regions and dominions with which Ferguson had concluded treaties. Ferguson's recommendation that the river be the frontier was based, in part, on information conveyed by a French colonial agent's map and Ferguson's concern for British interests in the region. He reasoned that French officer Louis Gustav Binger's 1890 map did not show France's influence extending beyond this river. Furthermore, he reported that France's proposed boundary would hinder access to rich gold deposits and would also cut off Britain's trade network with the "Mosi" kingdom that extended to Salaga. He closed by noting that various African communities had expressed their opposition to any type of division. Thus, while Ferguson considered colonial economic interests and the extent of proposed French and British territories he also registered local Africans' sentiments against dividing the region to support his recommendation for the British and French frontier.

Through his maps as well as his political treaties, Ferguson portrayed and facilitated the expansion of British colonialism. His maps illustrated his cartographic skills as well as his knowledge of political and colonial mapping techniques. Ferguson adopted the techniques used by contemporary European cartographers, and his use of lines, color, and lettering both documented the treaties he had secured and promoted British expansion beyond that explicitly negotiated by the individual treaties.[40] As Britain negotiated its colonial territory with other European powers, it is clear that the work of Ferguson, a Gold Coaster, provided solid evidence Britain could produce to document its colonial influence and claims.

FERGUSON AS AN INTERMEDIARY

Ferguson's mapping skills, political savvy, and allegiance to Britain are clear from his maps, reports, treaties, and letters. Less obvious are the significance and complications of his intermediary status, as an African employee of the British colonial state. Because of his African descent, Ferguson seemed better placed than British counterparts to carry out the work of colonial expansion. According to the Gold Coast's governor, Ferguson's knowledge of "native character and languages" facilitated his work.[41] Ferguson was able to communicate successfully with chiefs about their political interests and hesitancies to align with Britain. He drew on his language skills to negotiate treaties, and his knowledge of both regional and international geopolitics figured into these negotiations. He was likewise able to circumvent detection, as he was able to travel and maneuver without standing out as a foreign agent. Approximately thirty years after Ferguson's death, another African surveyor, Kweku Asante, working for the Department of Surveys in the Gold Coast, wrote a short biographical essay on Ferguson. Asante wrote glowingly of his predecessor: "Among his many qualifications one which influenced Government in selecting him for the various missions was that 'being a native he could travel with a small following and remain in the bush for long periods whereas the ordinary British Colonial Officer would have required a special escort, a doctor and interpreters.'"[42] Praise for Ferguson's works extended throughout the Gold Coast administration and abroad within the Colonial, Foreign, and War Offices. The British Government awarded Ferguson the Ashanti Star Decoration for his role with the 1895–96 Ashanti Expedition.[43] The Royal Geographic Society posthumously awarded him the Gill Memorial and a gold watch for his contributions to geographic knowledge.[44]

Despite these accolades, Ferguson occupied an inherently dangerous position as an intermediary for the British, as the circumstances of his death illustrate. While he was traveling in the northwestern regions of the Gold Coast in 1897, carrying out another expedition to secure territories and treaties on behalf of Britain, he encountered the army of the West African empire builder Samori Touré. Touré originally came from the Bissandugu area of present-day Guinea, where he began his own state-building efforts. Following clashes with French colonial forces in that region, he relocated to northern Côte d'Ivoire. Being closer to the Gold Coast, Touré sought to align himself with the powerful head of the Asante kingdom, the Asantehene. Touré and the Asantehene corresponded in 1895 about reestablishing their influence in the region.[45] The Asantehene sought Touré's assistance in "recover[ing] all the countries from Gaman to the coast which originally belong to Ashanti."[46] Gaman referred to the northern regions of Greater Asante at its height in the eighteenth century and which coincided to a great degree with the territory claimed by the British through Ferguson's treaties.[47] These joint interests of the Asantehene and Touré demonstrate political maneuvers and the level of coordination within Africa's own empires and among its leaders to thwart colonial advancement and secure their own interests.

Not surprisingly, the British administration feared the alliance of the Asantehene and Touré and sought to secure and protect its northern territorial claims from Touré as well as the French. In 1897, the governor sent F. H. Henderson, a traveling commissioner, along with Ferguson and members of the Gold Coast Constabulary, to secure this region as British territory and better document their claims for the upcoming Anglo-French negotiations. According to Henderson's report, Touré and his army initiated a series of attacks on them. Over a week of on-and-off fighting, Ferguson was shot in the leg and was unable to walk without assistance. Fearing their inability to retreat to a safer area, Henderson reported his willingness to meet with Samori Touré, despite Ferguson's protests against any such meeting. Trying to negotiate an end to the fight and not admitting any ill will toward Touré, Henderson argued that British interests were solely to stop French colonial expansion. Ferguson remained behind and was soon abandoned by his African carriers. Touré's army advanced, to find Ferguson alone. According to Henderson's report, the soldiers encouraged him to accompany them to their headquarters, but Ferguson refused and pointed an unloaded gun at them. The soldiers initially retreated, but returned to find Ferguson still alone and killed him, bringing his head to Henderson and Touré. Henderson and the carriers all survived this encounter, and Henderson recounted these events and exchanges later.

These final encounters demonstrate that Ferguson's intermediary status was not entirely defined by the duality of his allegiance to the British and his identity as a Gold Coaster. Complex political relationships formed among African empires, colonial powers, and power brokers across these fields of interest. Ferguson's role as intermediary was rendered unstable by this complexity and made him personally vulnerable. He was abandoned by his African carriers, who were supposed to carry heavy loads and support the expedition but were not prepared to escort him to a safe position far away from Touré's army. Ferguson actively supported British expansion, yet he feared for his life and therefore refused to meet with Touré, whereas the one Briton in this entourage, Henderson, met Touré and survived. The violent end of Ferguson's life and the display of his head to Henderson demonstrate that Samori Touré's army knew of Ferguson's status and that his death would be a significant loss to the Gold Coast administration and to Henderson. Moreover, Ferguson's murder reinforced Touré's reputation for fierceness, specifically in that his army had killed an African agent of British imperialism.

In the end, Ferguson was literally and figuratively trapped in a clash between all of these communities. Ferguson was nevertheless a key figure in both the establishment and early mapping of the Gold Coast. His cartography and his work as a diplomatic intermediary on behalf of British colonialism contributed significantly to the development of administrative hierarchies within the Gold Coast, to the geographic integrity of the colony, and to expansion of

British colonialism northward from the Gulf of Guinea. His work shows the importance of African intermediaries in, quite literally, charting the direction of British colonization in the Gold Coast. Coming from within the colonial system, trained and supported by high-level colonial administrators, Ferguson became a key political agent facilitating British colonial expansion. The British had repeatedly been thwarted by Asante's forces as they headed north, and Ferguson offered a way to sidestep Asante. He was able to travel through Asante's territories, reducing Asante's influence, and documenting and mapping its hinterlands. Finally, Ferguson's work began a trajectory of the colonial administration's engagement and training Gold Coasters in surveying practices. African surveyors' skills and contributions bolstered a relatively weak British team of bureaucrats and technicians, who lacked contextual knowledge of cultures and languages of the region. The next section considers how African engagement with colonial mapmaking became an established of feature colonial administration and development promoted by the bureaucratization of surveying practices.

COLONIAL ADMINISTRATION AND DEVELOPMENT, 1901–30

Between 1901 and 1930 government mapping and surveying in the Gold Coast expanded substantially to support the administrative needs and development agendas of the government. In 1901, colonial mapping bureaucracies were expressly established to support the extraction of gold, timber, and other natural resources and the administration of concessions. The Mines Survey Department quickly assumed additional surveying and mapping responsibilities and, in 1908, changed its name to the Survey Department. Its staff was responsible for documenting the colony's territories, towns, and people, and it launched a topographic framework for the colony, establishing a network of fixed points for the Gold Coast Colony initially, followed by Ashanti, and then the Northern Territories. There were certainly major challenges internationally and domestically during these years; however, the expansion of colonial administrative systems is notable, particularly during the 1920s.[48] Surveying and mapping the Gold Coast fit this trend, being practices that reinforced the unfolding of colonial infrastructures as well as resource development and exploitation. This section chronicles the expansion of surveying during this second phase of colonialism with particular attention to the involvement of Gold Coasters during this period. It considers major trends in the development of the profession and the bureaucratization of these practices within the colony, and where and how Africans were seen to contribute to and participate in mapping activities.

The evidence and sources for interpreting Gold Coasters' roles in colonial cartography change with the evolving organization of these practices. The main sources for this period are departmental annual reports, personnel files, and various administrative files, once held by the Colonial Secretary's Office or the governor. And unlike the previous period in which George Ferguson regularly communicated with the governor about his cartographic and political endeavors, the contributions of individual Gold Coasters are not well documented, as generally only the surveyor general or senior staff are listed on the maps. Thus, the specific contributions of individual surveyors are not evident for this period. This move toward increased anonymity reflects a normalization of cartographic practices that is characteristic of colonial governmentality.

AFRICANS AND THE SURVEY DEPARTMENT, 1901–20

At the turn of the twentieth century, there was great demand for colonial surveyors across British Africa but comparatively few qualified personnel available. To meet this demand, the British Colonial Office, in conjunction with local colonial administrations, established surveying departments across its African colonies.[49] The Gold Coast Mines Survey Department was founded in 1901. The demand for surveyors and draftsmen was particularly acute in the Gold Coast because of the boom in gold, timber, and other concessions that had begun with the expansion of British authority over Ashanti and the influx of prospectors. With this surge in concessions, there was considerable confusion between issuing leases and with coordination of the actual plots of land being leased. The Mines Survey Department and its surveyors were the key to the regulation of concessions and a new Concessions Bill. Department surveyors checked and validated plans, cut boundary lines, and conducted surveys of the leased lands. The department also licensed private surveyors, hired by mining and timber companies to produce surveys of their concessions.[50] Faced with such demand, both London-based offices and the Mines Survey noted the need for local staff.

The demand for African surveyors, draftsmen, and other assistants only partly stemmed from the extraordinary amount of survey work to be undertaken. The colonial argument for hiring Africans was also based on the economics of paying this African staff substantially less than it paid its metropolitan staff. The colonial administration, including both those based in Accra and London, did not expressly seek the political or cultural knowledge that an African staff could bring to the department. Rather, the colonial government recruited African staff to assist with the mundane tasks and demands for surveying and maps.

Both the Geographical Section of the British Association for the Advancement of Science and the Colonial Office asserted "the absolute necessity of resorting to native agency for its topography."[51] The Colonial Office circulated the recommendations of Thomas Holdich, chair of the Geographical Section of the British Association and author of *How Are We to Get Good Maps of Africa?* This 1901 pamphlet drew on his career in the Survey of India and his familiarity with the role of Indian surveyors. For Holdich, the need for African surveyors was based largely on economics, since Africans would be paid at a fraction of the salary of Europeans. Holdich saw Africans as providing the bulk of the work, as Indians had in the survey of India, whereas Europeans would serve more or less in supervisory positions. While Holdich's report was aimed at a continental scale he did take note of the contributions of George Ferguson in the Gold Coast. Holdich recommended that colonial officials identify other "natives of Africa who will exhibit the same peculiar aptitude for geographical map-making."[52] Holdich's reference to Ferguson's "peculiar aptitude" suggests that the recruitment of African surveyors was not simply an economic calculus. That said, Holdich did not explicitly acknowledge the full scope of Ferguson's contributions and the many ways that African surveyors might contribute to this second phase of the colonial project.

Within the Gold Coast's Mines Survey Department, A. E. Watherston, the first director, agreed in principle with Holdich's ideas and discussed options for training African staff. He regularly reported that the unit was understaffed. He went so far as to recruit a number of unpaid African staff to work in the department as assistant surveyors or laborers helping with chaining and traverse measurements. Watherston held racist views toward Africans. He wrote, for example, that Africans disliked physical work.[53] Given his prejudices, Watherson was disinclined to move beyond his minimal efforts to recruit African staff. This policy changed in 1905 when the Mines Survey came under new direction.

Under the leadership of F. G. Guggisberg, the Mines Survey hired four so-called native surveyors, including a draftsman, in 1905. This hiring marked a formal recognition of African professionals within the department. It also established a hierarchy based on race and professional training that regulated duties, supervision, salary scales, promotion grades, and other entitlements of

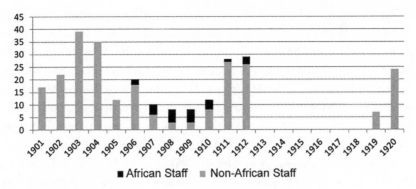

FIGURE 5.4. Staffing and NCOs of the Gold Coast Survey Department, 1901–20.

African and European staff. Guggisberg codified many of these practices in his reports and his 1911 *Handbook of the Southern Nigeria Survey: A Textbook of Topographical Surveying in Tropical Africa*.[54]

Of the four men employed at the Mines Surveys unit, one or possibly two of them were Gold Coasters, but all of them were listed as "natives." The men included E. J. Smith, a Gold Coaster, T. H. Vaughan, a West Indian surveyor, J. B. Essuman-Gwira, and a draftsman, Robert Josiah.[55] Guggisberg's 1906 report notes his willingness to hire more African professionals and that he received applications from "natives" who studied surveying in London; however, he was not satisfied with their skills.[56] In the absence of a local survey school, the number of African surveyors and draftsmen remained rather low.

In its first twenty years, the number of African surveyors and draftsmen employed in the department reached a maximum of three surveyors and one draftsman working at the department (see fig. 5.4).[57] By 1915, only two surveyors remained, as one surveyor left the Gold Coast and one of the men died. The African draftsman was promoted to a surveyor position.

The department had the option of sending students to Southern Nigeria for training at a survey school established there in 1908. However, from the record it appears that the department preferred to train African staff on the job. In addition to the four professional positions, many more Africans were employed by the department as laborers to assist in the surveying of the colony. The department closed for four years during the First World War as many of the European staff were dispersed to various war zones. Both the department's annual reports and other records were not maintained for some time as well. During this period, it appears that African staff were relocated to the Public Works Department.[58]

Annual reports and personnel files suggest that African surveyors worked on both town and topographic surveys and also helped establish the colony's topographic framework. The "native" staff was assigned to the town surveys of Accra and other large communities. A number of these town sheets were handled entirely by the African staff. Departmental reports indicated that the African surveyors were "very useful" in contributing to the first ten topographic sheets of the Gold Coast Colony, printed in 1907 and 1908. In his description of the topographical mapping of the Colony and Ashanti, Guggisberg noted that ten surveying parties were active and that each party was supported by fifty Africans. It is likely these African employees were predominantly laborers. But with only two or three Europeans in each party, Africans also fulfilled various technical roles—working as headmen, probationers, chainmen, sappers, and carriers. Guggisberg also explained in the same report that young Africans, who had just left government schools, had become good and "cheap" surveyors who were capable of filling in details on the maps between the framework and conducting compass surveys.[59] Thus, in addition to the four professional staff, many more Africans played supporting roles in the production of topographical maps of the Gold Coast.

Despite his stated interest in hiring and training Africans, Guggisberg's characterization of African surveyors was not always favorable. He considered the Africans to be less adept at cadastral mapping, noting that they did not grasp the mapping of "artificial features" such as concession boundaries. These concession boundaries were delimited based on negotiations between prospectors and local landowners and were regulated through colonial administrative offices, including judges, surveyors, and the deeds office. As a result, Guggisberg preferred to assign African surveyors to mapping the physical features of the landscape. It was in this way that Guggisberg himself drew a line between what African and European surveyors could and should measure and map.

EXPANDED OPPORTUNITIES DURING THE 1920S

Following World War I and the related hiatus of the Gold Coast's Survey Department, F. G. Guggisberg became governor of the colony in 1919. In his new role, he revitalized the Survey Department by supporting cadastral and topographic mapping and by funding new initiatives. For example, a special party was formed in the Survey Department to help handle the mapping of stool boundaries.[60] The department also compiled new maps to serve and educate the general public (including atlases and road maps). The government opened a publication office in Accra, which allowed for the local printing of

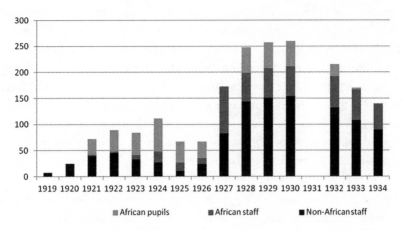

FIGURE 5.5. Staffing and students at the Gold Coast Survey Department, 1919–34. NAG ADM 5/1/96–106, Annual Reports 1919–30.

maps and other documents. Most important, Governor Guggisberg prioritized the economic development of the colony. The Survey Department became a key player and beneficiary of the governor's development plans.

The Survey Department's revitalization brought about many changes across the department, including new opportunities for its African surveyors and draftsmen. Due to an increasing demand for maps, there were both new positions and training opportunities. One of the most significant developments was the establishment of a new government-run survey school for African students. Students advancing through the program received practical training and apprenticeships as surveyors and mapmakers for the department.

Figure 5.5 provides an overview of the staffing of the Survey Department throughout much of the 1920s and into the mid-1930s.[61] African surveyors, draftsmen, and technical staff were found in all the department branches: the cadastral and topographic sections and records and reproduction. Africans were also posted to the newly formed provincial surveying units. The overall number of Africans employed by the Survey Department is most certainly underestimated. The annual reports only cite notable positions held by African personnel. It is likely that many of the lower-ranked positions that were either not associated with a title or not listed at all were held by African employees. While the highest-ranking staff are listed and named, many of the supporting assistant surveyors, draftsmen, clerks, and laborers are anonymous individuals within these reports. Thus, these documents probably underestimate the contributions of African staff to the surveying of the colony.

In addition to the growing number of African staff, there were new standards for assigning the rank of professional staff. In naming positions, the department implemented a four-tiered system for ranking surveyors. Promotion in the system was dependent on employees passing what were known as "efficiency bars." The reorganization enabled some African staff to hold a "European appointment," meaning they were paid on a scale that applied to European surveyors. Kweku Asante, hired as a chainman in the early 1910s, received multiple promotions in the early 1920s, including the rank of surveyor with a "European appointment." A. A. Young, a Nigerian cadastral and town surveyor, who worked in the department for eleven years, also held a "European appointment." With these appointments, the colonial administration modified its former practice of paying Africans according to a lower pay scale. The year of Asante and Young's promotion, 1922, the surveyor general, R. H. Rowe, wrote that further promotion was possible:

> If the two surveyors can continue to maintain their standards of faithful and loyal work, they may hope to rise still higher in their profession, and help by their example in the department to form that character and reliability so necessary in the African Surveyor before he can qualify for the higher appointments.
>
> To those African Surveyors who read this report I say clearly that, while high technical skill is essential and will be demanded of them, technical skill alone will not qualify them. Reliability, loyalty to their superiors, and such strength of character as to ensure proper control of their subordinates, are essential before recommendations for promotion will be made.[62]

The just compensation to Asante and Young for their work and the possibility of future promotions suggests a changing working environment in which European and African professional staff might be compensated more equitably. Yet no other surveyors received such promotions or held a "European appointment" in the years to come. The department's expansion in the 1920s also coincided with the starting of a survey school internal to the unit and creating a pipeline for Gold Coasters to enter the profession.

TRAINING AFRICAN SURVEYORS

The heads of the Gold Coast Survey Department knew that in order to establish a well-trained local staff, they would need to create a local training school for surveyors and draftsmen. The topic was repeatedly taken up by both Watherston and Guggisberg under their leadership of the Survey Department. The Colonial Office decided to support a surveying school in 1907 but located it in Southern Nigeria rather than the Gold Coast.[63] With Guggisberg's return as governor and his commitment to surveying and development, a second survey school in British West Africa was created in the Gold Coast in 1921. The Gold Coast Survey School first opened in Odumase and admitted twenty-three students in its first year.[64] The school regularly had more applications from students than it could accept, and applications generally increased over time as the school and profession gained a strong reputation. By 1927, 101 students had entered the training program and 26 had successfully graduated.[65] By 1930, 44 students had qualified as government surveyors.[66]

The training program entailed three years of instruction and practical training that took place during and after the formal instruction. Admission was based largely on successfully passing an exam, which encompassed arithmetic, elementary algebra, geometrical drawing, history, geography, English, and general knowledge. Exams were typically held once or twice a year in some of the larger urban centers in the Gold Coast Colony and Ashanti. Students also had to be at least sixteen years old.[67] Once admitted, students would sign a bond agreement, allowing them to receive some support for their training but also committing them to work for the department for four years following their successful completion.[68] According to its initial curriculum, the first year would focus on elementary surveying, math, and developing drafting skills. Students were introduced to topographical mapping—using rope and sound traverses and aneroid barometers. The second year, students would continue learning topographic skills, including leveling and plane-tabling, and prismatic compass traverses. They would also begin cadastral map training, learning leveling, large-scale plane-tabling, basic theodolite usage, and chaining.[69] Many of the annual reports specify some of the applied learning that students took part in, mapping missions and topography around Odumase. This practical training then culminated in the students' third year, when they were referred to as fourth-class "native" surveyors and were assigned to the provincial surveying units or other sections of the department for their practical training. Successful graduates would then begin their minimum of four years of service to the colonial administration.

For the Survey Department, the benefits of creating this program were multiple. First, it addressed the shortage of trained technical staff available to the department. Second, most African staff were paid on a lower salary scale than their European counterparts; thus, training local surveyors helped to keep costs lower. Salary, pensions, and allotments for field-based charges were lower for African surveyors and draftsmen, and transport costs to and from the Gold Coast were eliminated in the case of Gold Coasters. Another benefit was that students contributed to departmental initiatives as a part of their training. The head of the Cadastral Branch wrote, "2nd and 3rd year students have been employed for a period of several months in the field. They have made excellent progress. . . . The Survey School has well justified its existence, and without it, we would not be able to turn out the quantity of work that we are now capable of producing." A private letter from the head of the Topographical Branch indicated, "I never imagined that any of these fellows could become so efficient at the job. . . . Their sheets are quite up to the standard of those turned out by the European surveyors, and of course the costs are working out extraordinarily low."[70] Channeling students into government service enabled the Survey Department to be extremely productive in the 1920s. Further, the department knew that it would also have its most successful students enter as surveyors, working for the unit for at least four years. After these four years, Gold Coast surveyors and draftsmen could enter the private sector, if they chose to.

The survey school and the new opportunities within the department created employment opportunities for African students and surveyors. The survey school was an institution that could channel them into careers that engaged their talents in algebra, geometry, geography, and science and their general knowledge.[71] Retired Ghanaian surveyors who were trained during the colonial period reported that the surveying career also spoke to their sense of adventure. Over the course of the 1920s the number of African surveyors employed in the department reflected the growth of opportunities. However, for many of the students, their education had its limits. During the first ten years of the survey schools' existence, none of the students rose to the point of being compensated on a European pay scale. Additionally, by 1926, more than 40% of the students had been dismissed or transferred to other departments in the colonial government.[72]

As part of the students' mentoring, the surveyor general, R. H. Rowe, established a process for continued supervision of African surveyors. Within the Cadastral Branch, the 1923–24 annual report explained, "That the junior native surveyors have proved to be of considerable assistance does not mean that an efficient European supervising staff is no longer necessary. . . . Good European supervision will be essential. . . . The stage through which these junior surveyors are now passing is one in which they require constant supervision, very careful guidance and sympathetic help." These attitudes toward the African surveyors were not uncommon within the annual reports and departmental memos of its European staff. There was a tendency to offer praise to select and named professionals, as described below, and mentoring to promising junior surveyors. There was also some skepticism of African students, some staff, and certainly the unskilled laborers. Assessments reported in annual memos legitimized, first, the continued employment of European surveyors and experts. They also validated Europeans' holding higher ranks than the African staff. Such attitudes are unsurprising given the colonial context in which they operated. This narrative of continued patronage and mentoring was common in the period of colonial development.

MAPPING FOR ADMINISTRATION AND DEVELOPMENT

During the 1920s, the work of the department and its African surveyors resulted in considerable output in terms of the number of maps, the revision and expansion of the country's geodetic framework, and the planning and implementation of several development schemes. Africans were essential to the functioning of every departmental unit. By 1924, African surveyors led by Kweku Asante were assigned sole responsibility for the updating of a Cadastral Survey of Accra at a scale of 1:1,250.[73] Within the Topographical Branch, Africans were given greater responsibility, producing field sheets and significantly lowering costs of production. Major Bell, head of the Topographical Branch, calculated that a detailed topographical survey per one-inch sheet of 290 square miles cost £94 less in 1924 than it did the previous season. He also estimated that once a field camp was entirely made up of African surveyors and laborers, the cost would drop £294 below the 1923 amount. African surveyors and draftsmen were also advancing within the Reproduction and Records unit and at the survey school. F. O. Hanson was promoted as the senior African draftsman to fill a position vacated by a retiring European officer. Also, A. A. Young worked at the school as assistant instructor, and he was expected to be promoted to instructor once a vacancy opened up.[74]

As the employment of African personnel expanded over the 1920s, there was increasing emphasis on creating standards for techniques and practices. One of the first steps in this process was to establish a new topographic framework based on theodolite traverses and leveling tied to the transverse Mercator projection. With the framework, new beacons were established as points of departure for future surveys.[75] The Topographic Branch established its new standard scale as 1:62,500 (one inch to a mile), which let it update its small-scale maps published in 1907–8 and 1914. Also, most cadastral maps of cities and towns were completed at 1:6250 and 1:1250. Along with the scale changes, surveyors and draftsmen implemented standards for deriving their data and representing them. New tools were introduced during this period that enabled greater accuracy, including a steel tape for chaining and a wireless set that helped to establish longitude. The department established conventional signs for use on its topographic sheets and codified standards for orthography and place-names. In these ways, the department began to craft a more systematic approach to mapping the Gold Coast. African personnel were certainly involved in these processes, and thus were inculcated with the importance of European standards and accuracy.

The colonial administration needed African surveyors to carry out surveys and to support the new interest in town layouts, cadastral plots, and topographical sheets. Yet their contributions to these efforts became increasingly anonymous. Instead of crediting the individuals involved in compilation, published maps indicated only the surveyor general's name. This move occurred as mapping standards were established in the Gold Coast, and it included new standards of representation and scale replacing some of the more variable practices.

The period 1901 to 1930 was an important one for involving more Africans in the profession of surveying and mapmaking. The demand for skilled surveyors significantly increased during this period, when the administration sought to address its dependence on concession mapping by investing in the professional training of Africans to undertake this work at relatively low cost. For African surveyors, their options for training and advancement were initially limited. Situated within the institutions and power relations of British colonialism, Gold Coasters and other African staff had to fit into the subservient roles and hierarchies that the British had created for them. Despite these obstacles, African personnel played a prominent part in the Survey Department's activities. The increasing number of well-trained African surveyors and the accomplishments of certain individuals advanced the work and influence of the Survey Department. Higher-level administrators took note of some of these contributions and recognized them with promotions, in turn elevating the status and visibility of a select number of African surveyors.

COLONIAL CONSOLIDATION AND DECOLONIZATION, 1931–57

The visibility of African surveyors and their work declined during the Great Depression and World War II. These international crises constrained the ability of British authorities to allocate resources to the colonies and thus to the Survey Department. In addition, from the 1930s onward there was mounting pressure from the African press, student unions, trade unions, traders, and returning war veterans to end British colonialism through a series of strikes, marches, and other political actions. The official end of British colonialism in the Gold Coast came in March 1957.

For the Survey Department, the retrenchments of the 1930s reduced production and brought a temporary closing of the survey school. From 1931 to 1933, the Survey Department reduced both its European and African personnel substantially. It lost fifteen European positions, including two draftsmen, a lithographer, seven surveyors, and two supernumerary surveyors. Among its African personnel the department lost sixty staff and students. These numbers included twenty-four draftsmen, eight surveyors, and twenty-one pupil surveyors.[76] While the survey school reopened later in the 1930s, the number of students remained comparatively low. Staff numbers also remained lower than before the Depression, though exact numbers are unclear. When funding was in place for both training and implementation, new surveying technologies were introduced and facilitated production. As during the 1920s, the work of the department was cast as serving the development of the colony.

Training opportunities for African surveyors were cut for many years, but in the mid-1950s new opportunities arose for training and credentialing. Gold Coaster surveyors remained in lower-ranking positions in the department and were paid on different scales. Yet the credentialing options that opened up were important to the status and recognition of African surveyors. More generally, however, the training and involvement of African surveyors no longer received the attention that it had once had in Survey Department reports. Thus, in this period, African involvement, while still essential to the mapping of the colony, received less attention in reports, memos, or archival records from the department.

It is also important to note that during this late phase of British colonialism, the colonial archives and records of the Survey Department are less complete or sometimes nonexistent. With fewer Survey Department reports, reconstructing the involvement of Africans in the history of the department relies much more on the memories of African surveyors, as well as a few published reports and studies. In this section I highlight the training of these men and identify the priorities and interests of these surveyors.

TRAINING AND CREDENTIALING

Despite the retrenchments of the 1930s, most Africans working as government surveyors and draftsmen during the late colonial period continued to be trained in the government-run survey school in Accra. The training opportunities were generally very limited during the 1930s and 1940s but notably opened up for aspiring African surveyors in the 1950s.

During this period the survey school continued to be officially linked to the broader department, and its headmaster reported to the surveyor general. One notable change was that students were no longer identified as either African or native students, but, rather, they would begin their working careers as "pupil surveyors." Students were required to work for the Survey Department for a number of years after completing their education. This training institution and the students' attachment to the department helped inculcate the culture of the Survey Department in a cadre of surveyors and draftsmen.

In the mid-1950s, a few additional changes and opportunities opened up in the Gold Coast. Some departmental surveyors received additional training in the United Kingdom. This training gave them the opportunity to obtain credentials from the Royal Institution of Chartered Surveyors (RICS). In 1955, a branch of RICS was set up in Ghana, with the London office approving its draft constitution and regulations.[77] Also at this stage, the University of Science and Technology in Kumasi offered survey training courses. Several African students trained in surveying at the university level could earn the RICS certification, if they succeeded in the curriculum and exams.[78] The RICS training, while not fully standardized across the British Empire or Commonwealth, regulated syllabi and the series of examinations that students would need to pass to earn their certification.[79] Among the retired surveyors I interviewed, RICS requirements for credentials were thought to be rigorous but not always well suited to the needs and contexts of Ghanaian surveyors.

In the cases of both the RICS credentials and the Survey Department training programs, annual school exams were mechanisms for standardizing surveying practices. Established under a Gold Coast ordinance of 1928, the exams and "survey rules" created a set of expectations for all students and practitioners.[80] One retired surveyor explained that in the 1950s, when he was at the university, exams were even graded externally, suggesting even broader adherence to particular benchmarks for standardizing cartographic practices. Students had to pass through three grades of exams: professional, intermediate, and final.[81] Students who did not pass, he explained, could become technical officers, who were ranked beneath surveyors within the department. In this way, the Survey School, and later the university, became institutional mechanisms for establishing a core set of practices that students would have to master in order to be employed.

In addressing the broader political changes afoot in the Gold Coast in the 1940s and 1950s with the move toward independence, surveyors and draftsmen emphasized the apolitical aspects of their job. My informants noted that they remained committed to the impartiality of their practice. In response to questions on whether independence struggles led to new priorities or changes in the Survey Department, African surveyors reported that the neutrality of their practices supported the continuity of mapping practice. Their responses suggest that not only was there no change in practices and priorities with independence, but also that one should not expect to see any transformation based on changing political circumstances due to the objective nature of their work.

Interviews with retired employees of the Survey Department also clarified some of the ways that surveying standards were set within the department and the ways that objectivity and neutrality were achieved, as a matter of practice. I asked Alhaji Iddrisu Abu, the former director of surveys, about the role of Survey Department maps in creating a sense of national identity at the time of independence and afterward. He explained what he saw as the apolitical nature of surveyor's work: "Surveying has no national identity. It's a mathematical, factual situation. If something is a hill it's a hill. . . . A river is a river. Even if you fly a hundred miles from where you are standing it will not change if you are self-governing or somebody is governing you. . . . It has no racial, tribal or national identity. It's just facts."[82] Abu's assertion that surveying is an apolitical activity based on mathematically determined neutrality was echoed by all the retired surveyors and the cartographer with whom I spoke. Mathematical measurements, techniques, and calculations are core practices within surveyors' work. But in the department's role in surveying disputed property boundaries, stool boundaries, and international frontiers, underscoring neutrality and mathematical determination is a logical framing of their professional work. Asserting surveying's neutrality helped to secure the profession's role in politically contested decisions. For Abu or any surveyor to agree that the practices have a sociopolitical role in national identity creation would be to negate the value of surveying's impartiality.

The impartiality of surveying was asserted in a report to the Ghana Institution of Surveyors in 1991, where Abu detailed the state of surveying in Ghana and some of its historical origins. However, as the statement continues, he hints at the possibility of corrupting practice and the need to adhere to the professional standards. He wrote:

> The need for an impartial demarcation and redemarcation of land bound-aries is said to have brought about the professional called the Land Surveyor today. . . .
>
> From time immemorial the land surveyor's services was and still is impar-tial measurements. Trying always to find the "best fit" to each "environment" the surveyor, always allows his measurement, not sentiment to control his judgment.
>
> Fellow Surveyors, are we sure that we are living our professional lives to this standard? If yes then we are well equipped to look to the future.[83]

While all the retired surveyors note that they upheld their impartiality in their work, the last line in the quote above suggests that some surveyors might allow "sentiment" to cloud their measurements and decision making, despite the standards promoted by the department and the profession.

Despite the assertion about the mathematical neutrality of surveying, the retired surveyors were aware of how contentious boundary mapping could be, and they had multiple strategies for mapping socially constructed enti-ties, such as boundaries. Cadastral surveying for property or stool boundaries was described as dangerous by a number of surveyors, and reports of assaults or threats were also well known among them. Not only did the surveyors encounter and know of conflict at that scale, but A. H. Osei, who chaired the Joint Demarcation Commission that remapped the 285-mile boundary between Ghana and Burkina Faso in 1968–69, also referenced tensions that brewed between the two sides during the surveying.[84] Thus, in coping with violence or threats of violence, the surveyors drew on particular cultural prac-tices, higher authorities, and maxims and also acknowledged the limitations of their profession to help defuse these situations. For instance, during the map-ping of chieftaincy boundaries, several surveyors explained that they would arrange for representatives of both parties to be present in order to agree on the boundary line. S. W. Kuranchi noted that a particular plant, known as *ntornel*, was planted by people to mark the boundary in the past, and other surveyors noted that anthills, trees, or rivers might be other markers.[85] Such a plant would serve to mark the boundary on the ground, so long as the two parties still agreed to it. Further, a commonly stated mantra among several surveyors was that "chiefs know their boundaries" or "the people know their boundaries"; thus, it was not the role of a surveyor to weigh in on the decision. In cases of ongoing dispute among the land authorities, one retired surveyor advised the parties to take the matter to court rather than involving him to try to arrive at an agreed-upon property line. The mapping of socially constructed boundaries is one contradiction to the scientific neutrality of the map. Another contradiction is the valuing of some resources or landmarks over others in determining what gets surveyed and mapped.

A good example of the apparent neutrality of mapmaking is the adherence to using a standard set of symbols to represent cultural and geographic features on maps. These symbols were known "conventional signs" that Ghanaian surveyors spoke about at length when asked about them. The former cartographer and former head of the Cartography Section S. R. K. Loh indicated that these signs would designate houses, schools, roads, and so on, but the signs could also be adaptable to the cultural context. For instance, in northern Ghana, where many houses are round and built in circular compounds, the sign for homes and settlements is round. In maps of southern Ghana, where rectangular homes forms are common, the symbol for a home or settlement is rectangular. Loh indicated that these "conventional signs" were determined by the Survey Department, and copies were issued to all staff whether they were in the field or in the offices drafting maps. He further explained that the conventional signs were important because "we must all speak the same language." He explained that these symbols informed what was important to depict on a map and what data to collect when in the field. Loh also stated that ultimately the surveyor general had the authority to determine what should be depicted. Further, courses taught at the government survey school included lettering and conventional signs at different scales to ensure consistency.

These standards and conventional signs did not change substantially with the transition to independence. A copy of the key to conventional signs used prior to independence was marked up in the Survey Department library to show what would be changed in the postindependence topographic maps (see fig. 5.6). The new key no longer listed the location of the Chief Commissioner's House and instead indicated Preventative Service Stations (custom stations) occasionally mapped in colonial topographic sheets. The conventions established in colonial contexts largely remained in place in the postcolonial period. These "conventional signs" were not unique to the Gold Coast. They were adopted and implemented across Britain's African colonies. The valuing of political borders, post offices, and rest houses, for instance, was a means of supporting the colonial network and of planning new services in underserved

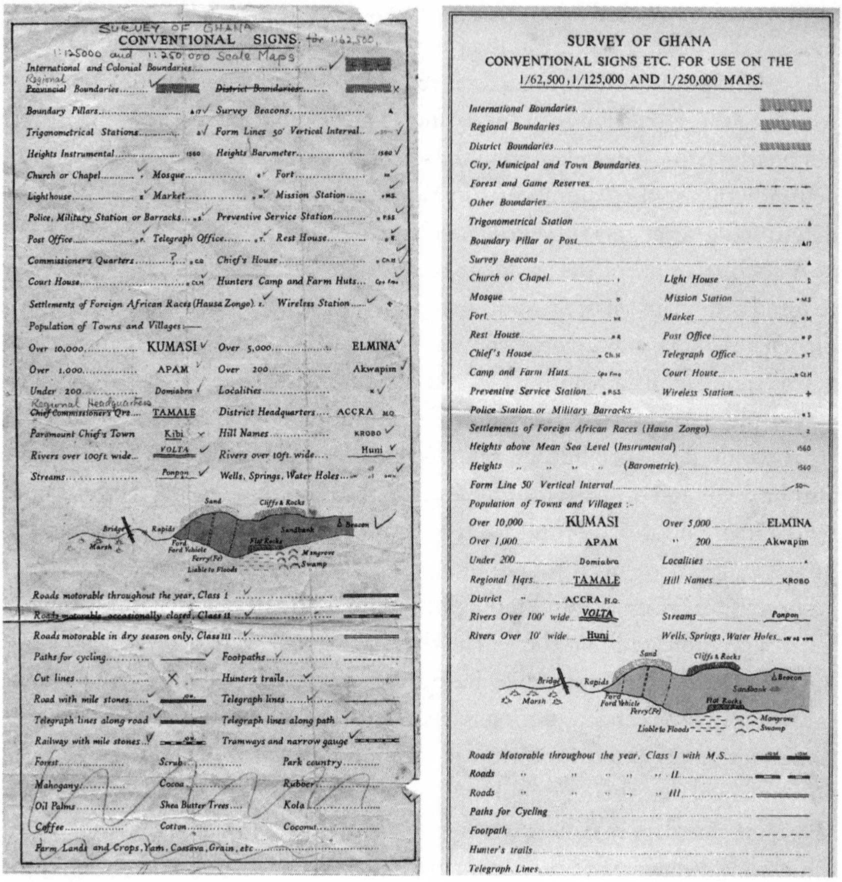

FIGURE 5.6. Survey of Ghana, conventional signs used before and after independence. National Archives of Ghana.

regions. However, the omission of information that might be deemed relevant in Ghanaian cultural contexts indicates that the maps' cultural construction was in fact still in accordance with British colonial rule.

In summary, this third period of African involvement in colonial mapmaking illustrates the emergence of a cartographic governmentality. Similar to Agrawal's notion of environmentality, in which local Indians incorporated the norms and best practices of Indian colonial forest councils, cartographic governmentality refers to the adoption of colonial cartographic standards and practices through training and credentialing that characterized the Gold Coast Survey School. With minor modifications, postcolonial maps of Ghana looked much like colonial-era maps as a result of this inculcation of colonial carto-graphic norms and practices.

AFRICANIZATION AND DEPARTMENTAL CHANGE

Ghana's independence movement did spur changes internal to the Survey Department. These changes came about in the context of political violence and mobilization. On February 28, 1948, a group of war veterans who had been denied benefits for their service to the British Empire during World War II marched to the seat of the British colonial government to submit a petition to secure those benefits. British police fired on the unarmed group, killing three ex-servicemen. This event spawned several days of violence across the country and fueled anticolonial political organization and action. The demonstration and its violent suppression helped give rise to the Convention People's Party (CPP) led by Kwame Nkrumah, who would become the first president of Ghana. Investigations into the violence and the intensification of the CPP's political mobilization pushed the British colonial administration to "Africanize" the public service staff. Africanization was a strategy of hiring and promoting qualified Africans into higher professional and administrative positions in the public service sector. It also provided increased training opportunities for Africans so that there would be more qualified people available for such openings. Changes did not take place over night. The case of the advancement of African surveyors in the Survey Department indicates that the Africanization policies were gradually implemented.

In 1949 African surveyors occupied few senior service appointments within the Survey Department. The number of African staff remained constant at four, and just one promotion took place within the senior echelons.[86] Several Ghanaians, however, were being trained overseas as part of the goal of enhancing their professional careers as explained below.

Africanization did not have an immediate effect on the staffing of the Survey Department, in large part due to lack of advanced training opportunities and credentialing. According to former surveyors, the training available to Ghanaian surveyors at the time was not seen as parallel to the training of foreigners. British surveyors were eligible to be credentialed through the Royal Institution of Chartered Surveyors; the surveying training available through the university in Kumasi did not offer this option until 1955. Informants noted that the educational opportunities available to Ghanaians were different, but so too were the appointments open to them. The racially tiered system enabled white surveyors to hold "professional appointments," whereas many African surveyors were assigned to lower-ranking staff and technical positions. While the title "African Surveyor" was no longer officially used informally to distinguish African from European appointments, the title continued to be used and still carried with it judgmental assessments. In an article on cadastral traverses, the surveyor general noted the heavy reliance on African staff in 1945. He wrote: "Owing to the smallness of the European establishment, practically all fieldwork must be done by junior African surveyors. Some of the African surveyors are extremely competent, but others show no great ability or desire to think for themselves. A junior African surveyor can safely be left to run a routine cadastral traverse with a minimum of supervision."[87] The report indicates a continuation of hierarchy and privilege assigned to white surveyors in the Gold Coast. None of the retired surveyors with whom I met spoke of a racist work environment. In fact, many of them acknowledged positive relationships with British personnel in the department. Yet many of the retired surveyors did talk about the new professional opportunities that resulted from the independence struggle and Africanization.

One retired surveyor, A. H. Osei, who began his training as a surveyor in Ghana in 1938, reported that Nkrumah's political mobilization aided surveyors' advancement. The Africanization order made it possible "for the training of local people to become professional men." He recounted the increasing availability of university training to various professions. He said that prior to this policy change, the hiring of outside, white, chartered surveyors limited the opportunities open to Africans. Osei explained that white surveyors, credentialed through the Royal Institution of Chartered Surveyors, were brought in and were one of the biggest reasons that Africans were held back "until these political boys came," referring broadly to the mobilization led by Nkrumah, J. B. Danquah, and others involved in the anticolonial struggle. Osei's reflections indicate that he benefited from the Africanization of the public service by virtue of the opening up of new training opportunities. After 1948, qualified Ghanaian surveyors were offered professional training and scholarships to University College in London, and Osei and two other men were among the first to benefit from these opportunities. Osei later became the deputy director of the Survey Department and a lecturer at the University of Science and Technology in Kumasi. Additionally, several years after independence one of his fellow scholarship recipients, R. J. Simpson, would become the first African to head Ghana's Survey Department. Africanization did ultimately change the staffing profile of the Survey Department.

While a number of retired surveyors highlighted the role of Africanization in creating new training opportunities and positions to Ghanaians, none of the men I interviewed indicated that the atmosphere was particularly jubilant when the first Ghanaian director was appointed. In fact, the retired surveyors felt that the appointment of an African as head of the Survey Department was part of the normal course of things following independence. Alhaji Iddrisu Abu, the former director of surveys, noted that while he was not on staff during the transition, he was employed by the Survey Department under Simpson and was familiar with the situation in which he was appointed. He explained:

> Before Simpson's time there was no Ghanaian or Gold Coaster who was of sufficient knowledge or luck to . . . hold a European post. A professional post they called a European post, you know because *they were the professionals* [said with a bit of humor]. The title accorded a certain authority and privilege. So to the extent that you were an African and you were a staff surveyor—you were called "a white man." You know, you're a "Black European."

Abu stated that by the time Simpson was appointed, most of the British had left the Survey Department. A few retired surveyors noted that some British surveyors had difficulties accepting staffing changes and still felt they were "a boss" even if they were a technical officer under a Ghanaian surveyor. Abu further observed that Simpson led the department both through his respect for others and by being self-disciplined himself. For example, he arrived at work each morning at 7:00 a.m.

Thus, during the struggle for independence, the mapping of Ghana did not seem to the surveyors and cartographers of the Survey Department to be pivotal in nationalist debates or concerns, or in the anticolonial struggle. While the appointment of Africans to higher professional positions of the Survey Department was personally significant to the surveyors I interviewed, they did not link the transition in leadership to changes in their mapping practices, specifically, their continued adherence to technical and mathematical standards that the department and profession valued and supported. Reflecting on the last twenty-seven years of colonialism and mapping, those who worked in the surveying and mapmaking institutions feel a strong sense of continuity in mapping practices. This continuity owes much to the surveying and cartographic culture and expertise established during the colonial era. Only minimal changes were noted by retired government surveyors. And, there is little apparent difference between the maps produced during the colonial and early postcolonial eras.

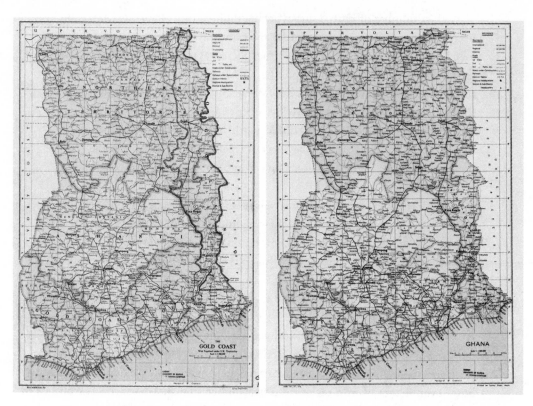

FIGURE 5.7. Survey Department administrative maps of the Gold Coast (1955) and Ghana (1957). Courtesy of the University of Illinois Champaign-Urbana Libraries.

The persistence of map practices and forms demonstrates the ways that the standardization of cartographic practices and training reinforced the trajectories of mapping for development, town planning, and boundary administration and regulation. As an example, the continuities of representations can be seen in the Survey Department's atlases—comparing the administrative maps in the 1955 edition issued before independence and the 1957 edition, produced after independence (see fig. 5.7). The administrative map of the country changed in these atlases only in so far as country's name changed, and the political configuration of the country changed only as new political regions were created and named. The blocks of color changed in correspondence to the creation of a new region, but otherwise, color choices were consistent between the two maps. Within the scope of the Survey Department's work, they maintain the same representation of hierarchies of boundaries, capitols, and transport networks. Map size, orientation, projection, and scale remain the same between these two editions. These continuities in the maps confirm, as the surveyors and draftsmen noted, that their practices changed very little with the move to independence.

Most importantly, retired surveyors and cartographers explained that they did not perceive a political shift with independence that changed their maps and cartographic practice, despite independence and nationhood being monumental and transformative within other realms. They noted some of the changes that did take place intradepartmentally, and many surveyors benefited from the opportunities that opened up. These departmental changes were mostly at the level of personnel changes and promotion opportunities.

CONCLUSION

Most studies of mapping in colonial contexts understate the role of indigenous populations in mapping the territory. This study of the Gold Coast experience argues otherwise, that Gold Coasters played a major role in the colony's cartography, though one that shifted with changing political, geopolitical, and economic circumstances.

During the first phase of colonial expansion, the colony's administration cultivated a close relationship with one African surveyor, George Ekem Ferguson, who made a significant, active contribution to the expansion of British colonialism in the Gold Coast and the cartographic construction of the colony. More concretely, Ferguson's mapping of the colony exemplifies important role of intermediaries, the ways in which colonialism engaged cartography, and the ways that standard cartographic practices were adapted to the needs of colonial cartography. Ferguson's case highlights the ways that early intermediaries occupied positions that were often highly influential, yet also subject to the multitude of interests and communities vying for power and influence. Ferguson was repeatedly asked to fulfill important missions that would expand British interests in the region, and he repeatedly succeeded in fulfilling his assignments. He also adopted particular practices of colonial mapping that illustrate the claim-making and administrative contexts of British colonialism, enabling Great Britain to demonstrate the extent of its claims and influence in the highly competitive geopolitical context of European colonial expansion. His role and his death also demonstrate the vulnerabilities of intermediaries, especially during the early phases of colonization—and his case is not uncommon among Africans working in the colonial service at the time.[88]

The second part of this study focuses on the period of the bureaucratization and standardization of surveying and mapping through the formation of the Survey Department and the expansion of African participation. This section highlights the ways that African professional staff and students greatly increased in numbers in the context of heightened demand for surveys and maps to fulfill an expansive colonial agenda. Employed by the colonial state, they worked to fulfill the governor's needs for administrative organization and development planning. Africans played central roles, especially in the preparation of the town and topographic surveys. However, African contributions to mapping also often were made anonymous by the reports and on the maps. Thus, unlike Ferguson's experience, the work of African surveyors was often hidden in this period. This occluding of African contributions was one sign of the bureaucratization of mapping and surveying.

Finally, this study's examination of the cartographic practices from the 1930s until Ghana's independence in 1957 reveals that there was significant continuity in the making of maps despite the political disruptions of the late colonial period. The Great Depression, World War II, and anticolonial political actions that spread across the colony in its last years led to increasing fractures within British colonial rule. For internal colonial units such as the Survey Department, regular reporting dwindled and disappeared, but African surveyors remained essential to the functioning of the Survey Department. In interviews, surveyors, first, expressed their belief in the objectivity of their practices, and that the training and the work of surveyors emphasized this objectivity, regardless of experiences that indicated otherwise. Second, they noted the move to professionalize their degrees and credentials; Royal Institute of Chartered Surveyor status was an important sign of their professionalism and service. Third, the surveyors noted that the Africanization that occurred across the administration and civil service ultimately opened up more positions to qualified Africans in surveying.

A history of Ghana's mapping, like any colonial mapping history, must engage the role of local participants, including surveyors, in order to represent the scope of mapping practices. Unlike past research that underplays local knowledge and involvement, this study's major contribution is to demonstrate the systematic involvement of local experts and surveyors throughout the colonial period. It shows that cartography was not strictly in the hands of colonial agents. Throughout the colonial period, British authorities were highly reliant on African mapmakers. And these African intermediaries were moreover essential to the continuity that prevailed in the mapping of an independent Ghana.

By exploring African involvement in colonial cartography, this study enriches the literature on the history of cartography. In particular, it speaks directly to Matthew Edney's predominantly Eurocentric focus on colonial cartography in *Mapping an Empire*. The mapping of British colonies, such as the Gold Coast, did more than legitimate British colonization, as Edney argues in the case of India. Based on their cartographic skills, Africans held influential and decisive positions in the determination of boundaries, borders, and regions, and Africans mapped these regions on behalf of the colonial state. Further, the mapping of the colony helped create a cartographic culture among African surveyors and cartographers, in which the objectivity and neutrality of their practice was valued. Mapping facilitated the exchange of information between colonists and colonized, in which both played active roles informing the depiction of the territory. The involvement of Africans also set in motion considerable continuities between the colonial and postcolonial periods, in relation to the mapped spaces and hierarchies of geographic information.

NOTES

1. Benedict Anderson, *Imagined Communities: Reflections on the Origin and Spread of Nationalism* (New York: Verso, 2006); Bernard Cohn, *Colonialism and Its Forms of Knowledge: The British in India* (Princeton: Princeton University Press, 1996); D. Graham Burnett, *Masters of All They Surveyed: Exploration, Geography, and a British El Dorado* (Chicago: University of Chicago Press, 2000).

2. Jeffrey C. Stone, "The District Map: An Episode in British Colonial Cartography in Africa, with Particular Reference to Northern Rhodesia," *Cartographic Journal* 19, no. 2 (1982): 104–28; Stone, "Imperialism, Colonialism and Cartography," *Transactions of the Institute of British Geographers* 13, no. 1 (1988): 57–64; Stone, *A Short History of the Cartography of Africa*, African Studies, vol. 39 (Lewiston, ME: E. Mellen Press, 1995).

3. Burnett, *Masters of All They Surveyed,* 2000.

4. Matthew H. Edney, *Mapping an Empire: The Geographical Construction of British India, 1765–1843* (Chicago: University of Chicago Press, 1997).

5. J. B. Harley, "New England Cartography and the Native Americans," in *The New Nature of Maps: Essays in the History of Cartography,* ed. Paul Laxton (reprint, Baltimore: Johns Hopkins University Press, 2001), 169–95.

6. K. H. Offen, "Creating Mosquitia: Mapping Amerindian Spatial Practices in Eastern Central America, 1629–1779," *Journal of Historical Geography* 33 (2007): 254–82.

7. T. J. Bassett, "Influenze Africane sulla cartografia europea dell'Africa nei secoli XIX e XX" [African influences on European mapping of Africa in the 19th and early 20th centuries], in *Culture dell'alterità: Il territorio Africano e le sue rappresentazioni,* ed. E. Casti and A. Turco (Bergamo: Edizioni Unicopli, 1998), 359–71; Bassett, "Indigenous Mapmaking in Intertropical Africa," in *The History of Cartography: Cartography in the Traditional African, American, Arctic, Australian, and Pacific Societies,* ed. D. Woodward and M. Lewis (Chicago: University of Chicago Press, 1998), 24–48.

8. O. Y. Balogun, "The Training of Nigerian Surveyors in the Colonial Era," *Surveying and*

Mapping 45, no. 2 (1985): 1159–67; Balogun, "The Native Surveyor: The Nigerian Surveyor under British Administration," *Surveying and Mapping* 48, no. 2 (1988): 81–87.

9. M. Foucault, "Governmentality," in *The Foucault Effect: Studies in Governmentality*, ed. G. Burchell, C. Gordon, and P. Miller (Chicago: University Of Chicago Press, 1991), 87–104.

10. Foucault, "Governmentality," 102.

11. G. Prakash, *Another Reason: Science and the Imagination of Modern India* (Princeton: Princeton University Press, 1999); M. Vaughan, *Curing Their Ills: Colonial Power and African Illness* (Stanford: Stanford University Press, 1991).

12. Prakash, *Another Reason,* 7.

13. Ibid., 4.

14. J. C. Scott, *Seeing like a State: How Certain Schemes to Improve the Human Condition Have Failed* (New Haven: Yale University Press, 1998).

15. A. Agarwal, *Environmentality: Technologies of Government and the Making of Subjects* (Durham: Duke University Press, 2005), 217.

16. Ibid., 223.

17. B. N. Lawrance, E. L. Osborn, and R. L. Roberts, *Intermediaries, Interpreters, and Clerks: African Employees in the Making of Colonial Africa* (Madison: University of Wisconsin Press, 2006); K. Arhin, *West African Colonial Civil Servants in the Nineteenth Century: African Participation in British Colonial Expansion in West Africa* (Leiden: African Studies Centre, 1985); A. A. Boahen, *African Perspectives on Colonialism* (Baltimore: Johns Hopkins University Press, 1987); Frederick Cooper, *Decolonization and African Society: The Labor Question in French and British Africa* (Cambridge: Cambridge University Press, 1996); L. Lindsay, "'No Need . . . to Think of Home'?: Masculinity and Domestic Life on the Nigerian Railway, c. 1940–61," *Journal of African History* 39, no. 3 (November 1998): 439.

18. Lawrance, Osborn, and Roberts, *Intermediaries*.

19. Boahen, *African Perspectives*.

20. D. Turnbull, "Boundary-Crossings, Cultural Encounters and Knowledge Spaces in Early Australia," in *The Brokered World: Go-Betweens and Global Intelligence, 1770–1820*, ed. S. Schaffer et al. (Sagamore Beach, MA: Science History Publications, 2009), 388.

21. Turnbull, "Boundary-Crossings," 402.

22. Arhin, *West African Colonial Civil Servants*; Elliott P. Skinner, *The Mossi of Burkina Faso: Chiefs, Politicians and Soldiers* (Prospect Heights, IL: Waveland Press, 1989).

23. Lawrance, Osborn, and Roberts, *Intermediaries*; A. Eckert, "Cultural Commuters: African Employees in Late Colonial Tanzania," in ibid.

24. Lawrance, Osborn, and Roberts, *Intermediaries*, 13.

25. Ibid.

26. Ibid., 14.

27. R. B. Bening, *Ghana: Regional Boundaries and National Integration* (Accra: Ghana Universities Press, 1999).

28. In this paper, I use "Asante" to refer to the people, and "Ashanti" to refer to the territory, following British nomenclature.

29. K. Arhin, *The Papers of George Ekem Ferguson: A Fanti Official of the Government of the Gold Coast, 1890–1897* (Leiden, 1974).

30. The area known as the Gold Coast Colony is not always synonymous with the Gold Coast. By 1901, the British colony known as the Gold Coast encompasses the Gold Coast Col-

ony, Ashanti, and the Northern Territories. If referring to the littoral region, I specify either "Gold Coast Colony" or "Colony," whereas, when discussing the entire British colony, I use either "Gold Coast" or "the colony."

31. See Arhin, *West African Colonial Civil Servants*; Roger G. Thomas, "George Ekem Ferguson: Civil Servant Extraordinary," *Transactions of the Historical Society of Ghana* 13, no. 2 (1972): 181–215; K. L. Korang, *Writing Ghana, Imagining Africa: Nation and African Modernity* (Rochester, NY : University of Rochester Press, 2003).

32. National Archives of Ghana (NAG) SC 24-1, *Gallery of Gold Coast Celebrities*, by Isaac S. Ephson (1969); NAG, *This Man Ferguson of Anomabu*, by M. J. Sampson (January 1956).

33. Ferguson's knowledge of the region depicted and therefore his contributions to this map are not articulated in the colonial records discussing the map. Instead, the divisions shown are noted to be based on the *Gold Coast Gazette*.

34. The National Archives (TNA): CO 96/159, Office of District Commissioner for the various districts.

35. The Crown Agents financially backed the map's printing, and it was the subject of much discussion by officials in the Colonial Office. These London-based discussions centered around the marking of the northern boundary of the Gold Coast Protectorate with a hatched line, and whether or not such a line might limit later northern expansion of British interests. After internal debate over this line, the Colonial Office concluded this boundary marking not to be a concern and allowed the copies to be circulated. The concerns with the map circulated mostly within the Colonial Office and the Department of State, and Ferguson was probably not privy to the discussions. See TNA CO 96/169, "New Map of the Gold Coast," discussed in Jamie McGowan, "Conventional Signs, Imperial Designs: Mapping the Gold Coast, 1874–1957" (Ph.D. dissertation, University of Illinois at Urbana-Champaign, 2013), chap. 2.

36. TNA FO 925/874.

37. Thomas, "George Ekem Ferguson," 181, citing TNA CO 96/200, Ferguson to Private Secretary, April 6, 1889.

38. Boahen, *African Perspectives* (n. 17 above).

39. TNA CO 879/37, Further Correspondence. Anglo-German Claims.

40. T. J. Bassett, "Cartography and Empire Building in Nineteenth-Century West Africa," *Geographical Review* 84, no. 3 (1994): 316–35.

41. TNA CO 96/223, Ferguson—Secret Mission, 1892.

42. Kweku Asante, "G. E. Ferguson: An African Pioneer," *Survey Review* 8 (1933): 102.

43. Ibid., 1933.

44. "The Monthly Record: The Society," *Geographical Journal* 3, no. 5 (May 1894): 420.

45. D. Owusu-Ansah, *Historical Dictionary of Ghana* (Lanham, MD: Scarecrow Press, 2005).

46. I. Wilks, *Asante in the Nineteenth Century: The Structure and Evolution of a Political Order* (London: Cambridge University Press, 1975), 304.

47. I. Wilks, *One Nation, Many Histories: Ghana Past and Present* (Accra: Anansesem Publications, 1996). In future research it is worth looking at whether Ferguson's successes in negotiating treaties throughout this region benefited from the legacy of Asante in the region.

48. G. E. Metcalfe, *Great Britain and Ghana: Documents of Ghana History, 1807–1957* (London: Thomas Nelsons and Sons, 1964).

49. Stone, "The District Map" (n. 2 above); C. G. C. Martin, *Maps and Surveys of Malawi: A History of Cartography and the Land Survey Profession, Exploration Methods of David Livingstone on Lake*

"Nyassa," Hydrographic Survey and International Boundaries, Geographical, Environmental and Land Registration Data in Central Africa (Rotterdam: A. A. Balkema, 1980).

50. Gold Coast Annual Reports, 1901, 1902, 1903.

51. T. H. Holdich, *How Are We to Get Good Maps of Africa?* (London, 1901); and TNA CO 96/396, Surveys.

52. Holdich, *How Are We to Get Good Maps of Africa?*

53. CO 96/396, Surveys.

54. F. G. Guggisberg, "Mapping the Gold Coast and Ashanti," *Transactions of the Liverpool Geographical Society* 19 (1911): 7–14.

55. In later years, Josiah changes his name to Kwantreng. NAG PF 3/54/104, Kwantreng.

56. NAG ADM 5/1/83, Annual Reports, "Report of the Mine Survey for the Year 1906."

57. It should be noted that the data featured in figure 4 are not entirely comprehensive; as not all nonprofessional positions are listed. They come from the departments' annual reports, which in some cases reported by name and title the people working in the department, but lower ranking positions were inconsistently listed. Undoubtedly many Africans served in the lower-ranked positions within Mines Survey and later the Survey Department.

58. Co 96/599, Survey Dept. Resuscitation.

59. Guggisberg, "Mapping the Gold Coast and Ashanti."

60. The stool is the symbol of Akan chieftaincy, and it likewise refers to the land and people over which the chief rules.

61. After the mid-1920s annual reports no longer included a full roster of positions and people.

62. NAG ADM 5/1/99, Annual Report Survey Department.

63. NAG ADM 5/1/83, Annual Reports, Report of the Mine Survey for the Year 1906; TNA CO 96/456, "Survey School for Natives"; NAG ADM 5/1/84, Annual Reports, Survey Department, 1907; Guggisberg, "Mapping the Gold Coast and Ashanti."

64. The school was within eleven miles of Akuse, where the Survey Department would also initiate its new topographic framework of the colony. Akuse would be the starting point for the primary chain. I. Curnow, "Topographical mapping in Africa," *Journal of Manchester Geographical Society* 41 (1925): 32–27.

65. NAG ADM 5/1/103, Annual Report 1926–27.

66. NAG ADM 5/1/106, Annual Report 1930–31.

67. NAG ADM 56/1/116, Survey Department.

68. NAG ADM 5/1/102, Annual Report 1925–26.

69. NAG ADM 5/1/100, Annual Report 1923.

70. NAG ADM 5/1/100, Annual Report 1923.

71. NAG ADM 56/1/116, Survey Department.

72. NAG ADM 5/1/103, Annual Report 1926–27.

73. "The Gold Coast Survey Department," *Geographic Journal* 68, no. 5 (1926): 451–52.

74. NAG ADM 5/1/98, Annual Report 1921.

75. NAG ADM 5/1/100 Annual Report, 1923–24.

76. NAG ADM 5/1/108 Annual Report, 1931–32.

77. Anon., "Extracts from Council Minutes," *Journal of the Royal Institution of Chartered Surveyors*, 1995, 34.

78. All of the retired government surveyors trained in the mid- to late 1950s received some, if not all, their training through this university, as opposed to the Survey School.

79. Anon., "Notice," 1965.

80. NAG PF 3/54/188 E. G. Smith.

81. Interview with Benedict A. Neequay, August 21, 2008.

82. Interview with Alhaji Iddrisu Abu, November 2008.

83. Iddrisu Abu, Ghana Institution of Surveyors Workshop, "State of Surveying and Mapping in Ghana," Land Surveying Day, July 25 and 26, 1991.

84. Interview with A. H. Osei, November 8, 2008.

85. Interview with S. W. Kuranchi; December 5, 2008.

86. NAG RG 5/1/396, Africanisation.

87. A. V. Lawes, "Cadastral Traverses in the Gold Coast," *Empire Survey Review* 8 (1945): 138.

88. Lawrance, Osborn, and Roberts, *Intermediaries* (n. 17 above).

CHAPTER SIX

MULTISCALAR NATIONS

CARTOGRAPHY AND COUNTERCARTOGRAPHY OF THE
EGYPTIAN NATION-STATE

Karen Culcasi

INTRODUCTION

As other chapters in this book have shown, much of the modern political world map, most especially in Africa, the Americas, and Asia, has imperial origins. The territorial entities that European imperialists constructed have had profound and enduring effects not only on the division of the world into discrete "nation-states," but also on the politics and economies of these supposed nation-states. As has been widely acknowledged, the boundaries that imperial leaders created reflected their own interests and were rarely reflective of indigenous or local geographies. Thus, during the mid-twentieth century, when many imperial and colonial territories gained their independence, these emerging nation-states were faced with the task of building national discourses and unity within state boundaries that had little meaning to their populations.

Egypt is something of a unique case in the study of postcolonial nation building, as it is one of the oldest societies on earth while also being a recent postcolonial construction. Even though it has a rich and long past, once Egypt achieved semi-independence from Great Britain in 1922 and embarked upon nation-building practices, its ancient past had little prominence in the burgeoning postcolonial national discourse. Egypt's ancient Pharaonic history and its Ottoman connections were dwarfed in importance in comparison with the more recent history of Egypt's leadership in anti-imperialist, pan-Arab, and pan-Islamist movements of the nineteenth and twentieth centuries (Goldschmidt 2004). These movements were indeed formative in Egypt's postcolonial national narrative, yet they are only part of what has formed Egypt's multiple and contested identities today.

In this chapter, I explore the complex discursive formations of an Egyptian national identity by examining how Egypt was cartographically constructed after its official, though nominal, independence in 1922. Through an examination of official and unofficial maps and atlases produced and used in Egypt, I provide a critical reading of the construction and the contestation of Egyptian national identity through a cartographic lens. The nation-building activities and discourses that both newly independent and well established nation-states employ are numerous and varied, yet cartography has played an inextricable and formative role in creating, sustaining, and even contesting the existence and legitimacy of nation-states. Most research concerned with postindependence nation building, whether cartographic or not, has examined these processes at the scale of the state. Indeed, the state is so prevalent in our thinking and framing of national identities that it has often limited our understandings of other ways in which the world is divided, ordered, and imagined (Agnew 1998). By examining the processes of nation building at multiple scales, we gain new insights into their postcolonial complexity. Scales are, arguably, the most elemental differentiation of geographic space and are key frameworks for the construction of place (Smith 1992a, 73). Scalar divisions, such as the urban, regional, national, supranational, and global, have traditionally been viewed as relatively stable geographic categories for ordering the world (Brenner 1998, 459–60). Debates since the late 1990s have challenged this traditional conception of scale, highlighting that scales do not have a natural or static existence, nor are they fixed on a hierarchical continuum, but instead they are the products of multiple, competing, and fluid discourses and processes (Paasi 2004; van Schendel 2002; Swyngedouw 1997; Brenner 2001; Marston 2000; Herod 1997; Agnew 1997; Delaney and Leitner 1997; Staeheli 1999; Smith 1992b; Leitner 1997). Transnational studies have also questioned the stability of the scale of the state and have highlighted how national identities often exist at intersecting supranational and local scales (Western 2007; Grewal and Kaplan 1994; Ong 1999). Regardless of the critiques of traditional notions of scale, and of the connections between nation and state, many studies on nations and national identities still frame the nation as existing at the scale of the state (Mountz 2013; Murphy 2013). In this paper, I draw on a multiscalar approach that underscores the tenuousness and complex intersections of scales. I move among the supranational, national, and local scales of analysis to highlight how geographic narratives at all these scales intersect to both support and at times contest an Egyptian national identity. My intention is not merely to change the scale of analysis (Herod 1997, 146–47) but, as Staeheli (1999, 55) argues, to consider the ways in which processes such as cartographic nation building operate at intersecting scales.

NATION BUILDING AND CARTOGRAPHY

The division of the world into independent nation-states became the dominant world order in the twentieth century as European empires disintegrated and the idea of self-determination was promoted. A "nation" is generally considered to be a community of people bound by a sense of shared history and culture, and often an attachment to particular territory. By contrast, a "state" is usually understood as a politically sovereign territory with a centralized government. Thus, a nation-state is the theoretical unification of a culturally homogeneous population residing within clearly defined political and administrative boundaries. But as it has been commonly argued, nation-states are not natural preexisting entities, but constructed and imagined political communities (Anderson 1991). Creating nations and nation-states is then a complex process that that includes a variety of different practices, activities, and discourses. The processes of constructing nation-states has varied immensely over space and time, but it often includes discourses and practices that seek to homogenize the nation and marginalize or assimilate minorities into the dominant group (Anderson 2001; Berger 2006; Baram 1990). Such national discourses are often initiated by the state, and fueled by educational institutions and the media. However, nation building is never linear, static, or entirely top down. Instead, it is a messy process that involves different actors and institutions, which also have the ability to alter and challenge dominant discourses of the nation. Moreover, the citizenry are not passive agents in the construction of nation-states either, as they often facilitate dominant discourses of the nation (Billig 1995), as well as challenge them.

Though nation-states are constructed and maintained through various practices and discourses, and by multitudes of people and groups, cartography is one of the most compelling ways that nations are made, legitimized, and ordered (Kashani-Sabet 1999; Craib 2002; Herb 2004; Zeigler 2002). A mapped territory, with clearly defined borders and meaningful place names, provides a powerful way to visualize otherwise abstract space. It serves to define who is included in the nation and who is excluded. And, as Anderson (1991) famously observed, a map can become a readily accepted and ubiquitous visual symbol or "logo" of "a territorial specific imagined reality." Thongchai argues that both British colonial powers and the Thai elite used "western" maps to create and signify the Thai nation. He asserts that the map's role is so powerful that the nation "is born in the map, and nowhere else" (1994, 174). The existence of a unified nation-state, as Clayton (2000, 338) summarizes, "depends on the advent of the state as a territorially defined entity and actor. Cartography was central to the arrival and presentation of states as nation-states. Maps were both instruments of state power and constitutive of the nation-state." Cartography's role in constructing nation-states, moreover, is not limited to the delineation of territory or serving as a logo. Cadastral, topographical, natural resource, administrative, or census maps have all been part of wider nation-state practices that sought to record and administer national territories. Such geographic knowledge makes the national territory and its citizenry knowable and controllable (Scott 1998; Pickles 2004).[1]

The role cartography plays in creating modern nation-states is part of longer processes of constructing European imperial space (Edney 1997; Heffernan 1995; Harley 1990). Maps and surveys were integral instruments of European colonial expansion and administration, and the lines Europeans drew on their maps of Africa, America, and Asia profoundly influenced the development of today's political map (Akerman 2009). Upon achieving independence, most former colonial territories inherited colonial boundaries that did not reflect local cultural and political divisions. Consequently, newly independent entities faced difficult challenges in building homogeneous and unified nation-states based on the lines that Europeans drew on a map.

In the remainder of this chapter, I examine how various maps and mapping projects in postcolonial Egypt created, supported, and complicated a nation-state narrative. Most of the maps and atlases produced immediately following independence stemmed from the preestablished British cartographic institutes in Cairo. But in the mid-twentieth century, the British neocolonial relationship weakened, and the maps that were subsequently produced altered Egypt's national narrative. Utilizing a multiscalar approach to examine the Egyptian national narrative since its semi-independence in 1922, I show some of the ways in which the supranational pan-Arab movement altered the mapping of Egypt and its neighboring Arab countries. Then, I discuss how the Copts and the

Nubians, who are marginalized groups in Egypt, have articulated and mapped a slightly different national narrative outside the state scale. Moving through these different mappings and scales highlights the diverse, fluid, and multiple national narratives in Egypt.

MULTIPLE IDENTITIES OF EGYPT'S PAST AND PRESENT

Egypt is a country of about 80 million diverse people. Most Egyptians ascribe to multiple identities that includes being Egyptian, but also Arab, Coptic, Berber, Bedouin, Caironese, Nubian, Christian, Muslim, and even Mediterranean, Pharaonic, Greek, and Armenian. What it means to be Egyptian has varied over time and space, but today it has primarily come to refer to someone who lives in the territorial state of Egypt (Goldschmidt 2004, 196).

Though Egypt is one of the oldest civilizations on earth, the Egyptian nation-state is also a postcolonial construction. The Kingdom of Egypt was unified around 3200 BCE as a monarchical state. During its height in the New Kingdom (approximately 1600–1000 BCE), Egypt's rule included the areas along the Nile, but it also stretched northeastward along the Mediterranean to include modern day Israel, the Occupied Palestinian Territories, and Lebanon. The last of the Egyptian dynasties fell to Roman rule in 30 BCE. By 642, Arab Islamic forces had conquered Egypt, and most Egyptians converted to Islam and adopted the Arabic language. From their capitals in Damascus and Baghdad, the Umayyad (661–750) and Abbasids (750–1258) Islamic empires controlled much of what we today delineate as Egypt. In 969, the smaller Fatimid Empire (909–1171) ruled Egypt and established Cairo as their capital. The Mamluks—a Turkish military slave class—came to power in 1250 and controlled Egypt until 1517, when the Ottoman Empire conquered Egypt. Egypt would remain a part of the Ottoman Empire for almost four hundred years.

Direct European imperialism began in 1798 when French troops under Napoleon Bonaparte invaded and occupied Egypt. Under Napoleon's brief rule, the twenty-volume *Descriptione del' Egypte*, which provided the first European attempt at creating a comprehensive scientific survey of Egyptian archaeology, history, and geography, was compiled. French and British colonizers continued to create knowledge of Egypt by ordering and structuring its cities, villages, citizens, and the economy (Gregory 1995; Mitchell 1991; Godlewska 1994). As European interests in Egypt grew, so too did European literature and stories of Egypt. Nineteenth-century travel writing about Egypt embodied common Orientalist stereotypes of the backward other (Said 1978), while also creating Egypt as an object of gaze and consumption for Europeans.

British intervention in Egypt in 1801 facilitated the collapse of French rule in Egypt. The brief but impactful French imperial rule over Egypt fueled the creation of an Egyptian nationalist movement. Muhammad (Mehmet) Ali (1769–1849), an Albanian officer of the Ottoman Empire who ruled over Egypt as an Ottoman subject from 1805 to 1848, strove and fought for an autonomous and modern Egypt. In the 1820s and 1830s he achieved high levels of Egyptian autonomy from Istanbul and fueled the beginning of Egyptian nationalism. However, with the opening of the Suez Canal in 1869 by the French Suez Canal Company, the British became deeply concerned about Egypt's geostrategic importance. Great Britain's interest in the canal was immense, because the canal reduced the seafaring voyage from Great Britian to its "crown jewel" of India by approximately 4,500 nautical miles. Thus, when the Egyptian monarchy fell into debt soon after the canal's opening, Great Britain purchased its shares and began to assert direct influence and control over the canal.

The expansion of British power in Egypt fueled the rise of Egyptian nationalism, which evolved into resistance and revolt against the British in 1879. Britain crushed the Egyptian rebellion in 1882, and in order to keep Egypt subdued Britain then occupied Egypt. Though it was still nominally an Ottoman province, Egypt was now a de facto colony of Great Britain. It was not until 1914, when Great Britain declared war on the Ottoman Empire, that Egypt ceased to be an Ottoman province and was officially recognized as a British territory (a "protectorate" to be specific) (MacMillan 2001, 411–12).

Egyptians continued to revolt and fight for their independence. During World War I, the slogan or chant "Egypt for Egyptians" became common in Cairo and Alexandria. Throughout urban and rural Egypt, there were well-organized demonstrations and revolts against British rule, which reached their height in 1919. Fearing that they would lose control of the Suez Canal, British forces suppressed the movement, but in 1922, to assuage the nationalist leaders, Britain established Egypt as an independent parliamentary monarchy. Though Egypt was technically an independent country as of 1922, Great Britain retained troops in Egypt and directly controlled Egypt's foreign affairs, its military and defense, communications, and the Suez Canal. Not until the 1952 Free Officers Revolution did Egyptians gain full control of their political system; and not until 1956 did Egypt gain control of the Suez Canal.

There was a strong sense of unity and nationalism in the Egyptian rebellions against the British (MacMillan 2001, 413; Fromkin 1989, 420). Both men and women from all classes and creeds demonstrated—including Christian Copts, Muslims, theologians, secularists, urban elites, and rural peasants. Egypt's ancient past had played very little role in the identities of most Egyptians prior to the World War I revolts, but during the anticolonial movements after World War I, the ancient history of Egypt was embraced by many leaders and organizers as a way to unify Egyptians around a rich and proud historical past (Fromkin 1989, 419–20).[2] Egypt was not unique in drawing on the past to legitimize its present. It is a common national practice to celebrate a nation's supposed age and history rather than its youth (Sparke 1998, 479).

The Egyptian national narrative that was growing in the early twentieth century also drew on Arab discourses, but not until the mid-twentieth century would the Arab discourses become central for Egyptian identity (Reid 2002; Mitchell 2002, 181–82; Gorman 2003, 62). Today, about 92% of Egypt's population identify themselves as Arab, and most Egyptians speak Arabic. But Egypt was not always considered part of the Arab World, nor had Egyptians always considered themselves Arabs (Goldschmidt 2004, 84; Gorman 2003, 62). Egypt did not participate in the Arab Revolt against the Ottoman Empire during World War I, nor did Egyptian leaders engage in the early twentieth-century pan-Arab movements for Arab unity (Loder 1923, 14). Though pan-Arabism was not an important discourse for Egyptians in the early part of the twentieth century, Egypt would soon become the leader of that movement. In 1944, Egypt supported the creation of the Arab League (Gershoni and Jankowski 1995, 14–15), and with the rise Gamal Abdul Nasser after the 1952 Free Officers Revolution, Egypt became the center of the Arab movement. Nasser was a charismatic leader whose political agenda merged nationalist, socialist, anti-Western, and pan-Arab ideologies. During his long presidency (1954–70), Nasser facilitated the rise of Arab nationalism as a powerful political ideology in the Arab states of North Africa and Southwest Asia. Though the pan-Arab movement was established in the early twentieth century as an anti-Ottoman and anti-imperialist movement, the creation of Israel in 1948 and its territorial expansion into Arab territories provided a powerful catalyst for the growth of the movement. In the 1956 Egyptian constitution, Egypt became the first state to officially proclaim that it was an Arab state and part of a wider Arab nation (Goldschmidt 2004, 127). In 1958, the pan-Arab movement reached its pinnacle when Syria and Egypt unified as the single state of the United Arab Republic (UAR). But less than three years later, a coup in Syria precipitated the demise of the UAR.

The UAR's demise signaled a weakening of Arab unity and an increased tendency for each state to protect its own independence (Sharabi 1966, 11–

12). The Arab nationalist movement continued to weaken during the 1960s. The defeat of the Arab armies by Israel in the Six-Day War of June 1967 was a devastating loss that humiliated the Arab armies and weakened the political cohesion of Arab states. As pan-Arabism weakened, the individual state nationalisms that had intersected with broader-scale Arab national discourses grew stronger. Egyptians, Iraqis, and Jordanians, for example, began to look more internally toward their own individual state nationalism, and less toward Arab unity and cohesion.

At the same time that national and supranational identities were growing, so too were Islamic movements (Gorman 2003, 62). In the latter part of the nineteenth century, Jamal al-Din al-Afghani (1838–97) promoted pan-Islamism and transformed it into a political ideology (Haim 1976, 10). In the early years of European colonization of predominantly Islamic areas, many Muslims resisted external Christian rule. Indeed, Islam was an important part of Egyptian life during, before, and after the pan-Arab movement; and it constituted an important part of the discourses of the Egyptian nation-state (Ajami 1978; Dawisha 2003).[3]

An Arab-Egyptian-Muslim national identity has dominated Egypt since the mid-twentieth century, but it has evolved over time and is replete with other contending identities. As I will discuss in the remainder of this chapter, maps produced and used in Egypt since its nominal independence in 1922 are part of wider national discursive formations. Examining Egyptian cartography produced at multiple scales that extend above and below the state scale provides an opportunity to examine the dominant and contending national narratives of Egypt, as well as the role of cartography in the construction and contestation of national identity.

CARTOGRAPHIC DISCOURSES OF THE EGYPTIAN NATION-STATE

Immediately following Egypt's nominal independence in 1922, cartography was utilized for functional and administrative purposes, as well as to symbolize Egypt's nationhood. But a careful reading of maps produced at multiple scales shows the variability and tenuousness of the Egyptian national discourse. This section examines the cartography produced in Egypt by two major state institutions, the Survey of Egypt and the Ministry of Education. It highlights the role of cartography in constructing, and at times contesting, an Egyptian national discourse. I focus on "official" state maps because they are the most common and numerous of all maps produced in Egypt; however, in the following subsection I will discuss a few countermappings of Egypt.[4]

THE ATLAS OF EGYPT AND THE TOPOGRAPHICAL ATLAS OF EGYPT

The leaders of quasi-independent Egypt utilized the cartographic facilities established by the British during their occupation. The most notable and productive of the former British institutions was the Survey of Egypt. It was established by Britain in 1898 (sixteen years after occupation). British leaders relinquished their control over the survey in 1922 with independence (Murray 1950).[5] After independence, the survey produced a comprehensive topographical series and a wide array of other reference maps at different scales. Though a few maps and atlases were made for general reference, most of the survey's projects were made for the administration and development of Egypt. Two of the largest and most circulated products that the survey produced were the *Atlas of Egypt* and the atlas *Collection of Topographical Maps of Egypt*, both of which are exemplars of national cartographic projects.

A national atlas can be used for general reference and administration and also as a symbol of nationhood, national unity, and national pride (Monmonier 1997). Not surprisingly, then, the first major cartographic project after Egypt gained independence was the *Atlas of Egypt: A Series of Maps and Diagrams with a Descriptive Text Illustrating the Orography, Geology, Meteorology, and Economic Conditions.* At the time of its publication in 1928, the *Atlas of Egypt* was the only one of its genre in the Middle East; even as late as 1960 there were no other national atlases in the Middle East except for Israel's (Monmonier 1997). Eugene Van Cleef, who wrote a review of the *Atlas of Egypt* the year after its production, referred to it as "a delightful piece of workmanship" (1929, 342). Produced by the Survey of Egypt in conjunction with Ministry of Public Works and Finance, it consists of thirty-one plates of color maps that span the entire country. The general map is at 1:2,000,000,[6] the orographic maps at 1:1,000,000, and a more detailed population map at 1:500,000. The atlas is divided into five sections, each created by a different department under different directors, and the style, scope, scale, symbols, and even place names and borders noticeably differ between sections.

This atlas exemplifies a national mapping project. The atlas discursively and symbolically asserts Egypt's independence as a clearly defined and demarcated national territory, and uses the official coat of arms of the Egyptian monarchy on the deep green and gold embossed cover. While the *Atlas of Egypt* is an example of a nation-building project, there is no nationalist or independence rhetoric within it. It does not include text or maps that allude to the greatness, history, or unity of Egypt, as many national atlases do. The completion of this cartographic project and its formal unveiling at the International Geographic Congress in Cambridge in 1928 worked instead to demonstrate that newly independent Egypt had the knowledge and capabilities to accomplish this extensive, scientific mapping project. However, these postindependence achievements highlight the ongoing strength of the neocolonial relationship that existed between Egypt and Britain in nominally postcolonial times. The text of the atlas is entirely in English, not Arabic, which is the official language of Egypt. Perhaps this is not surprising, given that English was the primary language of scientific discourse, but this nevertheless suggests that the atlas was not intended for the Arabic-speaking people of Egypt, but rather for non-Egyptians and the small elite English-speaking class of Egypt. Moreover, the editor of the atlas was Hussein Sirry Bey, an Egyptian who had been surveyor general for only one year, but the primary cartographers, including Dr. John Ball, W. F. Hume, O. H. Little, and L. J. Sutton, were British. The only other Egyptian recognized in the preface of the atlas as contributing to its production was Mohammed Amin Bahat Bay, the binder. Besides the fact that the *Atlas of Egypt* was produced in English with the help of British expertise, there are other cartographic signposts to an existing colonial relationship in the atlas. For example, the atlas indicates areas that are ripe for industrial development and shows the location of plantation agriculture like cotton cultivation, while neglecting other themes like water resources and local agriculture, which would have been of greater utility to Egyptian economic independence from Britain (Monmonier 1997, 369).

A second edition of the *Atlas of Egypt* was published in 1958, six years after the Free Officers Revolution. The maps in the 1958 edition atlas are exactly the same as in 1928.[7] The only change to the atlas is the complete erasure of any references to the monarchy, which was overthrown during the Free Officers Revolution of 1952. The monarchy's coat of arms no longer adorned the cover and was replaced by the official state seal of Egypt. Links to the monarchy were also *manually* removed from the atlas. The original date of publication and recognition of the role of the Survey of Egypt were literally scratched off each individual map. The introduction to the atlas, which previously had numerous references to King Fouad, was entirely removed, as was any recognition of the role of the British cartographers. The issuing of a second edition was not to update the maps, but to disassociate this national cartographic project from the fallen Egyptian monarchy, the Survey of Egypt, and the British cartographers who were integral to the first edition.

In 1929, one year after the publication of the *Atlas of Egypt,* the Survey of Egypt published a topographical atlas. The atlas *Collection of Topographical Maps of Egypt* does not cover the entire state but focuses primarily on the Nile Delta, the Suez Canal area, and the Mediterranean Coast, which are the locations of Egypt's densest populations and most important natural resources. Though this atlas does use English in its map titles and keys, unlike the *Atlas of Egypt,* it has most of its map text in Arabic. Produced during Egypt's semi independence, this topographic atlas and the 1928 edition of the *Atlas of Egypt* indicate a transition in Egyptian mapping. These atlases were not national icons for the consumption by the general public, but they did for the first time discursively construct the image and idea of an independent Egyptian nation. Following complete independence from Britain in 1952, however, the Egyptian government funded numerous school atlases and maps intended for a more general audience and with more prominent nationalist rhetoric. Perhaps unsurprisingly, different agencies of the Egyptian government (like the Ministry of Tourism) and private mapping companies and agencies (like the Remote Sensing Center) made maps and atlases that provided detailed geographic information about Egypt and were intended to foster national narratives of unity. For example, the Family Planning Office created the *Population Atlas* in Arabic in 1977. With seventy-two leaves of maps at a variety of scales and spatial scope, this atlas focused on Egypt's demographics as a national state and includes on the first page a logo map of the Egyptian nation. The Ministry of War has also produced maps of the Egyptian nation for general usage, such as the 1995 administrative map of *The Arab Egyptian Republic* at a scale of 1:2,300,000. This color map, about two feet by one, is in Arabic and shows the administrative units of Egypt within a clearly defined and delimited national territory. The Ministry of War, working in conjunction with the Survey of Egypt, also produced a series of topographical maps in the mid 1980s, spanning all of Egypt at a scale of 1:100,000.[8] The individual topographic maps from the Ministry of War were in color and entirely in Arabic. Though these topographic maps are not as evident an example of a logo map as the 1995 map or the *Population Atlas,* they provided information about Egypt that was critical for the government to know and to order the nation-state.

SCHOOL ATLASES AND MAPS

Perhaps the most influential map publisher for the building of the Egyptian national discourse was the Ministry of Education. Textbooks and school maps can play powerful roles in fostering national discourses. Geographer James

Blaut asserted that "textbooks are an important window into a culture; more than just books, they are semiofficial statements of what exactly the opinion-forming elite of the culture want the educated youth of that culture to believe to be true about the past and present world" (1993, 6). School atlases, more specifically, play a particularly important role in the construction and framing of a nation's territory by giving an imagined community a visual reality and then educating the nation's youth about the nation's geography (Gutsell 1972, 27; Monmonier 1997, 396).

According to G. W. Murray, a "technical expert" at the Survey of Egypt, the first *Arabic Elementary Atlas* for schools was issued in 1920. I have been unable to locate a copy of a 1920 edition, but have examined 1922 and 1930 editions. Prepared and published by the survey and financed by the Ministry of Public Information, the *Primary School Atlas of the World* is a thirty-five-page color atlas. The emphasis on Egypt in this world atlas is notable. Though this is a "world" atlas, it begins with a physical map of "North Africa, West," a region that includes Egypt. The scale of this map is 1:16,000,000, and it encompasses nearly the entire Nile watershed, a geographic feature that people as far back as Herodotus have claimed to be the ancient lifeline of Egypt. The second map in the atlas depicts the Egyptian nation (fig. 6.1). This map exemplifies the concept of a national "logo" map, a readily accepted and ubiquitous visual symbol of a national territory. It shows the nation-state's territory and high-lights its major natural and human geographic features, like the agricultural areas around the Nile and major urban areas. Indeed, the Nile River plays a prominent role in the graphic hierarchy of this map and indicates the central-ity of this river to the history, culture, and economy of Egypt. Interestingly, however, the boundaries on this map are not clearly demarcated. Two areas of this map that extend over the neatline are places that have been the focus of territorial and geopolitical disputes. First is the area of Al-Sallum, near the Libyan border, which was claimed by both Egypt and Libya until their brief war of 1977 settled the border. The missing border on the map is due to the fact that an agreement had not yet been made between Egypt and Italian forces that were occupying Libya.[9] The second segment of territory that extends over the neatline is the Hala'ib Triangle in the southeast. In 1899, the British drew the boundary between Egypt and Sudan at the 22nd parallel. However, in 1902 Egyptian and Sudanese representatives established an administrative boundary that extended northward into Egypt; the area between these two borders is known as the Hala'ib Triangle. Sudan was granted administrative control of this area because the people living there had stronger connections to Sudan than to Egypt. However, since the 1902 border was administrative, Egypt could

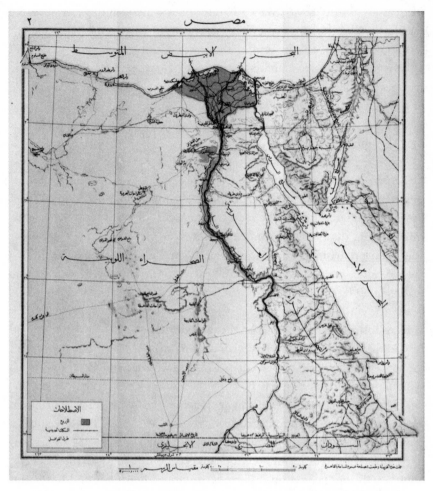

FIGURE 6.1. Page 2 of *Primary School Atlas of the World*, 1922. This map of Egypt was featured on the second page of this world atlas. It highlights the Nile River as a central and defining part of Egypt.

still assert its control and sovereignty over the area. The tenuous status of this region is evident in this map. Though both borders are drawn, the fact that the Hala'ib Triangle is drawn to extend over the neatline suggests that this territory is part of Egypt, not Sudan. As will be noted below, many maps produced in Egypt exclude the 1902 border altogether. The inclusion or exclusion of this 1902 border on maps is important if Egypt ever decides to claim full territorial sovereignty over the region. Tensions over the Hala'ib Triangle have heightened in recent years as speculation grows about the quantity and quality of natural resources (including water) in the area. Fortunately, conflict over the Hala'ib region has not been violent.[10]

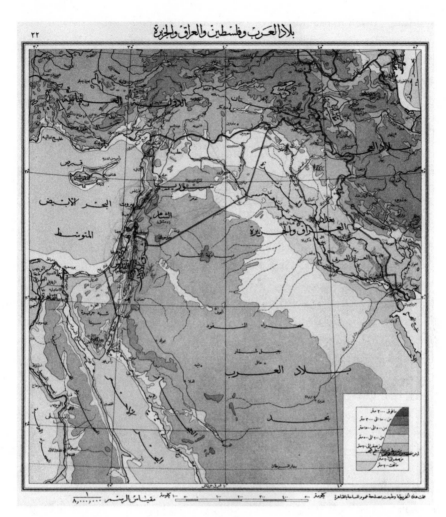

FIGURE 6.2. Page 22 of *Primary School Atlas of the World,* 1922. This maps shows the tentative boundaries of "Arab countries, Palestine and Iraq." It is notable that Egypt is not featured as an Arab country.

The title of this national logo map is the three-letter Arabic word that transliterates into English as "Misr." It is important to underscore that Egyptians refer to their country as Misr and themselves as Misreean and that the "Western" term "Egypt" is an exonym, or a toponym which is not readily used by local inhabitants, but instead imposed by outsiders.[11]

In this 1922 atlas there is one map that delineates a supranational Arab geographic entity. On page 22 in the Asia section of the atlas, there is a map of "Arab countries, Palestine and Iraq" (fig. 6.2). The lines on this map indicate the tentative borders of the British and French mandates of Syria, Palestine, and Iraq, as well as the borders of independent Iran and Turkey, which were still being negotiated after the fall of the Ottoman Empire during World War I. This map reflects the changing geopolitics of the region, which included the official entrenchment of British and French imperialism in the region. Interestingly, this map does not indicate tensions between Arab states and Iran. In this 1922 map, the water body between the Arabian Peninsula and Iran is labeled *al-'ajam*, which is translated as "Non-Arab" or "Persian." By the mid-twentieth century and particularly during the height of the pan-Arab movement, this waterway became universally known as the "Arabian Gulf" across the Arab countries. The naming of this gulf on maps has become a politically contentious point between Arab countries and Iran, as well as for commercial mapmakers across the globe (Abedin 2004).[12] Importantly, the scope of the Arab countries on this map does not include Egypt, which alludes to Egypt's peripheral status as an Arab entity before the rise of Nasser in the 1950s. Indeed, in other atlases produced in Egypt prior to the 1952 Free Officers Revolution,[13] there was similarly no connection between Egypt and its Arab neighbors.

After the 1952 revolution, the Ministry of Education published a series of state-funded Egyptian school atlases. These atlases reflect a dramatic change in Egypt's national and regional narrative. In 1965, thirteen years after the Free Officers Revolution and during the height of the pan-Arab movement, the *Arab Atlas* was first printed. I located and analyzed nine different editions of this official school atlas. These atlases, published between 1965 and 1986, all vary slightly in their content, but the style of maps, the ordering of pages, and the emphasis on the Arab world (as the title indicates), are remarkably similar. Each atlas contains eighty pages of color maps, and there is no text accompanying the maps.

The 1965 edition begins with two introductory pages on map projections, and then proceeds to pages on world flags, earth-sun relations, and time zones, which are then followed by physical and political world maps. The next two maps in the atlas are physical and political maps of an entity labeled the "Arab Homeland" (*al-watan al-arabe*) (fig. 6.3). This geographic entity is a powerful geopolitical construction (Culcasi 2011).[14] The political map clearly delineates the region first by highlighting in color the states that are included, and then, through the use of a thick red border, visibly separating the Arab Homeland from the rest of Africa and Asia. The individual boundaries of each state are marked with black dashed lines (though the pending borders of the southern Arabian Peninsula are not included), yet all the individual states are simultaneously united as part of the broader Arab Homeland. As part of the wider discourses of the pan-Arab movement, this map is an example of an emerging cartographic discourse that stressed the importance of Arab unity existing simultaneously with individual state entities. Perhaps unsurprisingly, Israel is not indicated on maps of the Arab Homeland; instead, the unified state of Palestine (*philistine*) is demarcated around what is commonly referred to as Israel and the Occupied Palestinian Territories today. The use of the place name "Palestine" and the placement of its borders are significant geopolitical symbols that cartographically assert Arab states' support of an independent Palestinian state and a rejection of Israeli encroachment on Arab lands. Moving a bit eastward, what was labeled the "non-Arab" or "Persian Gulf" in the pre-1952 atlases is consistently the "Arabian Gulf" in this and all other editions of the *Arab Atlas*.[15]

Egypt's position within the Arab world is central and firmly anchored on the map. Though Egypt is clearly an Arab state in this atlas, it is also Egyptian. The place name for Egypt on this Arab Homeland map is the United Arab Republic, not Egypt (or Misr as in the 1922 atlas). In a highly symbolic move, Egypt changed its name to the United Arab Republic in 1958 when it officially united with Syria, and Egypt retained the name UAR for nearly a decade after the demise of the UAR. It is also noteworthy that Egypt's contested boundaries are no longer ambiguous on this map. Indeed, Egypt's borders with Libya and Sudan are clearly marked in favor of Egypt.

The next two pages of this atlas contain six more thematic maps of the Arab Homeland (i.e., resources and land use). Then the atlas moves on to focus specifically on Egypt (or the United Arab Republic, as it is labeled here). There are two pages on the entire Nile watershed, followed by six pages of differently scaled maps of parts of Egypt, which include detailed maps of the Nile Delta and valley. The atlas then includes twenty-two pages of maps on the individual countries of the Arab Homeland (with Iraq being the next). Small-scaled maps of Africa, Asia, Europe, North and South America, and Australia follow, and then there are twelve more pages of various world-scale thematic maps (i.e., temperature maps and ocean currents). The large number of maps of the Arab Homeland and their placement in the front of the atlas are perhaps to be expected from the title of the atlas but are nevertheless telling that the Egyptian national narrative shifted to embrace a supranational discourse alongside a nation-state-based discourse.

The 1986 edition of the *Arab Atlas* was the last edition of the series. It is very similar to the one published twenty-one years earlier, but there are some notable changes. It begins as the 1965 edition did with flags of the world, two world maps, and then maps of the "Arab Homeland." However, in the 1986 edition, the homeland has grown to include Somalia, Djibouti, Mauritania,

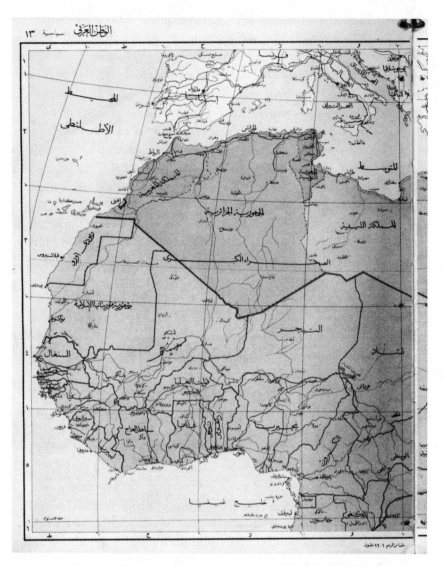

FIGURE 6.3. Pages 12–13 of the *Arab Atlas, 1965*. This map is titled "The Arab Homeland," and Egypt is featured prominently in this extensive Arab territory.

and (probably) the Western Sahara (fig. 6.4). These additions are probably due to the fact that Somalia, Djibouti, and Mauritania all joined the Arab League in the twenty-one years between the dates of publication of these two editions.[16] Another notable difference is the inclusion of Lake Nasser, which stretches across the border of Egypt and Sudan. In the 1960s, even before the construction of the highly controversial Aswan High Dam was complete, Lake Nasser

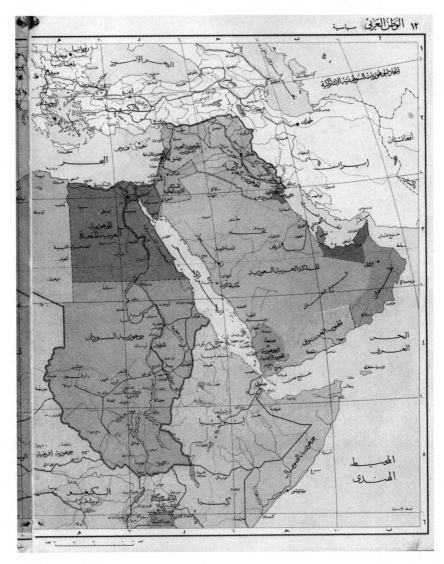

FIGURE 6.3. *Continued*

began to form and inundate Nubian lands in southern Egypt and northern Sudan. The official change in Egypt's name is also reflected in this 1986 atlas. In 1971, a year after Nasser died, and during a time when the pan-Arab movement was drastically weakened, Egypt's name was officially changed from the United Arab Republic to its current official name, the Arab Republic of Egypt. Thus, after thirteen years of stressing an Arab identity and disregarding an Egyptian

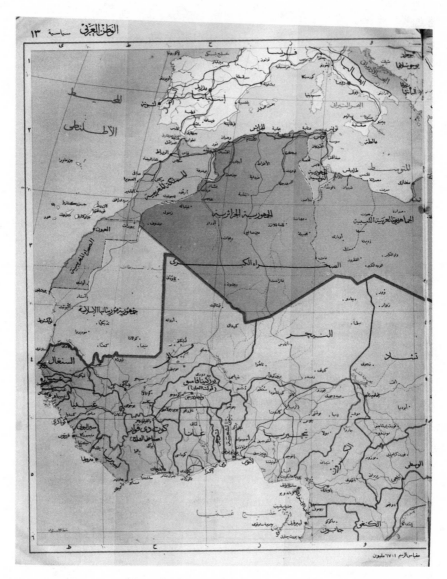

FIGURE 6.4. Pages 12–13 of the *Arab Atlas*, 1986. This map, a revision of the 1965 map above, is also titled "The Arab Homeland," but it is notable that the territorial extent of the "homeland" has grown since 1965.

one in its official name, Egypt's national identity was reemphasized while its Arab connections were maintained too.[17]

The *Arab Atlas* was not produced during the three-year unification of Syria and Egypt. However, I was able to locate two flat maps of the UAR, which were general reference and educational maps. One such map approved by

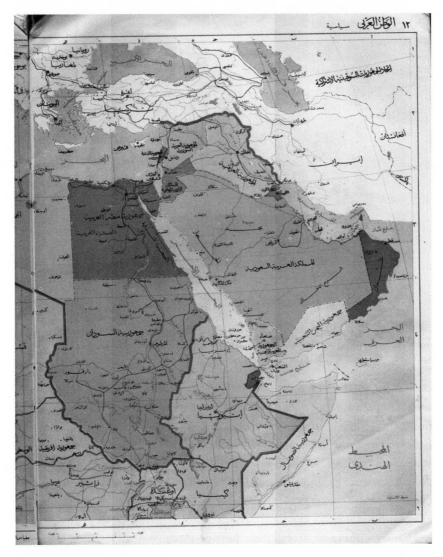

FIGURE 6.4. *Continued*

the Egyptian Ministry of Education (no date) shows the UAR at the scale of 1:3,000,000. This territorial state is colored light green with a darker green border and has the label UAR stretching across the two noncontiguous entities (jumping over Palestine and Jordan). A second map of the UAR, produced by the Modern Institute for Printing in 1959 at 1:3,000,000, clearly unifies these two provinces as well, while also showing the administrative boundaries for the Nile Delta and Syria in inset maps (fig. 6.5). Both these maps use the place

FIGURE 6.5. Excerpt of a map titled *The United Arab Republic,* 1959. The toponym "the United Arab Republic" and the coloration of territories on this map indicate the unity of Syria and Egypt.

names "Syrian Province" and "Egyptian Province" within the individual territories, as well as uniting them as the UAR.

Both state and private publishers made and released many other maps and atlases in Egypt. All the maps and atlases discussed above, as well as the many other maps that I examined but did not discuss in this chapter, indicate that an Egyptian identity intersects cartographically with an Arab identity. By examining both the national and supranational scales, it is clear that Egypt's identity is not only within a territorial state, but also supranational.

LOCAL OR COUNTERDISCOURSES WITHIN THE MODERN
EGYPTIAN NATION-STATE

In creating an image and idea of a unified Egyptian-Arab nation, local or minority groups have often been marginalized in both material and discursive ways (Gorman 2003). By focusing on scales other than the state, a modest counternarrative that questions the homogeneous Egyptian-Arab-Muslim national identity becomes evident. Though the maps discussed above included no direct or obvious signs of Muslim identity, Islam is a central part of everyday lives and politics in Egypt and a normative part of Egyptian identity for most Egyptians. There are many minority and marginalized groups in Egypt: the Berbers, a minority group located in the western desert of Egypt who speak Berber; the Bedouins, a scattered group generally considered seminomadic; small Armenian and Greek communities; Women's groups; and political parties like the Muslim Brotherhood.[18] All of these groups have had tense if not outright hostile relationships with the Egyptian government.[19] Marginalized groups across the globe have used various alternative or countermapping practices, whether to stake claim to lost territory or to assert the importance of conservation (Ramaswamy 2010; Crampton 2010; 123–27; Peluso 1995; Harris and Hazen 2009; Wood 2010, 111–55; Culcasi 2012; Hodgson and Schroeder 2002). Of the various minority groups in Egypt, two particular marginalized groups have created alternative mappings to stake claims to territories and assert a historical and cultural identity that diverges from the dominant Egyptian-Arab-Muslim discourse.

Nubians are a minority group located predominantly in southern Egypt and northern Sudan. They are usually referred to as an "ethnic group" who speak one of several different dialects of Nubian and consider the area of southern Egypt and northern Sudan along the Nile to be their ancestral homeland. This vague region has been referred to as "Nubia" since antiquity and is a central source of Nubian identity.[20] While many Nubians have adopted Arabic and Egyptian citizenship, they consider themselves Nubian (Poeschke 1996). But Nubians are not recognized as a distinct group in Egypt, so their population statistics are difficult to determine. Though Nubians have many grievances with the Egyptian government, including their lack of recognition, the status of their ancestral land in southern Egypt is their most central concern. They are not striving for autonomy or independence in their homeland, but recognition and compensation for the destruction of great swaths of their territory.

The building of the Aswan High Dam was controversial, both internationally and domestically. After Nasser failed to secure funds from the United Kingdom and the United States to build the dam, he accepted support from the USSR. Some of the money to build the dam was also garnered after Nasser took control of the Suez Canal from the British, which sparked the 1956 war between Egypt and Britain, France, and Israel. Domestically, the dam has also been controversial for several reasons. Though the dam has had benefits such as the control of floods and the creation of hydroelectric power, as is true of most dams, there is concern about its the long-term effects on soil fertility, salinization, and aquatic wildlife. The building of the dam also caused international and domestic alarm because it inundated several ancient Egyptian sites, as well as the homes and villages of approximately one hundred thousand Nubians. The construction of the dam in the 1960s created Lake Nasser, one of the world's largest human-made lakes, in the area Nubians call their home-land.[21] Most Nubians consider the construction of the Aswan High Dam and the flooding of their land an egregious and discriminatory act of the Egyptian government. Nubians were not involved in the decision-making processes or in the planning of the dam, and their resettlement sites were selected by Egyptian authorities in accordance with national economic and agricultural needs (Poeschke 1996). Many Nubians have continued to petition Cairo to recognize their plight and have formed the Nubian General Club to speak for Nubian interests.[22] Since many of the Nubian's concerns are over territory and their unique cultural identity, which is linked to a specific territory, they have at times used maps to illustrate their alternative discourse of history and identity.

UNESCO opened the Nubian Museum in Aswan on November 23, 1997, to showcase some of the ancient relics that were saved before the inundation of the area and to recognize the incredible achievements of the relocation of many ancient sites and relics (such as Abu Simbel and Philae). However, the museum also works to commemorate the lost land and some of the injustices that Nubians have experienced. The museum has attempted to portray and re-create a uniquely Nubian culture and Nubian territory in order to (at least partly) document, represent, and compensate those who have lost their land.[23] A textual introduction at the entrance of an exhibit titled "Nubia Submerged," which opened in 2000, states that the exhibit permits "Nubian people to revisit the villages where they cannot return anymore." The "Nubia Submerged" exhibit includes 152 black-and-white photos and descriptive, personal stories in both English and Arabic of the rescue of ancient relics as well as the history of the Nubian people. A diorama that exhibits life in Nubia before the dam showcases an idealized past, and much of the external structure of the museum was designed to resemble a traditional Nubian village. At the entrance of the museum, the visitor is immediately greeted with a towering statue (approximately fifteen feet tall) of Ramses the Great, a relic that was saved before the flooding. Below him, a large-scale (approximately twenty-five-foot-long) three-dimensional relief map highlights the sites of Nubia—including the Nile River, Lake Nasser, and the ancient monuments that were saved before inundation (fig. 6.6). The large three-dimensional map works as an introduction to the relics of Nubia and includes some brief explanatory text. Though there is nothing overtly geopolitical about this map, it provides a visual display to commemorate the lands lost to the building of the dam. This map, and the

FIGURE 6.6. This three-dimensional map of the Lake Nasser area is featured at the entrance of the Nubian Museum in Aswan, Egypt. This lake submerged much of Nubia in the 1970s. Photography by the author.

museum experience more broadly, were principally intended for cultural tourists and thus serve to educate people about the territorial struggles of Nubians within Egypt. The three-dimensional presentation of the map works within the context and structure of the museum, actively engaging visitors in their understanding of Nubia and the impact of the dam (della Dora 2009).

In another form, a now defunct website called Nubia Today disseminated a strongly pro-Nubia political discourse with maps. The website used historical maps to prove its territorial and historical legitimacy as a separate entity from Egypt. The map featured on the website's homepage was the first of several depicting "the Stolen land!" Using different shades of grey, the map shows that in 1800 Egypt and Nubia were two entirely distinct entities. Extensive explanatory text describes the lost lands and the Egyptian government's "ethnic cleansing against Nubian people." Six maps follow on subsequent pages, all of which depict Nubia and Egypt as separate entities and Nubians and Egyptians as different peoples. For example, "Johnson's Africa," from the 1863 American atlas *Johnson's New Illustrated (Steel Plate) Family Atlas*, delineates Egypt and Nubia as two entirely separate territories. Egypt is colored pale yellow and Nubia pink. The border between the two entities is clearly marked just north of the Tropic of Cancer, and both place names are written in the same boldfaced capitalized font on the map. Other maps included on this web site also attempt to assert that Egypt and Nubia have been distinct territories for centuries. Scottish artist and painter David Roberts toured the "Holy Land" and painted hundreds of images of Egypt in the mid-nineteenth century. Accompanying his art-work on the "Holy Land" was a reference map of his travels along the Nile River, which Nubia Today posted under the title "David Roberts in Egypt and Nubia." Unlike Johnson's map of Africa, the Roberts map showed no territorial distinction, but, nevertheless, the two separate place names in the title still facilitate an argument that there has been a long historical distinction between Egypt and Nubia.

The Copts, a Christian minority group in Egypt, have also experienced discrimination in Egypt (Gorman 2003, 147–50). Many have held positions of power, but, like the Nubians, they have had to struggle for equal treatment.[24] Copts are considered the descendants of the ancient Egyptians, while the majority of today's Arab Egyptians are descendants of migrants from the Arabian Peninsula. Though the Coptic language is still used in some of their churches, most Copts speak Arabic on a daily basis. They are hesitant nevertheless to refer to themselves as Arab out of concern that they may lose their distinct identity. Population estimates of Christian Copts vary among sources, but it is generally assumed that Copts constitute between 5 and 10% of the Egyptian population.[25]

Unlike the Nubians', the Copts' grievances are not territorial, and they have not utilized maps as a form of resistance the way that Nubians have. Nevertheless, Coptic organizations have made maps in order to highlight a different part of the Egyptian past than what is seen in official state mapping projects. More specifically, Coptic groups have used cartography to emphasize important Christian sites in a predominately Muslim country. For example, the Hanging Church, located in Coptic (Old) Cairo, has at its entrance a map encased in glass titled "Called My Son Out of Egypt" (it is in English) (fig. 6.7). In reference to the Holy Family's journey in Egypt, this map shows the various places that the Holy Family is likely to have traveled. Similarly, but in much greater detail, the Coptic Archeological Society (Société d'Archéologie Copte) published, in English, an atlas in 1962 titled *Atlas of Christian Sites in Egypt*. The six maps of this atlas are simple black and white maps that show the various sites visited by the Holy Family, as well as monasteries and churches scattered throughout Egypt. Also from the French Coptic Archeological Society is a folded map titled in French "Carte de l'Egypte Chretienne." This 1954 map shows a borderless area that centers on modern-day Egypt. It highlights the topography of the area with pictorial symbols, important Christian sites and communities, and important routes. Interestingly, on the right margin of the map there is a large yellow sun drawn in around Medina. Six beams of sunlight stretch westward across the entire map, stopping at the left margin of the map.

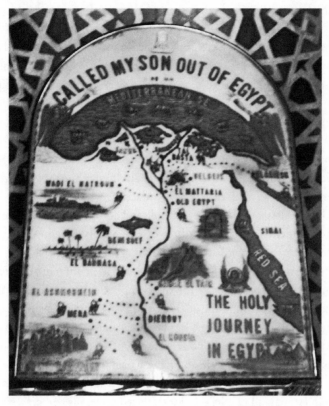

FIGURE 6.7. This pictorial map is displayed at the entrance of the Hanging Church, in Coptic (Old) Cairo. This map shows a subtly different cartography and territorial view of Egypt that emphasizes the Christian past. Photography by the author.

The sun was probably placed near Medina merely because it is the middle of the most eastward point of the map, while its light clearly penetrates the entire scope of the map. Even though this map depicts Christian sites in Egypt, its lack of borders suggests that religious sites and routes supersede the creation of modern borders of the Egyptian nation-state. These Coptic maps are not as politically charged as the maps on Nubia Today's web site. Nevertheless, they work as a subtle form of countercartography, telling an alternative story to the dominant Egyptian-Arab-Muslim one.

CONCLUSIONS

All of the maps reviewed here, whether they were part of the official state discourse or more marginalized ones, were made or used since Egypt achieved its independence from Great Britain, and thus might be expected to represent a postindependence and anti-imperialist cartography. However, the maps I examined in this chapter were entirely based on Western standards, and in many cases they were created within cartographic institutions that the British had established. I was unable to locate any modern examples of "indigenous" cartography (ancient Egyptian mapping notwithstanding). The colonization of mapping practices and Western representations of space are so ubiquitous that the influence of indigenous mapping today is feeble (Johnson, Louis, and Pramono 2006). Yet this conformity is not surprising: as Egypt began self-rule, it also became a part of a world that has been divided into sovereign nation-states, and that use Western cartographic practices.

As in all nation-states, there are multiple and competing national discourses in postindependent Egypt. Regardless of the widely accepted critique that nation-states are socially constructed and imagined entities, the nation-state remains the dominant manner in which the world is ordered and organized, and it continues to form a basis for the complex constructions of identity. But as I have shown through the analysis of Egypt's postindependence cartography, the official discourse of the Egyptian nation-state has emphasized both Egyptian and Arab identities. Though the Egyptian national narrative readily conforms to the dominant division of the world into nation-states, the mapping of the Arab Homeland does not. Maps showing the supranational entity of the Arab Homeland in part contest standard divisions and create a territorial entity that is not readily recognized outside of Arab states. Though subtle, the discourse of the Arab Homeland is a countercartography that invokes an anti-imperialist discourse, while simultaneously maintaining the geopolitical ordering of nation-states.

At smaller scales, marginalized groups have been largely silenced in both national maps of Egypt and supranational maps of the Arab Homeland. Yet minority groups like the Nubians and the Copts have embraced histories, politics, and cultural traits that differ from the dominant Egyptian-Arab-Muslim identity. Though neither of these groups has been prolific in its creation and use of maps, both have used maps to assert a different national narrative from the official or state discourse.

Cartographic discourses help construct and reflect postindependence nation building. However, moving among state, supranational, and local scales shows that national discourses are never merely state based. Ultimately, a nation-state like Egypt has complex and multiple identities, and though postcolonial mapping projects have often attempted to homogenize and unite newly independent nations, there is always complexity and diversity.

NOTES

I have drawn on some of the same data and made some similar arguments in papers published in *Political Geography* (Culcasi 2011) and the *Arab World Geographer* (Culcasi 2013).

1. Of course, maps have also played a key role in national defense and military strategies abroad.

2. There are conflicting ideas about and emphasis on the role of ancient Egypt in Egypt's modern national identity. For many Egyptians, though proud to be associated with one of the world's first civilization, Islamic and Arab heritage have and continue to be primary sources of identity construction (Reid 2002, 296–97).

3. Excerpts from the Doctrines of Muslim Brotherhood written by its founder Hasan al-Banna, articulate the levels of identity as to "their own particular nationalism, Egyptian nationalism." After that the Muslim Brotherhood supports Arab unity, since "the Arabs are the core and guardians of Islam," and "finally they strive for the Islamic League, which constitutes the perfect enclosure for the larger Islamic homeland" (Sharabi 1966, 11).

4. My sample includes a broad range of maps and atlases, which I collected during four months of fieldwork in Cairo. My primary locations for data collection included two different book markets (souks), the main library at the American University of Cairo, as well as its Rare Books Collection, and the Egyptian Royal Geographical Society's library and archives. I examined hundreds of maps and atlases looking for how different institutions mapped the Egyptian nation.

5. British leaders reasserted control of the survey during World War II but gave control back to Egyptian directors after the war.

6. The "General Map of Egypt," which begins the atlas, shows boundaries, relief, communication and transit lines, and the relative areas of cultivable land and deserts, the latter of which seem to dominate the image and geography of Egypt. The population of Egypt is concentrated along the Nile and in the delta, and because of this area's density, many maps and atlases focus on this specific region of Egypt.

7. I found only one copy of this 1958 edition, in the Rare Books Collection at the American University of Cairo.

8. I was able to locate and examine three of these topographical maps, of which two were at this scale and the third contained no scale information.

9. The southern portion of the Libyan (Tripoli)–Egyptian border was finalized in the 1925 Italian-Egyptian agreement, and the northernmost point of the border approaching the Mediterranean Sea was settled in 1977 between Egypt and Libya.

10. The next nine maps in this atlas are of Egypt's densely populated areas along the Nile and Mediterranean Sea. The atlas then moves on to include a few maps of Africa, Europe, Asian countries, and then North America, South America, and Australia. It ends with a few world maps, one of which is a political map on a Mercator projection, and an instructional page on scale and projections, latitude and longitude, and earth-sun relations.

11. The word "misr" means civilization, settlement, or a big city, in general terms. Egyptians are, of course, aware that the rest of the world refers to their country as Egypt.

12. See page 694 of the *Hans Wehr Dictionary of Modern Written Arabic* (Cowan 1994) and Mitchell (2002, 180–81).

13. The three other atlases include the 1926 *Historical Atlas* produced by the Ministry of Public Information and the the 1926 and 1939 editions of the *Primary School Atlas* produced by the Egyptian Ministry of Public Information, in consultation with George Philip & Son in London (a British map publisher).

14. Maps of the "Arab Homeland," as well as the UAR, subtly contest Western divisions of the world that recognize both Israel and the Persian Gulf. Further, the Arab Homeland also contests standard Western divisions of the globe into world regions that generally include "the Middle East" but exclude an "Arab Homeland."

15. In general, the cartographic construction of the Arab Homeland is not only very common in Egypt, but also in other Arab countries like Libya, Lebanon, Jordan, and Oman. Often the "Arab Homeland" is constructed as a uniform region with *no* internal nation-state boundaries.

16. Western Sahara is not recognized as a sovereign nation and thus does not have representation in the Arab League (or the United Nations).

17. In all *Arab Atlases* published from 1973 onward, this name change is implemented. This name change was globally recognized.

18. Interview with American University of Cairo sociology professor Saad Ibrahim in October 2005.

19. Though the Arab Spring had in great part served to unify Egyptians against their oppressive leaders, the ousting of Morsi in summer 2013 and the brutal suppression of the Muslim Brothers clearly indicates a lack of cohesiveness throughout Egypt.

20. The exact origin of the term "Nubia" is unclear, but around the fourth century the Noba people settled in this region, and the name derives from this group.

21. The name of the lake is also contentious. To most of the world it is Lake Nasser, but Nubians often refer to it as Lake Nubia (Poeschke 1996, 33, 111). The dam was not completed until 1971, but inundation was planned and began earlier.

22. They have offices in Abdin and Cairo. Interview with anonymous member of club on Thursday, December 1, 2005.

23. As paraphrased from the museum's web site http://www.numibia.net/nubia/doc_center .htm.

24. In 2006, Free Copts published their first edition of a magazine titled *Independent Copts*, in which "independent" meant free in speech and thought. http://english.freecopts.net/english //index.php?option=com_frontpage&Itemid=1.

25. 3.5 million, or 6% of the population, according to Boustani and Fargues 1990, 29. See also Gorman 2003, 147–48, which states that 10% of the population is "probably not wildly inaccurate."

REFERENCES

Abedin, M. 2004. All at Sea over "the Gulf." *Asia Times*, December 9.

Agnew, J. 1997. The Dramaturgy of Horizons: Geographical Scale in the "Reconstruction of Italy" by the New Italian Political Parties, 1992–95. *Political Geography* 16 (2): 99–121.

———. 1998. *Geopolitics: Re-visioning World Politics*. New York: Routledge.

Ajami, F. 1978. The End of Pan-Arabism. *Foreign Affairs* 57 (5): 355–73.

Akerman, J., ed. 2009. *The Imperial Map: Cartography and the Mastery of Empire*. Chicago: University of Chicago Press.

Anderson, B. 1991. *Imagined Communities*. 2nd ed. London: Verso.

———. 2001. Writing the Nation: Textbooks of the Hashemite Kingdom of Jordan. *Comparative Studies of South Asia, Africa and the Middle East* 21 (1 and 2): 5–14.

Baram, A. 1990. Territorial Nationalism in the Middle East. *Middle Eastern Studies* 26 (4): 425–48.

Berger, M. 2006. Nation Building to State Building: The Geopolitics of Development, the Nation-State System and the Changing Global Order. *Third World Quarterly* 27 (1): 5–25.

Billig, M. 1995. *Banal Nationalism*. Thousand Oaks, CA: Sage.

Blaut, J. M. 1993. *The Colonizer's Model of the World*. New York, Guilford Press.

Boustani, R., and P. Fargues 1990. *The Atlas of the Arab World: Geopolitics and Society*. New York: Facts on File.

Brenner, N. 1998. Between Fixity and Motion: Accumulation, Territorial Organization and the Historical Geography of Spatial Scales. *Environment and Planning D* 16:459–81.

———. 2001. The Limits of Scale? Methodological Reflections on Scalar Structuration. *Progress in Human Geography* 25 (4): 591–614.

Clayton, D. 2000. The Creation of Imperial Space in the Pacific Northwest. *Journal of Historical Geography* 26 (3): 327–50.

Cowan, J. M., ed. 1994. *Arabic-English Dictionary: The Hans Wehr Dictionary of Modern Written Arabic*. 4th ed. Urbana, IL: Spoken Language Services.

Craib, R. B. 2002. A National Metaphysics: State Fixations, National Maps, and the Geohistorical Imagination in Nineteenth-Century Mexico. *Hispanic American Historical Review* 82 (1): 33–68.

Crampton, J. 2010. *Mapping: A Critical Introduction to Cartography and GIS*. Malden, MA: Wiley-Blackwell.

Culcasi, K. 2011. Cartographies of Supranationalism: Creating and Silencing Territories in the "Arab Homeland." *Political Geography* 30:417–28.

———. 2012. Mapping the Middle East from Within: (Counter) Cartographies of an Imperialist Construction. *Antipode* 44 (4): 1099–1118.

———. 2013. Cartographies of Nation Building: Creating and Contesting the Egyptian Geobody. *Arab World Geographer* 16 (1): 30–53.

Dawisha, A. 2003. *Arab Nationalism in the Twentieth Century: From Triumph to Despair*. Princeton: Princeton University Press.

Delaney, D., and H. Leitner. 1997. The Political Construction of Scale. *Political Geography* 16 (2): 93–97.

della Dora, V. 2009. Performative Atlases: Memory, Materiality, and (Co-)Authorship. *Cartographica* 44 (4): 240–55.

Edney, M. H. 1997. *Mapping an Empire: The Geographical Construction of British India, 1765–1843*. Chicago: University of Chicago Press.

Fromkin, D. 1989. *A Peace to End All Peace*. New York: Henry Holt and Co.

Gershoni, I., and J. Jankowski. 1995. *Redefining the Egyptian Nation*. Cambridge: Cambridge University Press.

Godlewska, A. 1994. Napolean's Geographers (1797–1815): Imperialists and Soldiers of Modernity. In *Geography and Empire*, edited by A. Godlewska and N. Smith, 31–53. Cambridge, MA: Blackwell.

Goldschmidt, A. 2004. *Modern Egypt: The Formation of a Nation-State*. Boulder, CO: Westview Press.

Gorman, A. 2003. *Historians, State, and Politics in Twentieth Century Egypt*. New York: Routledge.

Gregory, D. 1995. Between the Book and the Lamp: Imaginative Geographies of Egypt, 1849–50. *Transactions of the Institute British Geographical Society*, n.s., 20:29–57.

Grewal, I., and C. Kaplan. 1994. Introduction: Transnational Feminist Practices and Questions of Postmodernity. In *Scattered Hegemonies: Post Modernity and Transnational Feminist Practices*, edited by I. Grewal and C. Kaplan, 1–33. Minneapolis: University of Minnesota Press.

Gutsell, B. 1972. National Atlases: Their History, Analysis, and Ways to Improvment and Standardization. Monograph 4, *Cartographica*.

Haim, S. 1976. *Arab Nationalism: An Anthology*. Berkeley: University of California Press.

Harley, J. B. 1990. *Maps and the Columbian Encounter: An Interpretive Guide to the Traveling Exhibition*. Milwaukee: University of Wisconsin Press.

Harris, L., and H. Hazen. 2009. Rethinking Maps from a More Than Human Perspective: Nature-Society, Mapping and Conservation Territories. In *Rethinking Maps: New Frontiers in Cartographic Theory*, edited by M. Dodge, R. Kitchin, and C. Perkins, 50–67. New York: Routledge.

Heffernan, M. 1995. The Spoils of War: The Société de Géographie de Paris and the French Empire, 1914–1919. In *Geography and Imperialism, 1820–1940*, edited by M. Bell, R. Butlin, and M. Heffernan, 221–64. New York: St Martin's Press.

Herb, G. H. 2004. Double Vision: Territorial Strategies in the Construction of National Identities in Germany, 1949–1979. *Annals of the Association of American Geographers* 94:140–64.

Herod, A. 1997. Labor's Spatial Praxis and the Geography of Contract Bargaining in the US East Coast Longshore Industry, 1953–89. *Political Geography* 16 (2): 145–69.

Hodgson, D., and R. Schroeder. 2002. Dilemmas of Counter-mapping Community Resources in Tanzania. *Development and Change* 33:79–100.

Johnson, J. T., R. P. Louis, and A. H. Pramono. 2006. Facing the Future: Encouraging Critical Cartographic Literacies in Indigenous Communities. *ACME: An International E-Journal for Critical Geographies* 4 (1): 80–98.

Kashani-Sabet, F. 1999. *Frontier Fictions: Shaping the Iranian Nation, 1804–1946*. Princeton: Princeton University Press.

Leitner, H. 1997. Reconfiguring the Spatiality of Power: The Construction of a Supranational Migration Framework of the European Union. *Political Geography* 16 (2): 123–43.

Loder, J. d. V. 1923. *The Truth about Mesopotamia, Palestine and Syria*. London: George Allen & Unwin.

MacMillan, M. 2001. *Peacemakers: The Paris Peace Conference of 1919 and Its Attempt to End War*. London: John Murray.

Marston, S. A. 2000. The Social Construction of Scale. *Progress in Human Geography* 24 (2): 219–42.

Mitchell, T. 1991. *Colonizing Egypt*. Berkeley: University of California Press.

———. 2002. *Rule of Experts: Egypt, Techno-Politics, and Modernity*. Berkeley: University of California.

Monmonier, M. 1997. The Rise of the National Atlas. In *Images of the World: The Atlas through History*, edited by J. Wolter and R. Grim, 369–99. New York: McGraw-Hill.

Mountz, A. 2013. Political Geography I: Reconfiguring Geographies of Sovereignty. *Progress in Human Geography* 37 (6): 829–41.

Murphy, A. 2013. Territory's Continuing Allure. *Annals of the Association of American Geographers* 103 (5): 1212–26.

Murray, G. W. 1950. *The Survey of Egypt, 1989–1948*. Cairo: Ministry of Finance, Survey of Egypt.

Ong, A. 1999. *Flexible Citizenship: The Cultural Logics of Transnationality*. Durham: Duke University Press.

Paasi, A. 2004. Place and Region: Looking through the Prism of Scale. *Progress in Human Geography* 28 (4): 536–46.

Peluso, N. 1995. Whose Woods Are These? Counter-mapping Forest Territories in Kalimantan, Indonesia. *Antipode* 27 (4): 383–406.

Pickles, J. 2004. *A History of Spaces: Cartographic Reason, Mapping and the Geo-coded World*. New York: Routledge.

Poeschke, R. 1996. *Nubians in Egypt and Sudan: Constraints and Coping Strategies*. Saarbrucken: Verlag fu Entwicklungspolitik.

Ramaswamy, S. 2010. *The Goddess and the Nation: Mapping Mother India*. Durham: Duke University Press.

Reid, D. M. 2002. *Whose Pharaohs? Archeology, Museums, and Egyptian National Identity from Napolean to World War I*. Berkeley: University of California Press.

Said, E. 1978. *Orientalism*. New York: Vintage Books.

Scott, J. C. 1998. *Seeing like a State: How Certain Schemes to Improve the Human Condition Have Failed*. New Haven: Yale University Press.

Sharabi, H. 1966. *Nationalism and Revolution in the Arab World*. Princeton: D. Van Norstrand Company.

Smith, N. 1992a. Geography, Difference, and the Politics of Scale. In *Postmodernism and the Social Sciences*, edited by J. Doherty, E. Graham, and M. Malek, 57–79. New York: St Martin's Press.

———. 1992b. Contours of a Spatialized Politics: Homeless Vehicles and the Production of Geographic Scale. *Social Text* 33:54–81.

Sparke, M. 1998. A Map that Roared and an Original Atlas: Canada, Cartography, and the Narration of Nation. *Annals of the Association of American Geography* 88:463–95.

Staeheli, L. 1999. Globalization and the Scales of Citizenship. *Geography Research Forum* 19: 60–77.

Swyngedouw, E. 1997. Neither Global nor Local: "Glocalization" and the Politics of Scale. In *The Global and the Local*, edited by K. Cox, 137–66. New York: Guilford.

Thongchai Winichakul. 1994. *Siam Mapped: A History of the Geo-body of a Nation*. Honolulu: University of Hawaii Press.

Van Cleef, E. 1929. The Atlas of Egypt (Book Review). *Geographical Review* 19 (2): 342–43.

van Schendel, W. 2002. Geographies of Knowing, Geographies of Ignorance: Jumping Scales in Southeast Asia. *Environment and Planning D* 20:647–68.

Western, J. 2007. Neighbors or Strangers? Binational and Transnational Identities in Strasbourg. *Annals of the Association of American Geographers* 97 (1): 158–81.

Wood, D. 2010. *Rethinking the Power of Maps*. New York: Guilford Press.

Zeigler, D. J. 2002. Post-communist Eastern Europe and the Cartography of Independence. *Political Geography* 21:671–86.

———

CHAPTER SEVEN

ART ON THE LINE

CARTOGRAPHY AND CREATIVITY IN A DIVIDED WORLD

Sumathi Ramaswamy

We have lived within the lines we have traced, and been made the subjects we have become.[1]

Is there an inescapable hegemony of the cartographic line ushered in by the modern science of mapping? This is the question that provokes my reflections here on one specific cartographic line that was legislated into existence with the formal end of British rule on the South Asian subcontinent on August 15, 1947. While the drawing of the so-called Radcliffe Line inaugurated the arrival of India and Pakistan as independent nations on the world stage, it also set in motion decades of tense confrontation, war, and violence that have remained with us unresolved to this day.[2]

In the dash and in the line, Gunnar Olsson suggests, lies the history of cartography. In the dash and in the line lies also the history of modern imperialism and the nation-state.[3] The cartographic impulse begins in the drawing and interpreting of lines that transform the world in which we dwell into a "geocoded" realm, making us subjects of these lines, as the epigraph for this essay reminds us.[4] In the words of the late J. Brian Harley, lines on maps become the new "dictators" of representation, disciplining land and territory through their "ethic and virtue of ever more precise definition."[5]

Yet does everyone capitulate—and in the same manner—to what anthropologist-turned-novelist Amitav Ghosh once called the "enchant-ment" of mapped lines?[6] In this essay, I explore the work of several artists of the South Asian subcontinent who have attempted to take on the statist line-making project to create alternative contours of belonging. There is now important scholarship that has moved the spotlight away from the high drama of state politics and the "transfer of power" from British to South Asian agents in August 1947, to focus instead on the everyday tragedies of 10 to 12 million people who were displaced, separated from families, abducted, mutilated, raped, or killed in the months leading up to and following the announcement of the so-called Radcliffe Award. This scholarship has turned to a rich archive of imagination constituted by fiction and film to write histories "from below" that have brought in the leavening perspective of the everyday, the gendered, and the subaltern to the work of a few men huddled over maps, census docu-ments, gazetteers, and other paper products of a bureaucratic state that was in the process of simultaneously coming apart and reconstituting itself anew.[7] And yet curiously outside the realm of film studies, there are few analyses of the visual and artistic responses to the Partition of India, and of the manner in which visual artists on both sides of the new border—elite and subaltern, working in numerous media ranging from oil on canvas to mass-produced imagery—began to contend with the new lines of power (but also of trag-edy) that came to transform and circumscribe their lives.[8] In exploring how the artist's practice grapples with the cartographer's, I ask how and whether artistic visual labors favor fuzzy contours and blurred boundaries over the unyielding lines of state and scientific cartography. Does the work of artists offer an aesthetic and ethic of inclusion and coexistence in distinction to stat-ist claims of exclusive proprietorship and bounded singular sovereignty? Or, are artists as well caught up—as "map-minded" moderns, even patriots—in the seeming hegemony, even "dictatorship," of the cartographic line?[9] These are some of the main questions that drive the reflections in this essay.

The boundary is the imaginary line that draws attention to itself by violence.[10]

Inspired by anthropologist Michael Taussig's call to allow the image to billow
out into our driving concept and to power the engines of our analyses, I begin
my exploration of these questions by turning to a striking illustration printed
on the cover of *Time* magazine on October 27, 1947, a little over two months
after the British formally withdrew from their Indian empire (fig. 7.1).[11] It is
titled *India: Liberty and Death*, crucially rephrasing the famous slogan (which
US readers of the magazine would have recognized immediately) attributed to
Patrick Henry from nearly two centuries earlier in the context of the loss of
Britain's North American colonies in the first age of empire. The print features
a demonic four-armed naked female figure, skulls adorning her head, wielding
a bloody dagger, which she plunges into her own right breast, which the other
hand clutches; her breast in turn is placed over that part of the map of India that
a cartographically literate reader would identify as Punjab, onto which drips
vivid red blood from the mutilated organ. While the outline map of "India"
(out of which the demonic figure seemingly erupts) is colored yellow, and
Kashmir is unambiguously part of it, "Pakistan" (also left unnamed) is green,
its eastern wing's outline conforming more to the "notional" rather than the
actual boundary eventually awarded by Radcliffe's Boundary Commissions.[12]

The print is signed by Boris Artzybasheff, active between 1941 and 1965 as one
of *Time*'s most important illustrators.[13] The work, however, might reflect not
only the Ukrainian-born Artzybasheff's artistic predilections for rich colors,
bold design, and imaginative symbolism, but also the spirit of the accompanying cover story that it helped illustrate and was tellingly titled "India-Pakistan:
The Trial of Kali."[14] For the brutal ongoing slaughter that the subcontinent
was witness to in those hot, heady months of 1947, the cover story put Kali
herself on trial. Introducing her to the magazine's reader as "goddess of death
and catastrophe, wife-conqueror of the eternal Siva, the dancer," the authors
of the essay insisted, "Kali, the Black One, could stand as symbol (or perhaps as
scapegoat) for the horror that had walked hand in hand with bright liberty into
India." The essay concluded, "If India could descend to the depths, it could
also look up to moral Himalayas. Its recent sin was great, but not unique, especially not unique in origin. It sprang from Kali, from the dark and universal
fear which rests in the slime on the blind sea-bottom of biology."[15]

OCTOBER 27, 1947

In this issue—TIME's
CURRENT AFFAIRS TEST!

TIME

THE WEEKLY NEWSMAGAZINE

INDIA
Liberty and death.

Artzybasheff

FIGURE 7.1. Boris Artzybasheff, *India: Liberty and Death*, October 27, 1947. Cover page illustration, *Time: The Weekly Magazine* (New York). Image courtesy: Time Inc.

As many scholars have argued, there has been a morbid fascination in the modern West with Kali, a fascination that has its roots in the British colonial preoccupation with this figure variously described by anxious and fearful administrators as "the goddess of death and destruction," and a "terrible goddess" whose worship appealed to "the grossest and the most cruel superstition of the masses."[16] The importance of the Kali figure in the Hindu-Indic religious imagination notwithstanding, the *Time* magazine image is highly idiosyncratic both in associating the divine form of this specific goddess with the modern geographic form of India and in the visual innuendo that the goddess had turned from attacking cosmic demons to destroying the map of the country. In other words, there is no precedent that I know of in the Indian pictorial archive for this particular image, and that itself makes it singular, but also problematic in its proposition that the catastrophic violence that accompanied the partition of the subcontinent had nothing to do with the British as they beat a hasty retreat from the subcontinent. Instead, some primordial force ("black grace"), as embodied by this ferocious goddess, had erupted from the very soil of India to destroy its people.[17]

This was not the only image of a violated map of India that *Time* published. A little over a year earlier, on April 22, 1946, it printed another cover with the provocative title *Mohamed Ali Jinnah: His Moslem tiger wants to eat the Hindu cow* (fig. 7.2).[18] Created by another prolific illustrator for the magazine, Boris Chaliapin, this image shows two tigers attacking the map of India, their claws tearing away at the areas we know of as Punjab and Bengal, while a (British?) lion watches from the margins—possibly a sign that the empire was already retreating from the scene of impending violence. A sinister-looking Muhammad Ali Jinnah is in the foreground of the image. Jinnah's appearance on the cover with its inflammatory title indelibly associates him with the macabre scene in the background: he was all that stood between "India and Independence," in the words of the accompanying cover story, which offered a largely unflattering portrayal of the avowed Father of Pakistan.

Unlike the Artybasheff image, which to the best of my knowledge was not appropriated by the Indian media, the Chaliapin cover was republished soon after in July 1946 in *Chitramayi Jagat*, a Marathi-language newspaper of Hindu nationalist inclination, with a title that translates as "From the Perspective of America: The Question of Hindustan."[19] In this new context, the image fed into a growing (Hindu and Indian) nationalist discourse that I have discussed elsewhere as the "martyrdom" of the Indian map.[20] It also resonates with a dense discourse saturated with "surgical metaphors" of amputation, dismemberment, and scarring to which Joya Chatterji has alerted us.[21] Indeed, it is worth recalling that the apostle of nonviolence Mohandas Gandhi frequently referred to the impending division of his beloved India as "vivisection," leading Jinnah to caustically remark, "It is amazing that men like Mr. Gandhi and Mr. Rajagopalachariar [a Gandhi ally] should talk about the Lahore Resolution [that announced Jinnah's decision] in such terms as 'vivisection of India'

FIGURE 7.2. Boris Chaliapin, *Mohamed Ali Jinnah: His Moslem Tiger Wants to Eat the Hindu Cow*, April 22, 1946. Cover page illustration, *Time: The Weekly Magazine* (New York). Image courtesy: Time Inc.

and 'cutting the baby into two halves.' . . . Muslim India and Hindu India exist on the physical map of India. I fail to see why there is this hue and cry. Where is the country that is being divided?"[22] Even Jawaharlal Nehru, the future first Prime Minister of an independent India, whose secular-socialist vocabulary did not usually resort to the somatic idiom, lamented, "But above all, what was broken up which was of the highest importance was something very vital and that was the body of India."[23] And in fact, from at least the early years of the twentieth century, the territory variously called "Hindoostan," "British India," or just "India" on colonial maps and laid out as empty cartographic space within a lattice of latitudes and longitudes was also contrarily imagined as the body of Mother India, or Bharat Mata, and gloriously pictured in all manner of visual media ranging from oils and acrylics to chromolithographs and cinema. Elsewhere, I have documented the rush to claim cartographic space delineated on imperial maps for the body of this new goddess of territory who comes to supplement it in many different ways: occupying and filling up the map (for example, fig. 7.12); merging partially with it; perched, seated, or standing on it; and most destabilizing of all, dispensing with it entirely and standing instead for the geo-body of India (for example, fig. 7.11). Ironically, the map of India, the proud artifact of a secular colonizing state and its sciences, is in fact a necessary guarantor of Mother India's persona as deity of *national* territory. In turn, and once again ironically, the scientific map form itself comes to be anthropomorphized, its secular space filled with the sensuous, feminine, and very Hindu presence of Bharat Mata, dominantly imagined as benign and benevolent (albeit occasionally, with elements of Kali surfacing in her picturing), in need of help from her children, especially her sons, to protect and cherish her, break her shackles, and restore her to power and dignity in a free India.[24]

BRINGING THE MAP BACK IN

[They] told me they wanted a line before or on 15th August. So I drew them a line.[25]

Although the Partition of India was arguably precipitated by nongeographic and extraterritorial factors, the actual determination and drawing of the new boundary lines was a cartographic act, even if one principally dictated by the avowed imperatives of religious demographics.[26] Indeed, it would not be far-fetched to say that the cartographic wars—politics by other means—that India and Pakistan wage to this day had their birth in the events of 1947.[27] Yet, with the exception of a few (which are cited and discussed here), the most influential and invoked scholarly works on Partition largely ignore the role played by maps, and cartographic practice more generally.[28] What then can we learn by bringing maps (back) into the frames of our analyses of this most foundational of events of modern South Asia?

Even the least map-minded of scholars do not fail to invoke a 1966 poem by W. H. Auden titled "Partition":

> Unbiased at least he was when he arrived on his mission,
> Having never set eyes in this land he was called to partition
> Between two peoples fantastically at odds
> With their different diets and incompatible gods.
> "Time," they had briefed him in London, "is short. It's too late
> for mutual reconciliation or rational debate.
> The only solution now lies in separation . . ."
> . . .
> Shut up in a lonely mansion, with police night and day
> Patrolling the gardens to keep assassins away,
> He got down to work, to the task of settling the fate
> of millions. The maps at his disposal were out of date
> And the Census Returns almost certainly incorrect,
> But there was no time to check them, no time to inspect
> Contested areas. The weather was frightfully hot,
> And a bout of dysentery kept him constantly on the trot,
> But in seven weeks it was done, the frontiers decided
> A continent for better or worse divided.
>
> The next day he sailed for England, where he quickly forgot
> The case, as a good lawyer must. Return he would not,
> Afraid, as he told his Club, that he might get shot.[29]

A work of imagination, Auden's poem captures in spirit the air of confusion, haste, and melancholy that hangs over the work of Sir Cyril Radcliffe and the two Boundary Commissions—one for the western Indian territory of Punjab and the other in the east for Bengal—that he was appointed to chair, soon after the critical announcement on June 3, 1947, that British India was indeed to be divided. That announcement (made by Prime Minister Clement Atlee in the

House of Commons in London, and broadcast by All India Radio at 7:00 p.m. Indian Standard Time) was importantly not accompanied by any territorial specifics, nor indeed was the "talismanic" word "Pakistan" even mentioned by Atlee in London, or Viceroy Mountbatten in New Delhi.[30] Nehru confusingly declared, "The India of geography, of history and traditions, the India of our minds and hearts cannot change," while Jinnah's official statement ended with "Pakistan Zindabad" (Long Live Pakistan).[31] Indeed, it was not until August 17—three days after Pakistan was born, and two days after independent India was created—that the "award" of the Boundary Commissions was made public to the people whose lives were catastrophically transformed by this very act of inscription.[32] Between June 3 and August 17 and in the immediate aftermath, the map as artifact flickers in and out of official records and public discourse, at times a concrete object over which men pored and pondered, at other times, a spectral presence, sometimes even a virtual nonentity. If all of us have become convinced, especially under the influence of J. B. Harley, that maps anticipate and enable modern empire, the Indian summer of 1947 could perhaps persuade us of the contrary truth, namely, that accurate and adequate maps may not be all that relevant for the dismantling of empires, or for at least this one.

Much has been made of the fact that Radcliffe, "India's mapmaker," was a career lawyer with no firsthand knowledge or experience of the subcontinent.[33] Before his arrival in Delhi on July 8, his formal "Indian education" might have been largely limited to a briefing in London prior to his departure, "a thirty minute session over a large-scale map with the Permanent Under Secretary at the India Office."[34] Contemporaries in the colonial administration did not deem this apparent lack of familiarity or hands-on experience to be a problem, and in fact it was even seen as a virtue because it conferred upon him an aura of impartiality.[35] Radcliffe was neither a cartographer nor a geographer with any technical expertise in boundary making. Similarly, like him, the South Asian members of the two Boundary Commissions were lawyers by training and high court judges by profession. Although we can presume that modern schooling would have given all these men some modicum of cartographic and geographic knowledge, this does not mean that they were necessarily adept or literate in reading and interpreting complicated maps over the course of a mere six weeks.[36] A highly reticent man by all accounts, Radcliffe reportedly burned all his papers, so we have only indirect reporting on his views of what transpired in those confused few weeks he spent in India, and even less of what he made of the momentous boundary-making project with which he was entrusted. Scholarly interpretations of the available record have long insisted that the maps placed at Radcliffe's disposal were inaccurate, inappropriate, or, in some cases, plain unavailable, although recently Lucy Chester has questioned this conclusion.[37] Leonard Mosley, a foreign correspondent whose reminiscences of this fraught period have been used by many historians, claimed that one of Radcliffe's principal worries was finding a map of suitable scale to carry out his task. "It seems extraordinary that when you have to decide the fate of 28,000,000 people you are not even given the right map to do it with." Mosley goes on to observe, "With the slings and arrows of importunate Muslims, Hindus and Sikhs whistling about his ears, Radcliffe took up the largest contour map he could find and began to draw."[38] The French journalist Dominique Lapierre quotes Radcliffe as saying, "The equipment I had at my disposal was totally inadequate. I had no very large-scale maps." It was not just a matter of adequacy, but also of accuracy. "The information provided on those [official maps] I did have sometimes proved to be wrong. I noticed the Punjab's five rivers had an awkward tendency to run several miles away from the beds officially assigned to them by the survey department."[39] Locked away in his "lonely mansion," Radcliffe had little direct exposure to either the land he divided up or the people who lived on it, and neither he nor members of his two committees had time (or perhaps even the inclination) to undertake local surveys. So, as Yasmin Khan eloquently writes, in the end "they retreated behind closed doors, working from maps using pen and paper, rather than walking the land and grasping for themselves the ways in which vast rivers, forests and administrative districts interlocked and could best be separated."[40]

If the mighty map-minded British Raj in its last days allowed nongeographic considerations (or "other factors") to have the final word, its colonial subjects on the contrary demonstrated a surprising flair for the deployment of maps and placed unusual faith in cartographic efficacy in the immediate years leading up to Partition and especially in the months before and after August 1947. I say unusual and surprising because there is very little evidence that prior to this time that most Indians, even the educated among them who had been exposed through colonial schooling to cartographic artifacts, routinely drew upon maps in their daily or professional lives.[41] With Partition imminent, however, all of a sudden maps began to be invoked and used in very revealing ways. Penderel Moon, a senior administrator, writes of attending a meeting convened by the Muslim League in Lahore in late June in a large private home: "On the floor and on a big table a number of maps of the Punjab were strewn about, variously coloured and chequered so as to show the distribution of the population by communities. *We all fell to poring over these maps.* It became plain in a very few minutes that no one had any definite idea where we should claim that the dividing line should run."[42] In turn, one of the Muslim League members of the

Punjab Boundary Commission, Muhammad Munir, wrote in deep suspicion of Radcliffe's secretary Christopher Beaumont's "distinctly pro-Hindu leanings." Thus, "whenever I went to his office, I found him poring over a large map and was surrounded by Hindus."[43] The surveyor general of India, G. F. Heaney, recalled receiving a request in October 1947 from a young Sikh officer for a large number of maps. "I glanced at the list and it was at once apparent to me that it included most of the Punjab, now part of Pakistan, in addition to much of northern India and adjoining part of the UP [United Provinces]." On orders from his superiors, who rightly feared the use of such maps in the ongoing cross-border and cross-community genocide, Heaney only gave him a single copy of each map, but he subsequently heard that the Sikh officer "tried, without success, to get further copies."[44] Earlier in July, the Boundary Commissions in Bengal and Punjab were inundated with petitions that resorted to maps as objects of persuasion. To be sure, the presence of maps in these memorials was a response to the official request that all memoranda submitted by interested parties "should be accompanied by such maps as may indicate the proposed line of demarcation between the two new Provinces."[45] Nonetheless, the fact that each of the interested parties—the Congress, the Muslim League, and the Sikhs—rallied around and created maps is important to note, as is the fact that in the deliberations before the commissions, they debated each other cartographically on the placement of various dots and dashes and lines. It is unclear who actually drew these maps, but it is revealing that in producing such artifacts, these men did not have access to official Survey of India maps, whose circulation was highly restricted.[46] Not least, the various contending parties also published aspirational maps in local newspapers and other media.[47]

We do not yet know of the precise mechanisms by which ordinary Indians or Pakistanis learned of the precise shape and contour of their newly formed countries, given the confusion that prevailed for several weeks on both sides of the border following independence.[48] On the one hand, Hindu nationalist magazines such as the *Organiser* published an inflammatory image of map and mother being destroyed by men like Nehru and Jinnah.[49] On the other, other mass media played it safe by continuing to use maps of undivided India to usher in the new nation-states. Lucy Chester draws our attention, for example, to an advertisement for Bata Footwear published in the *Times of India* on August 15, 1947, in Calcutta's *Morning News* and even in Jinnah's own *Dawn* that used a map of undivided India (fig. 7.3), as did advertisements of other ventures like Lilaram and Sons (a silk merchant and tailor), the Punjab National Bank, and the United Commercial Bank.[50] There might well have been commercial imperatives for wanting to appeal to the largest numbers of consumers and not

FIGURE 7.3. "The Dawn of Freedom."
Ad for Bata Footwear, published in *Times
of India*, August 15, 1947. Image courtesy:
Lucy Chester

wishing to lose existing networks that were now severed by new boundaries.
But it is also the case that such maps continued to recur after the lines of separation had been determined and drawn because of a prevailing fantasy that
these newfangled borders might well be soon undone and the land restored
to its former wholeness.[51] Be that as it may, there is also no doubt of the sheer
absence—in the public domain and for a while—of actual knowledge of the
legally announced border: both newly formed states were so overwhelmed

in the months following independence with the violence and mayhem that accompanied Partition that educating their new citizenry about the shape of their new homelands appears to have been not high on their agendas.

Learning the new shape of their nation must have surely come as a shock to Pakistanis in particular, for their country indeed more or less turned out, as Jinnah famously feared in 1944, "maimed, mutilated, and moth-eaten," divided in two parts, with hundreds of miles of hostile territory separating them.[52] Until recently, scholars have tended to argue that not least of the conundrums with which to contend in the years leading up to Partition was that the principal sponsors of the idea of "Pakistan," including (and especially) Jinnah, were strategically vague about the shape and form of their homeland. Thus, Ayesha Jalal argues that formulations regarding "bordered separation" were themselves fairly chaotic and shot through with ambiguity.[53] Even the historic Lahore Resolution of March 1940—which most scholars agree began to concretize the project of Pakistan—only demanded, "The areas in which Muslims are numerically in the majority, as in the North-Western and Eastern zones of India should be grouped to constitute Independent States in which the constituent units shall be autonomous and sovereign."[54] So, David Gilmartin concludes, "The two-nation theory, the basis for the Muslim League's Pakistan demand, was a fundamentally non-territorial vision of nationality, and for most Muslims, the meaning of Pakistan did not hinge primarily on its association with a specific territory."[55]

There are several striking exceptions, however, to what has been deemed a general attitude of geographic and territorial "indifference."[56] In his address to the Muslim League annual meeting in December 1930, poet and thinker Muhammad Iqbal first voiced the idea of a separate Muslim state to occupy what was then the northwestern part of British India. This vision was subsequently cartographically materialized in the numerous publications of Choudhary Rahmat Ali, a student (and aspiring lawyer) at Cambridge University who is credited with coining the name, "Pakistan," for the imagined nation. On January 28, 1933, Rahmat Ali released an appeal titled *Now or Never: Are We to Live or Perish for Ever?* in which he proposed that the Muslim homeland was to consist of the "northern units of India, viz., **P**unjab, North-west Frontier Provinces (**A**fghan Province), **K**ashmir, **S**ind and Baluchi**stan**."[57] In 1935, a two-page letter addressed to the House of Lords (who were then considering the Government of India Bill) included a header image which was a map of British India, the parts colored green and covering Baluchistan, Sindh, North West Frontier, Kashmir, and Punjab, named "Pakistan," the rest marked as India.[58] Over the next decade, Rahmat Ali published several maps that clearly point to the cartographic imperative at work in modern nationalist imagina-

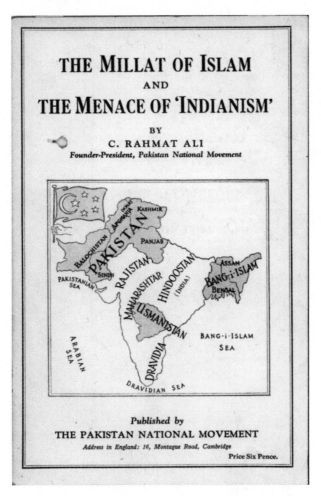

THE MILLAT OF ISLAM
AND
THE MENACE OF 'INDIANISM'
BY
C. RAHMAT ALI
Founder-President, Pakistan National Movement

Published by
THE PAKISTAN NATIONAL MOVEMENT
Address in England: 16, Montague Road, Cambridge
Price Six Pence.

FIGURE 7.4. Cover page for Choudhary Rahmat Ali, *The Millat of Islam and the Menace of "Indianism."*
Cambridge: Pakistan National Movement, 1940. © The British Library Board, IOR L/P&/J/8/689 f105

tions, even for those for which territorial clarity and certitude are problematic. It is clear from such images that Rahmat Ali's future "Pakistan" would be a bounded territory that lay mostly in the northwestern part of southern Asia, although its boundaries shifted over time across these maps, and new territories began to be cartographically affiliated with the core Muslim homeland. For example, in a map published in 1940 and reissued over the next few years, two new "geo-bodies" make their appearance to join "the northern units of India" singled out in 1933 as constituting Pakistan. These are named "Bang-i-Islam" (roughly coinciding with Bengal and Assam) and "Usmanistan," roughly the princely state of Hyderabad (fig. 7.4). Over the years, as he elaborated on his

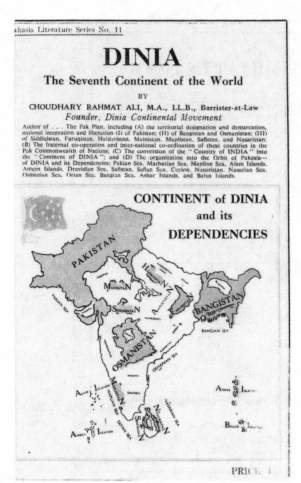

FIGURE 7.5. Cover page for Choudhary Rahmat Ali, *Continent of Dinia and Its Dependencies*. Cambridge: Dinia Continental Movement, 1946.

"prolific, even manic territorial" vision, other geo-bodies scattered across the subcontinent (and islands in the Indian Ocean, parts of which were also appropriately renamed) were added, their principal qualification some sort of historic or demographic connection to Islam and Muslims.[59] These would constitute a "Pakistan Commonwealth of Nations," subsequently expanded into a continental federation called "Dinia," a clever anagram of "India" that also resonated with the Arabic word for religion, *din* (fig. 7.5).[60] In a book first published in 1935 that went into several editions over the next few years, Rahmat Ali also charted the "national story" through a series of maps that begins with "Pakistan in Geological Times" and ends in the present. Several of these maps had the words "Pak Empire" boldly inscribed across the entire subcontinent. "It is important to remind the reader at the outset that the history of Pakistan is not that of a country which will some day be carved out of 'India.' On the contrary, it is the history of a country which, though at present incorporated in India, has always existed in its own right, and on whose life seventy centuries look down from the lofty peaks of the Jabaliya and the legendary passes of the Khaibar; and to whose future, no human power can set any limits."[61]

Rahmat Ali was not alone in generating such aspirational maps, reminding us of Thongchai Winichakul's insistence that "a modern nation-state *must* be imaginable in mapped form."[62] Yasmin Khan notes that in the months leading up to August 1947, Jinnah was inundated by fan mail from across the subcontinent which included "different maps of Pakistan carved in wood."[63] Even prior to 1947, images of wildly different geo-bodies (variously named) begin to proliferate in the public domain, all of them nevertheless resorting to the protocols of scientific cartography and the use of lines to create bounded spaces as well as taking the mapped form of British India as the starting point for their imaginations.[64] It is not clear whether the creators of these maps—many of whose names have become obscured over time—necessarily saw the new homeland as wholly separate from India, the lines and hatchings of cartography giving a sense of the certitude of unambiguous belonging and territorial homogeneity that existed perhaps only on paper.[65] Thus, in 1938, Syed Abdul Latif, a retired university professor of English at Osmania University in Hyderabad, published a pamphlet titled *A Federation of Cultural Zones for India* whose cover image was a map titled *Cultural Distribution of India* (fig. 7.6). The colored "Muslim zone" included the following regions (many of which had substantial, consequential, or majority Muslim populations): Hyderabad, East Bengal and Assam, the areas around Delhi and Agra (although the region has not been specifically named), Sind, Baluchi states, British Baluchi States and North West Frontier Province, Punjab (which includes also half of western Kashmir), and Bahawalpur.[66] Around the same time, under the patronage of Nawab Sir Muhammad Shah Nawaz Khan of Mamdot, a map showed a "Quinquepartite Confederacy," made up of various federated units including the "Indusstan Federation," that roughly occupies the place marked "Pakistan" in today's maps along with Kashmir.[67] Neither of these maps, however, invokes the name "Pakistan," notwithstanding its existence on Rahmat Ali's maps from slightly earlier in the decade.

The word "Pakistan" did appear on a very interesting bilingual map (in Urdu and English), first published in 1939 and reissued in 1945 (fig. 7.7).[68] On the map, a territory called "Pakistan Caliphate" stretches from the western edges of British India to the east, claiming much of the Hindi/Hindustani belt of northern India; the rest of the territory is given the name of God's earth or

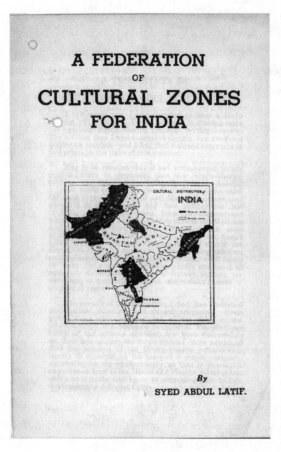

FIGURE 7.6. *Cultural Distribution of India*. Cover of Syed Abdul Latif, *A Federation of Cultural Zones for India*. Hyderabad, 1938. © The British Library Board, IOR L/P&/J/8/689 f105.

territory inhabited by non-Muslims but under the protection of Islam pending their imminent conversion. Rahmat Ali himself has also been credited with another map on which the words "Pakistani Empire" are boldly inscribed across the entire subcontinent (painted in a dark hue), and the body of water adjoining its west coast correspondingly labeled "Pakistani Sea."[69] Indeed, such varied imaginings fed into the propaganda machine of the rival Congress, some of whose members suggested that the party should recirculate such "posters and flyers around the streets with a suggested map of Pakistan under the caption 'Are you reading to leave your house, land, property and everything and go to [such a] Pakistan?'"[70]

In retrospect, it is all too easy to dismiss these varied visualizations as fantasies not worthy of our attention, and indeed little critical attention has been paid to such wishful maps in much of the existing scholarship on the idea of Pakistan (with some exceptions whose work I have invoked here). Nations,

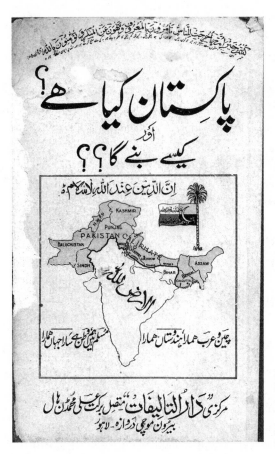

FIGURE 7.7. Cover image of Urdu pamphlet *Pakistan Kya He Aur Kaise Banega* [What is Pakistan and how will it be created], 1945 (originally printed in 1939). Image courtesy of David Gilmartin.

however, yearn for territorial form, and these mapped wishes are symptomatic of that yearning, but are also revelatory of what I have called the hegemony of the cartographic line. Ayesha Jalal has argued that Jinnah and his core followers almost until the very end generally operated with the notion of shared sovereignty, "which seemed the best way of tackling the dilemma posed by the absence of any neat equation between Muslim identity and territory." Thus, the boundaries between the proposed Muslim homeland and "India" had to be "permeable and flexible, not impenetrable and absolute."[71] Inevitably, though, as Muslim nationalists rushed to materialize their contending visions of "Pakistan" on paper, the flexible and fluid imagination of shared sovereignty began to be undone by the firm and unyielding dashes and lines of (printed) cartographic practice. To recall Bruno Latour in this regard, "There is nothing that man is truly capable of dominating: for man everything is immediately too large or too small, too mixed or composed of successive layers that dissimulate in view what he would like to observe. Surely! Yet one thing and one thing alone is dominated by the gaze: it's a sheet of paper extended on a table or tacked onto a wall. The history of science and technology to a large degree tells of ruses that allow the world to be brought onto this paper surface."[72]

ARTFUL MAPPING IN BAZAAR INDIA

> I was struck with wonder that there had really been a time, not so long ago, when people, sensible people, of good intention, had thought that all maps were the same, that there was a special enchantment in lines.[73]

In the bazaars and streets of independent India, in and after August 1947, nonofficial maps began to appear, delineating the new boundaries of the new nation(s). Such maps were frequently the work of artists whom I have called "barefoot cartographers," some of whose practices I have traced back to the closing decades of the nineteenth century.[74] Even with no obvious training in the use of maps, in the science of their production, or the aesthetics of their creation, these men play no small role in popularizing what Benedict Anderson refers to as the "logo" form of the national geo-body so that it becomes recognizable and familiar, not needing the crutch of naming or identification.[75] If we follow John Pickles, such men are exemplary of "the legion of mapmakers and map users that is not part of the professional cadre of expert cartographers."[76] Although they might have an inexpert, undisciplined, and informal relationship to the science of cartography and its mathematized products, to which they turn for various purposes, their lack of specialist cartographic knowledge should not, however, be read to mean that they are naive or apolitical. On the contrary, I suggest that such "artful" mapmakers of the street and the bazaar have a critical and constitutive role to play in producing and disseminating knowledge about the form of the nation among the citizenry. I would even propose that it is through the labors of barefoot cartography and such artful mapmakers—more so arguably than through the highly specialized operations of the state or science—that many become familiar with the shape of national territory that they came to inhabit as citizen-subjects.

Consider figure 7.8, titled *New India*, possibly published sometime after Indian independence and the so-called "accession" and "integration" of princely states between August 1947 and 1951[77] and before the process called "states reorganization," which began in 1953.[78] The lion capital (the newly installed national emblem), and the tricolor national flag at the top of the print

No. 400   NEW INDIA

FIGURE 7.8. *New India*. Print bearing signature of Banshi, circa 1950. Priya Paul Collection.

suggest that it was possibly meant to celebrate the arrival on the political landscape of the Republic of India on January 26, 1950. Although "India" is not
named as such, and is instead cartographically depicted as a proliferation of
numerous constitutive units, all meticulously delineated and named, the newly
created Pakistan is identified, the color green reserved as in many such maps

and prints for the new Muslim-majority country. This print adheres to the terms of state cartography in its general conformity to national boundaries as these began to be inscribed in official maps after August 1947. And yet, what sets *New India* apart from such normative and official maps of the country and makes this an instance of barefoot cartography, as I have defined it, is the inclusion of the portraits of the leaders of the nation—the "big men" of India—arranged in roundels around its borders. It is almost as if the newly won national territory cannot be merely shown as empty cartographic space, marked off by geometric lines and blocks of hues, and instead needs the legitimizing presence of these figures, left unnamed but well known to any patriotic citizen as the men who had led their country to freedom. These familiar faces then appear to introduce the recently configured national territory (the nation's "geo-body") to the citizen-subject, lending their recognizable presence to the new spatial reality that had come to fundamentally alter the lives of everyone on the subcontinent after August 15, 1947.

There are other examples from the dawn of Indian independence that resemble a prolific genre of popular imagery that is called the school or educational chart. In figure 7.9, also titled *New India*, the emphasis is certainly on distinguishing Pakistan (in deep green) from the "new" India, but the artist—whose name might well be R. S. Mukherjee, as printed on the bottom right—also appears to be keen on showing the continuing presence of the so-called "princely" states, which are set off in bright yellow within Indian national territory—not yet divided up into the fourteen new administrative units—colored red. Gandhi beams down on the newly created nation-state. His haloed presence possibly dates this print to after his death in January 1948, although by that time the vast majority of these princely states had merged into India or Pakistan (some rather contentiously).[79] In *New India No. 2* (fig. 7.10), such big men are displaced by the Everyman, tilling the soil of the nation to yield a rich harvest, while Gandhi smiles down on vignettes of the patriotic-bucolic (although one suspects that he might not have entirely approved of the presence of the industrial-scale technology in the fields of Nehruvian India!).

In such prints, although national borders show varying degrees of conformity to principles of cartographic accuracy, and are sometimes partly or wholly erased (as in fig. 7.10), there is a general commitment to lines and boundaries, which are drawn with clarity and firmness that speaks to state cartography's preoccupation with the limits of territorial sovereignty. Nonetheless, we also see that in such productions, there is a rush to fill in the empty cartographic space of statist maps with all manner of activities and bodies, most often the torsos and heads of the new big men of India: Gandhi, Nehru, and others.

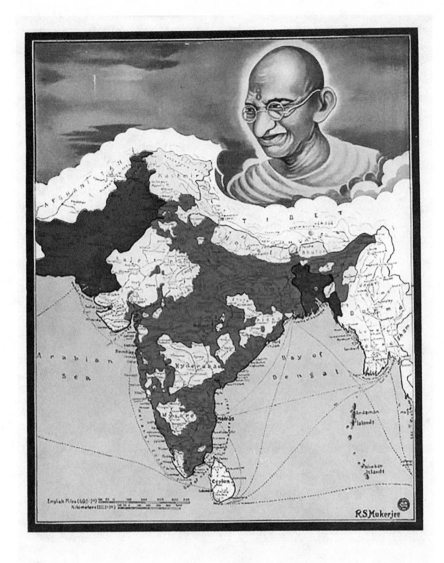

NEW INDIA
No. 688

FIGURE 7.9. *New India*. Print bearing signature of R. S. Mukerjee. Published possibly by Empire Calendar Manufacturing Co., circa 1947–48. Priya Paul Collection.

NEW INDIA No. 2
No. 700

FIGURE 7.10. *New India No. 2*. Print bearing signature of Sushil Das. Published possibly by Empire Calendar Manufacturing Co., circa 1950s–1960s. Priya Paul Collection.

This is because, as I have suggested elsewhere, barefoot cartography in India, even while cheekily reliant on the state's cartographic productions, also routinely disrupts them by injecting the anthropomorphic, the devotional, or the maternal into the spaces of secular science. It thus has an affective and worshipful, even idolatrous, investment in national territory in contrast to command cartography's geometrical grids of certitude and lines of power. This is most apparent when we turn to images where Mother India is shown occupying the map of India. Soon after independence, P. S. Ramachandra Rao—who had produced other such images in the past—painted a bodyscape in which Mother India's sari, clad in the new national tricolor, is arranged to suggestively approximate the shape of India in a manner that appears to leave out the new national territory of Pakistan, east and west (fig. 7.11).[80] More recently, an artist by the name of Appu painted a bodyscape where Mother India, also clad in the national tricolor, stands against a silhouette of a map of India, in which both Pakistan and Bangladesh have not been incorporated.[81] All the same, such images are in contrast to many others where, clearly, the historical fact of Partition is denied, erased, or occluded, and in which Mother India's body is deployed in various ingenious ways to claim lands that no longer legally or geopolitically are part of independent India. For instance, consider an illustration that appeared in 1955 in a Tamil textbook from southern India, a part of the new nation-state relatively less directly affected by the demographic catastrophe that accompanied the drawing of the Radcliffe Line. Nevertheless, the new borders are totally erased, as Mother India is shown claiming the whole map of an undivided India (fig. 7.12). The imperialist manner in which her body is deployed here is further heightened by the fact that her scarf—significantly colored green, the color of Pakistan—gracefully and seemingly innocently reaches out into Pakistani national territory.

Map of India in figure 7.13 is a reminder that such prints may not be as benign as they look at first glance: the print shows the newly delineated map of India in the company of Jawaharlal Nehru, who looks directly out at us with a slightly weary look.[82] Possibly the most striking aspect of this print is the Indian tricolor flying triumphantly over the nation's territory, the staff bearing it firmly planted on Kashmir.[83] Indeed, the most revealing acts of deletion, occlusion, or incorporation in all such productions of barefoot cartography become apparent in the pictorial fate of the contested territory of Kashmir.

In July 1947, when Radcliffe arrived in India to begin his work of division, his terms of reference did not extend to Kashmir, which was then a princely state and hence not a part of British India. It was then under the rule of the Hindu Dogra dynasty but comprised largely Muslim subjects with territo-

Published by:—
R. ETHIRAJIAH & SONS,
Pictures & Glass Merchants,
102, Devaraja Mudaly St.,
P. T. Madras-3.

Copyright Reserved by the Publishers

Printed by
Associated Printers, (Madras) Ltd.,
Madras & Bangalore.

भारत देवी

THE SPLENDOUR THAT IS INDIA
BHARATH DEVI

FIGURE 7.11. *The Splendour that is India: Bharath Devi.* Print bearing signature of P. S. R. Rao published by P. Ethirajiah and Sons, Madras, circa 1947. Author's collection.

FIGURE 7.12. Bharat Mata. Frontispiece to Tamil schoolbook by V. Lakshmanan, *Putiya Aarampakkalvi Tamil (Moonram Puttakam)* [New elementary Tamil: Book 3]. Mannargudi: Shri Shanmugha Publishing House, 1958. Image courtesy of Tamil Nadu State Archives, Chennai.

FIGURE 7.13. *Map of India*. Print circa August 1947, publisher unknown. Priya Paul Collection.

ries adjacent to the soon-to-be-created India and Pakistan. Its Maharaja, Hari Singh, fantasized about striking out on his own in August 1947, but such fantasies were cut short when, in the autumn of 1947, tribesmen from Pakistan crossed over into the northern areas of his kingdom and occupied parts of it in solidarity with those who had rebelled in the western part of the state to establish Azad Kashmir (Independent Kashmir). In response, Hari Singh turned to Nehru's government for military help, and on October 26 signed Kashmir's accession to the Indian Union. India in turn airlifted troops into Srinagar, driving the insurgents out.[84]

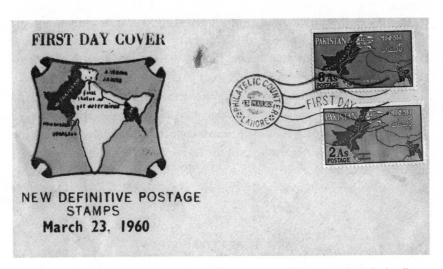

FIGURE 7.14. First Day Cover issued by Pakistan Post Department, March 23, 1960. Author's collection.

From then until today, the stalemate over Kashmir has continued, with neither side ready to give up on its territorial claims. "Cartographic wars" have accompanied real and bloody wars on the ground. Kashmir is thus identified as "disputed territory" on official maps of Pakistan, and as Ananya Kabir has noted, the north-eastern edge of the country "remains ostentatiously unbounded—an astounding rebuttal of the universal dependence of national maps on borders."[85] Or, in some maps, the rough outline of Kashmir is highlighted with the words, "Final status not yet determined" (fig. 7.14).[86] By contrast, official maps of India unequivocally and without question subsume the entirety of Kashmir, disavowing the Line of Control (LOC), the de facto border that separates Indian-administered Jammu and Kashmir from Pakistan-administered Azad Kashmir and the Northern Areas. On the eastern side as well, Aksai Chin remains another no-man's land, claimed since the 1950s by both India and China, a claim over which they both fought an unsuccessful war in 1962: here as well, a so-called Line of Actual Control constitutes the de facto border, which state cartography ignores. International maps showing otherwise are stamped with the statuary comment, "The external boundaries of India as depicted on this map are neither correct nor authentic." State cartography has typically resorted thus to words to back up the authority of the line. Barefoot cartography, by contrast, ingeniously places Mother India's head on Kashmir, the halo and the crown claiming all of the territory without question for the cause of India (see, for example, fig. 7.12). In "solving" thus the problem of Kashmir somatically, such prints also reveal that symbolically the loss of Kashmir would amount to a decapitation of Mother India herself and is worth sacrificing one's own life and limb to prevent. Anthropologist Christopher Pinney has observed that "the trauma of Partition was never visually represented by the commercial picture production industry" of India. I amend this argument to propose instead that the trauma is countered by at least some sections of this industry by an aggressive countercartography in which the new lines and boundaries that carved the body of Mother India into bits and pieces on statist maps are dissolved, and her wholeness is restored to her in the placement of her limbs, the swirl of her sari, the flow of her hair, and so on.[87]

ART IN THE AFTERMATH[88]

All I desired was to walk upon such an earth that had no maps.[89]

In contrast to these proliferating mass-produced images in which the map of India puts in an "artful" appearance in the aftermath of the drawing of the Radcliffe Line, gallery artists in both India and Pakistan seem to have largely ignored the maps of their new nation-states until very recently.[90] For instance, India's most well-known and famous modernist, Maqbool Fida Husain—a Muslim by birth who lived through Partition—seems to have adopted an off-hand attitude toward borders and boundaries: "For me, India's humanity is what is important, not its borders." In his autobiography he writes, "Man may create borders and LOCs [lines of control], but the sky remains vast and free." Although from the mid-1970s until the eve of his death in 2011, the map of India puts in an important appearance in his works—the fraught borders with neighboring nation-states undone by the placement of Mother India's body—his seeming casualness about the all-important boundary, whose drawing compelled him and his family to make a painful choice in August 1947, is striking.[91]

In Pakistan also, as Virginia Whiles observed as recently as 2010, artistic responses to the Partition were rare, as she quotes from an interview with the artist Ahsen Jamal, "There's so much confusion still due to Partition and because of the division on religious grounds . . . should it have been a secular state? Should the divisions have been on [a] geographical basis, was the alternative a system of federated states? I just don't know and it's all still so emotional."[92] And as in India, the map of the nation has not figured in artworks until recently, although the reasons for its absence might be different. As art historian Iftikhar Dadi speculates, "The map of Pakistan is not primordial" for its artists; it was after all a newfangled entity drawn up in 1947. "Moreover, the loss of East Pakistan in 1971 and the Kashmir dispute continue to haunt the map of Pakistan."[93]

One artist on the Indian side of the new border who did address the trauma of 1947 early on is Satish Gujral, whose compelling Partition Series (1950–57) elicited the following comment from one of the foremost critics of his times, John Berger, when he saw some of the works on display in London in 1955:

> If one stands in the middle of the gallery, one looks around at figure after figure wrapped in cloaks and draperies that hang in a rhythm of an undulating wail; and against these are sharp, jagged shapes of red, orange, and bruised purple, and hard green. When the draperies are white they seem to give relief to the flaring furnace of color around them—the whites of hospital sheets; but then this relief is immediately counteracted by their shape—they become hospital sheets torn into shreds. Bodies are huddled on the ground. Mouths are loose with cries. Even in the straightforward portraits there is a rawness that precludes all ease. Gujral paints in the wake, as it were, of a heavy terrible blade. To believe him one has to risk blunting the imagination which our security has allowed us to refine.[94]

Gujral indeed painted in the wake of a heavy terrible blade, but it has to be noted that none of these amazing works incorporate the dividing line that splintered a former shared homeland, and that compelled the artist's family to undertake a painful migration from their beloved Lahore into the new independent India.

Across the border in Pakistan and around the same time, Ustad Allah Buksh, who had formerly made a living in Bombay as a commercial artist, painted two canvases, *Anthropomorphic Landscape I* and *Anthropomorphic Landscape II: Partition*, neither of which, despite their suggestive titles, shows the contentious border that divided two landscapes and the peoples who lived on them.[95] Like Gujral, Allah Buksh evaded the reality of the Radcliffe Line in these canvases, although in 1947 its very drawing completely transformed his life and made impossible a return to his former homeland.

Beginning in late 1969 and for a few years, the Bombay-based Gujarati artist Tyeb Mehta produced a number of paintings as part of his so-called Diagonal Series that incorporate a jagged line that ran from the upper right to the lower left of the canvas, fragmented bodies, limbs, and torsos of humans and animals torn asunder scattered on either side (fig. 7.15). It has been argued that the trauma attending the Partition remained "the structuring element" of

FIGURE 7.15. Tyeb Mehta, *Diagonal*, 1969. Oil on canvas, 176 × 264 cm. Image courtesy of Vadehra Art Gallery, New Delhi.

this Muslim artist's entire productive career.[96] In the words of art critic Ranjit Hoskote, "As the archetypal sign of scission, the diagonal is the most prominent expression of the psychology of the schism that has haunted Tyeb. . . . The echo of the Partition resides in this slashing arbitrary gesture that changes space, memory, and the future forever."[97] All the same, although the diagonal in these powerful paintings visually "echoes" the Radcliffe Line, it is striking that, at least in his published interviews over the years, the artist himself abstained from overtly connecting it back to the violent border-making acts and activities of 1947, and instead preferred to trace its inspiration to his important encounter in 1968 with the US American artist Barnett Newman's "zip" paintings at the Museum of Modern Art in New York.[98]

CONTOURS OF YEARNING, FORMS OF MOURNING

The border is the line where there is nothing to see.[99]

It appears that we have had to wait at least half a century after Partition for the modernist imagination in the subcontinent to turn explicitly cartographic, almost as if the artists on either side of the border were waiting for a momentous occasion, such as the fiftieth anniversary of decolonization, to contend head-on with what I am calling the imperialism of the line underwritten by statist cartography to generate alternate contours of mourning and forms of yearning. Many such works have been recently brought together in an ongoing project titled *Lines of Control*.[100] In the words of the cocurator, Hammad Nasar, "The term 'Line of Control,' or its three-letter acronym LOC, is embedded in the vernacular language and imagination of the subcontinent, referring to unfinished cartographic business in the disputed border region of Kashmir. More generally it refers to the messy legacy of decolonization."[101] The project itself was initiated at a symposium held at the Royal Geographical Society (RGS) in London in August of 2007 to mark the sixtieth anniversary of decolonization, Partition, and the birth of two new nation-states. The exquisite irony here is hard to miss, given what we know of the RGS's implication in colonial enterprises over the course of the nineteenth century, and in the territorial and cartographic division of the world under British imperial rule. Equally ironically, in the absence of any sort of official or public memorials on either side of these fraught and contested borders, the Radcliffe Line itself has come to serve as a memorial of sorts that the creative imaginaries of the subcontinent have made their own.[102] As Amitava Kumar wrote after a recent visit to Wagah (the only point at which citizens of India and Pakistan can officially cross by foot or road over to the other side), "At the white line that divides the two countries, it is impossible not to think of Radcliffe."[103]

So how does this line come to figure in the art practices of a new generation of artists—many of whom have not experienced the trauma of 1947, at least directly? Consider, for instance, the Lahore-based Pakistani artist Farida Batool's holographic lenticular print mounted on board, *Line of Control* (2004).[104] The work shows two nude bodies pressed up against each other, from torso to thigh. Commenting on this work, Nasar writes, "The image's focal point is the line where the bodies meet. Shift in any direction and the illusory bodies move with you, but the line stays almost static."[105] The bodies locked thus in carnal embrace are brown; the line between them, a jagged dark crease. The erotic charge that animates, literally, this lenticular print is also a reminder that the border that now separates India and Pakistan as geopolitical entities also inhibits sexual ties and intimacy between the citizens of both countries. On the other hand, "Batool's portrayal of brown-toned, naked bodies in amorous embrace suggests the potential for human acts of defiance, the ability to resist the colonial lines of division."[106]

Another striking work in the *Lines of Control* exhibit is revealingly titled *River/Disease* (1999; reconfigured in 2009). It is the work of the New Delhi–

based Anita Dube (fig. 7.16). This multimedia work reminds us that for some, including Radcliffe in 1947, "water was the key."[107] In its later reiteration, this work reproduces the cartography of the Indus system, which as well was partitioned by the drawing of the Radcliffe Line, despite Sir Cyril's keenness to maintain the system intact (although the artist curiously occludes the Beas, the only river whose flow is limited to India). The "rivers" themselves are made up of hundreds of mass-produced ceramic "eyes" of various sizes. These eyes typically adorn the faces of deities in temples, and given that rivers are generally viewed as divine (and feminine) in Hindu-Indic thought, it is perhaps fitting that the artist has anthropomorphized these flowing bodies of water thus.[108] Dube herself observes in an interview with art historians Yasodha Dalmia and Salima Hashmi, "The eyes are like people for me and this could speak of large migrations in history. It could be from Kosovo or the migration from Pakistan to India. The sheer vulnerability and the futility of these migrations is expressed by the eyes."[109] Curator Hammad Nasar adds, "The image of the river overflowing its banks, suggesting the madness of crowds, gains added poignancy from the religious origin of Dube's material, and tension over territorial water rights adds contemporary frisson to fraught historical context."[110] At the same time, more than fifty years after Partition, this work disavows the critical partitioning of these waters in 1947 and recuperates a time before the Radcliffe Line was drawn across the land through which these rivers have meandered since time immemorial. Eerily, this artwork in fact ends up resurrecting Radcliffe's avowed hope of maintaining the Indus system intact, a desire that was summarily rejected by the leaders of both sides.

The signature work of the *Lines of Control* project is a mixed-media installation titled *Bloodlines*. It is the cross-border collaboration of Nalini Malani and Iftikhar Dadi, begun first in 1997 on the fiftieth anniversary of decolonization and recently updated in 2009 (fig. 7.17). Malani was born in Karachi in what was then British India but is now based in Mumbai, where her family moved after Partition. Dadi's parents were born and raised in British India but moved to Karachi in 1947, which is where he was born and raised. In contrast to many artists before their time, neither of them has shied away from taking on the Radcliffe Line that divided up their former familial homes and dislocated their families. In consciously collaborating across this contentious border (which imposed restrictions on travel and face-to-face meetings), the artists present their work as "a protest against the present situation, yet also concerned with the urgent possibility of looking beyond."[111] Across sixteen panels, they use thousands of gold, blue, and crimson sequins to dramatically materialize the 1947 border that created two nation-states (and the subsequent 1971 renaming

FIGURE 7.16. Anita Dube, *River/Disease*, 1999, reconfigured 2009. Enamel, copper, and Blu-tack, 305 cm × variable width. Reproduced with permission from the artist.

of East Pakistan as Bangladesh), the red of Radcliffe's pen, and also the red of the blood that has flowed across this terrain since that foundational act. In Dadi's words, "*Bloodlines* is a Martian landscape, mapped with detached scientific objectivity by the Radcliffe Commission, an arbitrary line of demarcation soaked with blood."[112] By resignifying the color red thus, Malani and Dadi's work reminds us of cartography's implication in the bloody violence that followed the drawing of the Radcliffe Line.

In contrast to its appearance as bloody and red, the line appears in Zarina (Hashmi)'s woodcut print *Dividing Line* (2001), not as a firm red stroke of a pen but as a bleak black jag searing across a cream-colored handmade Indian paper, a work of "staggering economy" (fig. 7.18).[113] Born in Aligarh in British India in 1937, Zarina studied printmaking in Paris in the 1950s and learned how to work with woodblocks in Japan in 1974. Of the many contemporary artists discussed in this essay, Zarina (who has lived in New York for the past several decades) is arguably the most self-conscious in her use of maps and cartographic lines in the body of her work: "Making maps was a natural consequence for the life of a traveler. When maps were not available, I would draw my own from books at the library. Maps also became a necessity to chart my route and find my destination." She also recalls, "Studying maps, I became aware of borders. The first border I drew was the border between India and Pakistan, the dividing line that split families, homes and the fabric of life of millions of people."[114]

One of Zarina's important interlocutors, the art historian Mary-Ann Lutzker-Milford, observes, "The harshness of the dense black cut strokes [in *Dividing Line*] against the slivered textured surface lends a certain ominous tone of apprehension to the print. It is innocent in its aesthetic abstraction yet offers potent reminders of divisive political decisions."[115] She has also likened "its convoluted, wayward passage" to "an umbilical cord that floats, waiting to be claimed and buried, in order once again to bring wholeness to the subcontinent."[116] Speaking to Ranu Samantrai in 2001–2, Zarina referred to many living on both sides of the border "who feel that the dividing line goes through the heart."[117] Contrarily, in an interview that the artist gave after *Dividing Line* was exhibited in Mumbai in 2004, she observed, "The line is just in everyone's head. . . . Our generation has come to peace with it a long time ago."[118] And yet the hard—if ragged—firmness of the line in her work (carved and gouged, rather than drawn) is a reminder that the Radcliffe Commissions' hasty handiwork from more than sixty years ago has an enduring materiality and affective consequences that are (possibly) impossible to overcome. At the same time, as Ranu Samantrai perceptively observes, "the print does not name the two nations, or even explain that the divide in question is geo-political. Absent

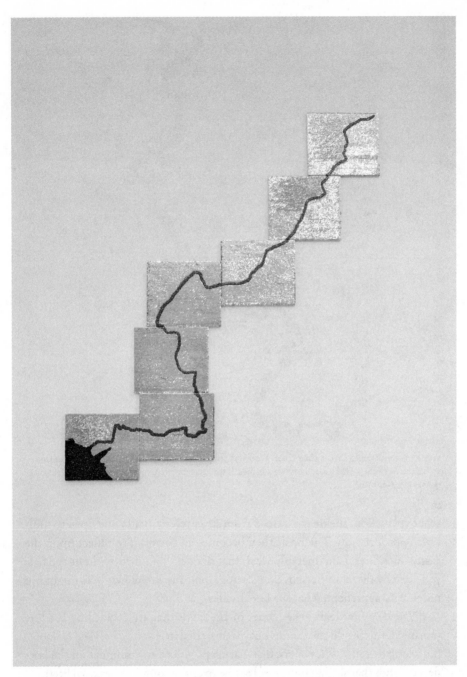

FIGURE 7.17. Iftikhar Dadi and Nalini Malani, *Bloodlines*, 1997/2009 (detail). Mixed media (sequins and thread on cloth). Reproduced with permission from the artists.

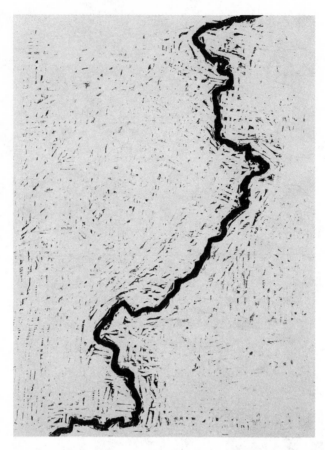

FIGURE 7.18. Zarina, *Dividing Line*, 2001. Woodcut printed in black on handmade Indian paper from an edition of twenty. Sheet size: 25.5 × 19.5 inches. Reproduced with permission of artist and Luhring Augustine, New York.

that explanation, the image is freed from strict referentiality and open to multiple appropriation. The print then becomes an occasion to reflect upon the many lines, literal and metaphorical, that divide."[119] In fact, without the title provided by the artist herself, even cartographically literate viewers might not necessarily apprehend what the line divides.

If *Dividing Line* leaves the names of the lands that are divided by the line unnamed, this productive anonymity is unsettled in *Atlas of My World IV* (2001) (fig. 7.19), one in a series of six that Zarina produced to comment on the borders of lands that she has journeyed across over the course of her own lifetime. In this print, *Dividing Line*'s jagged line separates the countries labeled (in Urdu) as Hindustan (India) and Pakistan, even as the line "appears to float above the physical map extending beyond the edges of the print."[120] In "A Conversation

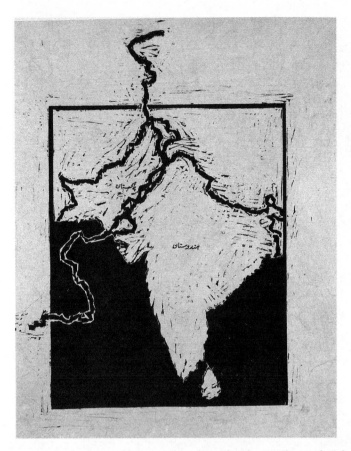

FIGURE 7.19. Zarina, *Atlas of My World IV*, 2001. From a portfolio of six woodcuts with Urdu text printed in black on handmade Indian paper, mounted on Archies cover white paper. Sheet size: 25.5 × 19.5 inches. Reproduced with permission of artist and Luhring Augustine, New York.

with My Self," Zarina writes, "I have often been questioned about the map I used to draw the border. Perhaps I distributed territory incorrectly. I didn't have to look at the map; *that line is drawn on my heart.*"[121] Once again, the affective erupts to leaven the geopolitical imperatives of state and scientific cartography.

I end this discussion by considering the work of an artist whose work is not included in existing iterations of the *Lines of Control* project, Nilima Sheikh's luminous *Firdaus IV: Farewell* (2004), which I read as another reminder that the nation's yearning for form—itself a highly fraught process—can and does get entangled with the individual citizen-artists's yearnings for particular places (fig. 7.20).[122] *Farewell* is part of a serial project that Sheikh has been involved with since 2002 on the troubled "territory of desire" that is Kashmir.[123] In ana-

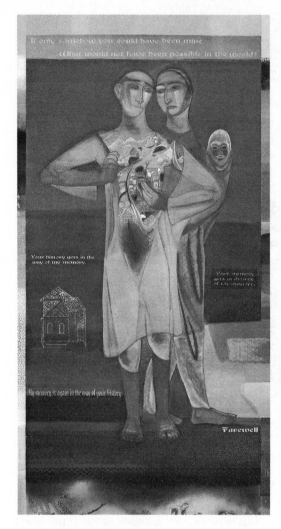

If only somehow you could have been mine...
What would not have been possible in the world?

Your history gets in the way of my memory.

Your memory gets in the way of my memory.

My memory is again in the way of your history.

Farewell

FIGURE 7.20. Nilima Sheikh, *Firdaus IV: Farewell*, 2004 (detail). Painted and stenciled on both sides of hanging canvas scroll in casein tempera, 25.4 (h) × 15.24 cm (w). Reproduced with permission from the artist.

lyzing this series, which she reads as producing "new maps of longing," the postcolonial critic Ananya Kabir asks, "How does one leave the map, when on the other side seems only yet another map of longing?"[124] When Radcliffe was charged with dividing up the subcontinent, the kingdom of Kashmir fell outside his mandate and hence outside the immediate reach of his line. Nonetheless, Kashmir did not escape the consequences of 1947, and indeed, as I have already observed, it remains a flashpoint in statist cartography on both sides of the border. I have also flagged the manner in which barefoot cartography in India attempts to somatically solve the "problem" of Kashmir on paper by placing Bharat Mata's head in such a manner as to occlude realities on the ground and claim the territory for India.

Sheikh offers a different response. In *Farewell*, the artist has engaged with the verses from the poem with the same title by the late expatriate Kashmiri poet Agha Shahid Ali, a poem (in English) that is "at one—but only one—level . . . a plaintive love letter from a Kashmiri Muslim to a Kashmiri Pandit (the indigenous Hindus of Kashmir are called Pandits)."[125] Stenciled across Sheikh's scroll are phrases from Ali's haunting poem, such as "They make a desolation and call it peace"; "Your history gets in the way of my memory"; "My memory is again in the way of your history"; and, most poignantly of all, across the top of the central panel, "If only somehow you could have been mine, what would not have been possible in the world?" These words appear to be articulated by the central figure in the panel—who in the artist's imagination stands in for the poet himself—who is shown gently opening his robe (the Kashmiri "pheran") to reveal a beauteous pastoral landscape, possibly tattooed or painted on the surface of his skin.[126] Like Kabir, I cannot resist seeing the peninsular form of the subcontinent in this act of gentle revelation. Is this central figure suggesting that the Muslim citizen is just as attached to the mapped form of India as his addressee, the Kashmiri Pandit (and by extension all Hindu Indians)?[127]

To a Hindu religious viewer, this act of revelation might recall a widely known image, popularized by twentieth-century mass-produced lithographs, of the monkey-god Hanuman, who tears open his chest to reveal his devotion to the Lord Rama and his wife Sita, permanently installed in his heart, as it were. Patriotic bazaar artists appropriated this religious imagery and "nationalized" it from the 1930s, and in their artworks, nationalists who tear open their chests reveal the torsos of martyrs who gave up their lives for the cause of the nation; in the 1960s, H. R. Raja, a Muslim artist who produced widely for the mass-market, painted a dramatic image of one of these martyrs, who tears open his chest to show that Mother India herself has taken up residence there.[128] In contrast, though, Sheikh's scroll eschews such violent imagery, and instead—like the words scrolled across its surface—suggests a spirit of reconciliation, a painted "love letter" from one estranged citizen to another. To recall Aga Shahid Ali's words, "If only somehow you could have been mine, what would not have been possible in the world?"

At another level, in the peninsular form of the nation painted on the (estranged) citizen's body, Sheikh appears to reject the imperialism of the cartographic line that has become such a part of South Asian geopolitics since British surveys of India began in the late eighteenth century, and since independence and the fraught birth of two nation-states in 1947. Instead, she opts for a persuasive contour, as well as a suggestion that these lands that have become so contested well have the hope to become paradise again, pastoral and peaceful.[129] In the gentle and delicate opening of the robe to reveal a painted peninsular form, *Farewell* also is at a radical remove from the image with which I began this essay, Artzybasheff's *India: Liberty and Death* (fig. 7.1), in which the female breast is subject to violent self-mutilation. Freedom can indeed coexist with peace—and not with violence and death—possibly if we give up our enslavement to lines that have divided and separated us. Thus, elaborately inscribed on the back of the scroll are the following words (in English):

> That failure of the subconscious was the border. The line of control did not run through 576 kilometres of militarised mountains. *It ran through our souls, our hearts, and our minds.* It ran through everything a Kashmiri, an Indian, and a Pakistani said, wrote, and did. It ran through the fingers of editors writing newspaper and magazine editorials, it ran through the eyes of reporters, it ran through the reels of Bollywood coming to life in dark theatres, it ran through conversations in coffee shops and TV screens showing cricket matches, it ran through families and dinner talk, it ran through the whispers of lovers. And it ran through our grief, our anger, our tears, and our silences.

> The buses carrying the passengers from Muzaffarabad traveled under a drizzling grey sky to Srinagar. It is a road that has been deserted after dusk for a decade and a half. I watched thousands of women, men, and children stand along the much soldiered road, waving hands and umbrellas, welcoming the ones who had stepped across the line. There was no fear that evening. There were only hands reaching out of the bus windows, waving in the air, *as if each wave would erase the lines of control.* I raised my hand and waved.

> Curfewed Night

> Basharat Peer[130]

Ananya Kabir has argued that the artist is the most effective vigilante of national desire, and that it is the artwork that compels viewers to revisit traumas, realize silences, and empathize with suffering.[131] It is also the case that a large majority of the artworks that I have discussed in this closing section, and that are part of the *Lines of Control* project, are by female artists, many of them among the most prominent of their times and contexts.[132] These works thus alert us to the work of gender in cartographic practice, a work that has been largely ignored by scholars and historians of cartography until fairly recently. Historically up until this day, mapmaking has been overwhelmingly a masculinist enterprise, imposing upon our earth the dictatorship of the cartographic line, transforming unbounded space, quite often violently and with the force of arms, into a geocoded world of which we are then rendered subjects, as John Pickles reminds us in the epigraph that I invoke at the start of this essay. Many years ago, Brian Harley noted that because everywhere in the modern world the state has been the principal patron of cartographic activity, "maps are preeminently a language of power, not of protest."[133] Given the nexus between state power and the exercise of militaristic masculinity in scientific mapmaking, it is perhaps not a surprise that it falls upon the female artist—operating outside the realms of science and state, and armed only with her tools of trade and her creative spirit—to crystallize so sharply and consistently an alternative vision (of desire, of yearning and mourning, but also of protest) that encourages citizen-viewers—and others—to free themselves from the hegemony of map-made lands, even teaching us to become enchanted with other forms, other contours, other lines.

ACKNOWLEDGMENTS

With gratitude to Lucy Chester, Iftikhar Dadi, Venkat Dhulipala, Rich Freeman, and David Gilmartin for their comments on earlier drafts of this paper; to Iftikhar Dadi, Zarina Hashmi and Nilima Sheikh for detailed conversations regarding their artworks discussed here; to Alyssa Ayres for responding to queries regarding Rahmat Ali's maps; and to audiences at various venues where I have presented earlier drafts of this paper: the Newberry Library, North Carolina State University, Cornell, Duke, the University of Michigan, and the University of Vermont. Not least, very many thanks to Jim Akerman for his kind invitation to participate in the Nebenzahl Lecture series, and for his support of this project. A much-abbreviated version of this essay appears in Ramaswamy 2011.

NOTES

1. Pickles 2004, 3.

2. For the most detailed analysis to date of the Radcliffe Award as it affected Punjab, see Chester 2009. For the drawing of the new border in Bengal, see Chatterji 1999; 2007; 19–60; and

van Schendel 2005. Although the term Radcliffe Line is used loosely to designate the entirety of the India-Pakistan boundary, the actual line "awarded" by the commission in Punjab "ran from the border of the Kashmir state in the north to the border of [the princely state] Bahawalpur. . . . Further south, the existing boundary between the province of Sindh and the region of Rajputana became the international boundary" (Chester 2009, 109). In Bengal, as van Schendel notes, "the new border was anything but a straight line; it snaked through the countryside in a wacky zigzag pattern" (van Schendel 2005, 54).

3. I borrow here from Pickles 2004, 17–18. Drawing on the writings of the nineteenth-century German philosopher Johann Fichte, Martin Brückner writes, "The nation-state always began and ended with the cartographic line demarcating political boundaries. . . . With the strike of a map-maker's pen, the line became a cognitive and disciplinary tool transposing local into national identities" (Brückner 2006, 119). For a perceptive analysis of the inexorable shift from "zonal frontiers" to "linear boundaries" that has accompanied the emergence of the modern nation-state system, with a specific emphasis on British India, see Embree 1989.

4. Pickles 2004, 5–9.

5. Harley 2001, 62.

6. Ghosh 1988, 256–57.

7. For key overviews, see especially Brasted and Bridge 1994; Gilmartin 1998; Tan and Kudaisya 2000; and Gilmartin 2015. These important reviews do not focus on cartographic issues at stake in this vital boundary-making enterprise.

8. I borrow the phrase "line of power" from Gunnar Olson (Pickles 2004, 4).

9. For the concept of the "map-minded," see Harvey 1993, 17.

10. Attributed to Italian sculptor Gilberto Zorio (Flood and Morris 2001, 327).

11. Taussig 1993. I thank Robert del Bonta for alerting me to this image.

12. The notional boundary was the working line used by the commissions as specified in the Indian Independence Act of July 18, 1947. The inclusion of Kashmir in the map of India is remarkable in itself, given that at the time of the creation of this image its fate was quite unclear.

13. The published scholarship on this artist is very limited, and my comments are based on research in Special Collections, Syracuse University Library (which houses his papers), and communications in February 2012 with and from Bill Hooper, Archivist, Time, Inc. to whom I am very grateful for his time and thoughts.

14. http://www.time.com/time/magazine/article/0,9171,854810,00.html (last accessed on March 12, 2011). Time Inc. archival papers that I was given access to unfortunately do not contain any specific information on the commissioning of Artzybasheff's print.

15. The magazine printed some readers' responses to this cover story in its issue dated November 17, 1947. One reader from Ohio praised Time for "this masterpiece of reporting," which enabled "the people of America [to] certainly have a better understanding of the age-old problems facing present-day India." Another respondent from Los Angeles declared the article "a classic," "impressive from every standpoint, literary, philosophical, humanitarian." Only one reader from Brooklyn—with the South Asian name of Sondhi—criticized the magazine for its "superficial one-sided analysis of the blood-soaked scene."

16. Ramaswamy 2010a, 108–10.

17. Artzybasheff's image might well speak to a dominant perception until the 1980s among those scholars who wrote largely within the framework of imperial historiography "to attribute the transfer of power to India to circumstantial forces outside of Britain's control" (Brasted and

Bridge 1994, 99). Lucy Chester also reminds us that Mohandas ("Mahatma") Gandhi had been pushing for an unconditional withdrawal by the British, "with uncompromising calls to leave India to God. If that is too much, then leave her to anarchy" (Chester 2009, 16).

18. I thank Lucy Chester for drawing my attention to this print.

19. I thank Lee Schlesinger for help with translating the Marathi title. The Marathi iteration of this image has been reproduced and discussed (albeit without the connection to Chaliapin's work) in Kaur 2003, 239–42.

20. Ramaswamy 2010a, 231–33.

21. Chatterji 1999, 185–86, 242.

22. Quoted in Devji 2013, 26–27. Borrowing the terminology from Gandhi and others, the Bengali newsmagazine *Millat* on April 11, 1947, likened the Congress party in that province to the matricidal maniac Parashuram (a revered figure in Hindu mythology) out to "slice Mother into two" (quoted in Bose and Jalal 1997, 184). To this day, Hindu nationalist discourse in India continues to invoke the trope of vivisection.

23. Quoted by Krishna 1996, 195. It is not Indians alone who used such somatically charged metaphors. Soon after the Radcliffe Award was announced, the Pakistani newspaper *Dawn* declared that "territorial murder" had been committed (Chester 2009, 110).

24. Ramaswamy 2010a, esp. chap. 1.

25. Cyril Radcliffe, quoted in Chester 2009, 182.

26. I am indebted to David Gilmartin for his discussion of this point with me. "If religion was the principle for division (which it was), then how could reasoned lines ever have been drawn? And of course, they weren't " (email communication, July 27, 2011). See also Chester 2009, 99.

27. I borrow the phrase "cartographic war" from Lal 2012, 74.

28. To be fair, this is also because of the immense difficulties faced by scholars in getting access to official cartographic data more generally for the period. As Ranabir Samaddar writes, "Maps are a barred subject" (quoted in van Schendel 2005, 13).

29. By the time he published these lines fairly late in his life and career, Auden, known for his leftist poetry, had moved to the United States, given up his British citizenship, and also become highly skeptical about Cold War politics especially as these impinged on decolonization movements. "Auden's poem appeared at a political moment when the question of international partitions resurfaced in the public sphere (for example, the partition of North and South Vietnam in 1954, and the erection of the Berlin Wall in 1961). By highlighting the restiveness of the final days, 'Partition' punctures the façade of order that the departing British empire struggled to present to the world " (J. Menon 2012, 29).

30. Only Jinnah invoked Pakistan at the time of this historic announcement (Khan 2007, 3). On the "talismanic" power of the word "Pakistan," see ibid., 44.

31. Sherwani 1969, 235–36.

32. The confusion over the new boundaries is poignantly reflected in what is perhaps the most well known of Urdu short stories centered on this foundational event, Saadat Hasan Manto's "Toba Tek Singh," first published in Urdu in 1955 (http://www.columbia.edu/itc/mealac /pritchett/oourdu/tobateksingh/translation.html [accessed on April 7, 2011]). Among other things, the story reflects "the fundamental absurdity of maps and nations" (Kumar 2001, 48).

33. Khilnani 1997, 7.

34. This is information based on accounts by journalists Leonard Mosley and Dominique

Lapierre (Chester 2009, 76). Chester's detailed research does not confirm this account. In fact, she suggests that his wartime work with the government would have provided Radcliffe with considerable background knowledge on Indian politics (ibid., 44–49).

35. Chatterji 1999, 186–87.

36. For brief biographies of the men who served on these committees, see Chester 2009, 56–58; and Chatterji 2007, 25. In Punjab, one concerned group hired the Australian geographer Oskar Spate to help make its claims, and in Bengal, the noted Bengali geographer S. P. Chatterjee was similarly recruited.

37. Arguing that politics and other imperatives took precedence in the border-making act of 1947, Chester notes that the cartographic information available to Radcliffe was not "significantly out of date" and that maps have been a "convenient scapegoat" (Chester 2009, 99).

38. Quoted in Chester 2009, 85.

39. Quoted in Chester 2009, 86. Similarly, on the problem of using Bengal's "volatile" rivers as natural borders to which Radcliffe had not given "any thought," see Chatterji 1999, 221–24.

40. Khan 2007, 106; see also 88. Chester reminds us of the persistence into the postcolonial period of existing administrative divisions (at the *zillah* and *tehsil* level, and even village boundaries) created by two hundred years of colonial rule in the commissions' line-making work (Chester 2009, 78–79). On the other hand, the new "lines of power" did destroy the integrity of two provinces, Punjab and Bengal, also mostly products of British rule.

41. For an alternate reading that suggests that elite Indians were indeed using maps from as early as the 1850s to settle property disputes, see Bayly 1996, 161, 314.

42. Quoted in Khan 2007, 104; emphasis mine.

43. Quoted in Chester 2009, 119. In an interview with Lucy Chester in February 2000, Beaumont vehemently denied such accusations.

44. Quoted in Chester 2009, 136–37.

45. Quoted in Chester 2009, 66. See also Leonard Mosley's comment that the delegations that showed up before the commissions "arrived armed with their maps" (ibid., 84).

46. Chester 2009, 66. Their utility, given that they may not have been all that up to date, is also worth noting (ibid., 88). For difficulty of access in Bengal to maps and even up-to-date census data, see Chatterji 2007, 29–30.

47. Chester 2009, 66–69; Chatterji 1999, 198–204, 207–10; Chatterji 2007, 31–32, 37–38, 42–43, 46–48; and Dhulipala 2015, 158, 194, 337.

48. Chester notes, for instance, that the maps of the new countries printed in newspapers in Pakistan on August 19, 1947, were largely illegible (Chester 2009, 135). Similarly, Yasmine Khan draws attention to the hastily sketched and hurriedly distributed maps through which vast numbers learned of the contours of new homelands (Khan 2007, 3, 97, 125).

49. Ramaswamy 2010a, 232. This image resonates with a cartoon published a few weeks earlier on July 9, 1947, in the British-owned *Pioneer* newspaper, which showed Mother India (encased in a coffin) being sawed with a giant blade wielded by Jinnah and Nehru, while an Englishman looks on and laments, "I only hope nothing goes wrong Madam."

50. Chester 2009, 128. See also Ata-ur-Rehman 1998, 212; and Khan 2007, 124. The map published in Ata-ur-Rehman's work is particularly revealing. This map, whose provenance is not known, is captioned with the following words: "TO EVERY MUSSALMAN MAN WOMAN & CHILD WHO FOUGHT SUFFERED & WON THE FIRST BATTLE FOR PAKISTAN THROUGH THE PUNJAB MUSLIM LEAGUE 1947." Notably, although celebrating "the battle for Pakistan," the

map shows an undivided India. For a fantastic account of an attempt in April 1946 by supporters of the idea of Pakistan to slash at a map of undivided India during the screening of a film titled *Forty Crores* in Bombay, see Dhulipala 2015, 22.

51. This was particularly true in Hindu nationalist publications. For example, the title page of *We or Our Nationhood Defined*, originally published in 1939 and then reprinted (as a 4th edition) in 1947 shows a map of undivided India, faint white lines in the west and in the east hinting that all was not well (Golwalkar [1939] 1947). For the continued use of maps of undivided India in Hindu nationalist symbolic activities into our times, see especially Lal 2012, 72–73.

52. Jalal 2000, 422. This important study has a useful discussion of the various attempts or "schemes" from the 1930s to territorially imagine "Pakistan" but unfortunately does not analyze the role maps played in such imaginations (ibid., 327–34; 386–409).

53. Jalal 2000, 327. Jalal thus writes, the primary defect of the spate of schemes that began to be articulated especially from the late 1930s "was in identifying areas over which territorial sovereignty could realistically be asserted" (388). In his recent work, Venkat Dhulipala has, however, quite convincingly demonstrated that in the immediate years leading up to Partition, both supporters and opponents began to concretize—with the help of the map—the idea of Pakistan (Dhulipala 2015).

54. Pirzada 1995, 220. For Jinnah's speech in Lahore on the eve of the adoption of the resolution that invokes "the British map of India," see ibid., 212. For Jinnah's assertions that demonstrate a well-developed geographic and cartographic sensibility, see Dhulipala 2015, 179-182, 227-229.

55. Gilmartin 1998, 1081. See also Jalal 2000, 397–98.

56. Devji 2013, 27–28.

57. This four-page leaflet has been reprinted in Aziz 1978, 1–10; bold emphasis in original. This appeal (probably precipitated by the deliberations of the recently concluded Round Table Conferences in London, in which the future of India had been debated) was cosigned by three others although authored by Rahmat Ali. For the background to the letter, see ibid., xvii–xix. In the original text, Rahmat Ali used the word "Pakstan," and thereafter changed it to "Pakistan" (ibid., xxiii). For a recent analysis of Rahmat Ali's spatial and cartographic ideas, see Ayres 2009, 25–27, 105–23. Ayres notes that Rahmat Ali's doctoral dissertation (from the University of Paris, titled "Contribution a l'Etude du Conflict Hindou-Musulman" and dated to 1933, did not include maps (email communication, April 24, 2011).

58. British Library, India Office Records, L/P&J/8/689, fols. 494–95. The letter begins, "Dear Sir, May I venture to address this appeal to you on behalf of the people of Pakistan at this critical hour, when Parliament is giving final shape to the Government of India Bill, for your valued sympathy and support in our fateful struggle against the ruthless coercion of PAKISTAN into the proposed Indian Federation." At the top of the first page of the document, inscribed in pencil are the following words: Secretary, P & J (S) Dept: "Are you interested in this great movement?" and a response, "I fear I am not excited" (signature unfortunately undecipherable). I thank Leena Mitford for this reference. This letter has been reprinted in Aziz 1978, 23–27, without the penciled comments or the header image.

59. Ayres 2009, 25. Ayres draws our attention to the circulation in Lahore in 1939 of one of Rahmat Ali's maps of Pakistan on the letterhead of Majlis-e-Kabir Pakistan, and notes that his "Bang-i-Islam" presciently anticipates the 1971 creation of Bangladesh (ibid., 25; 119).

60. Citizenship in Dinia would be extended to all those who were united against "the men-

ace of Indianism," itself characterized as based on upper-caste Hinduism and backed by British power.

61. Rahmat Ali 1947, 173. When he finally returned to Pakistan in 1948 on the eve of his death, Rahmat Ali was reportedly heartbroken over the partition of Punjab and Bengal, a "betrayal" he partly attributed "to the machinations of Radcliffe" (Aziz 1978, lvi).

62. Thongchai 1996, 76.

63. Khan 2007, 43. For examples from earlier years of the deployment of maps of Pakistan in public meetings and fora, even Moharram processions, see Dhulipala 2015, 231, 257, 445.

64. For a popular print labeled by Akbar Ahmed as "Map of a projected Pakistan in the mid-1940s which includes Punjab and Bengal as full provinces in Pakistan," see Ahmed 1997, 112. In this print (for which unfortunately Ahmed does not furnish further details), a flag bearing a crescent moon and star flies over the territory marked "Pakistan" (in English) on the map of an unnamed India, while the face of a youthful-looking Jinnah is placed in the skies above the map. It is important to note that the words "Calcutta" and "Zindabad" (Victory)—rather than "Pakistan"—are inscribed on the territory known as Bengal, introducing an element of ambiguity on the latter's status. Such "aspirational" maps were drawn not just by Muslim nationalists, but also men like B. R. Ambedkar, the future primary drafter of the Constitution for Independent India, and the Gandhian politician C. Rajagopalachari (Dhulipala 2015).

65. Jalal valuably observes that schemes that were developed in Muslim-majority regions were more likely to have a more clearly articulated secessionist agenda than others (Jalal 2000, 394–97).

66. Interestingly, the "Hindu zones" are identified according to languages spoken: Tamil, Malayalam, Canarese, Maharati, Andhra, Oriya, Bengali, Hindi, Gujrati, Rajastani, Punjabi. Thanks to Leena Mitford at the British Library, who alerted me to this publication (India Office Records, L/P&J/8/689, fols. 422–27). On Latif, see Pirzada 1995, 154–59.

67. The map was also published in a book titled *Confederacy of India*, published in Lahore in 1939. Mamdot, an influential landlord, was the president of the provincial Muslim League party in Punjab (Gilmartin 1988, 182). I thank David Gilmartin for alerting me to the map and discussing it with me.

68. The 1939 map was printed by the Himayat-i-Islam Press in Lahore in a text titled *Khilafat-i-Pakistan Scheme*, written by Mohammad Abdus Sattar Niazi, a leader of the Pakistan Muslim Students Federation (Jalal 2000, 475). Jalal does not mention the map, but see Khurshid 1977, 119–12. Niazi was also coauthor of *Pakistan Kya He Aur Kaise Banega* (What is Pakistan and how will it be created), which reprinted this map on its cover in 1945 (Gilmartin 1988, 207–13). I thank David Gilmartin for bringing this map to my attention and discussing it with me.

69. Ata-ur-Rehman 1998, 103. I have been unable to find out more about this map, including its provenance and date.

70. Quoted in Khan 2007, 45.

71. Jalal 2000, 400. See also Gilmartin 1998, 1081–83.

72. Quoted in Jacob 2006, xxv.

73. Ghosh 1988, 256–57.

74. Ramaswamy 2010a.

75. Anderson 1991, 175.

76. Pickles 2004, 60.

77. Covering about one-third of British India in area, and 562 in number with varying legal

arrangements and degrees of sovereignty and privileges, the autonomous princely states more or less lost hope by June 1947 of striking out on their own and had to choose between joining either India or Pakistan. By August 15, 1947, most had joined India, and those that had not (such as Junagadh, Hyderabad, and Kashmir) were compelled to do so over the course of the next few months, and subsequently either merged with the "provinces" of India over the course of the next couple of years, or formed autonomous "states unions." The classic "eyewitness" treatment of this process can be found in V. P. Menon 1956. For scholarly analyses, see especially Copland 1997, 229–87; Ramusack 2004, 245–74; and Guha 2007, 51–96. None of these studies, however, discuss the role that cartographic knowledge and maps obviously played in this complicated and contentious "endgame of empire." A document issued by the Government of India in 1951 proudly noted, "On the eve of Independence the map of India was studded with as many as 562 States. . . . These yellow patches on the map of India have now disappeared. Sovereignty and power have been transferred to the people. The edifice of new India has arisen on the foundation of the true patriotism of the Princes and the people" (Government of India 1951). For further discussion of *New India*, see http://tasveergharindia.net/cmsdesk/viewgallery.aspx?id=93&EId =116&ImageId=2.

78. Although the demand for internal reorganization of provinces conforming to linguistic (and ethnic) identity goes back to the 1920s, it was only in 1953 with the carving of Andhra Pradesh out of Madras and the passage of the States Reorganization Act in 1956 that this desire translated into a geo-political and cartographic reality. The creation of Maharashtra and Gujarat in 1960, and the reordering of Punjab in 1966, further altered the map of postcolonial India, a cartographic realignment that continues to this day (Guha 2007, 189–208). Despite the transformative importance of the mid-twentieth century "states reorganization," the role of cartographic knowledge in the process remains underdocumented.

79. Such maps also appear in the many languages of India. For examples in Hindi and Bengali, see figures 4–5 in http://tasveergharindia.net/cmsdesk/essay/116/index.html.

80. Kashmir, however, is claimed for India, as it is in many such productions.

81. This has been reproduced in Ramaswamy 2010a, 51.

82. The presence of the Indian national tricolor suggests that this is a map of independent India, separated from the territories colored deep green. We have to look hard, though, to find the word "Pakistan" inscribed across the green mass to the west, while the eastern section of the new nation is still inscribed with the word "Bengal" across it. Yet a clearly delineated green line separates the new nations of Pakistan and India as announced in mid-August 1947. The map in this print possibly alludes to the state of affairs in the months immediately following Partition and independence, when princely states such as Baluchistan, Khairpur, and Bahawalpur had yet to formally accede to Pakistan. Bahawalpur and Khairpur joined Pakistan in October 1947, and the various Baluchi states (Makran, Kharan, Las Bela and Kalat) did so in early 1948 (Wilcox 1963, 68–85).

83. For another example, see http://tasveergharindia.net/cmsdesk/essay/116/index.html, figure 8.

84. Zutshi 2004, 299–322.

85. Kabir 2009b, 8–9. For an argument regarding the "overwrought" cartography in Kashmir that has resulted in "a tangle of thick, thin and broken lines," see also Kabir 2009a.

86. Note that more than a decade after it was absorbed into India, the former princely state of Junagadh is still claimed as part of Pakistan (see also Khan 2007, 98).

87. Pinney 2004, 146. In amending this argument, I rethink as well my own argument in Ramaswamy 2010a, 341 n. 59.

88. I have adapted this phrase from Sinha 2006, 9–15.

89. Oondatje 1993, 261.

90. This disavowal is striking especially given that modernist art of the West is one of the reference points for these artists. For examples from Europe and the United States of such engagements with the national map form, see especially Cosgrove 2005; and Harmon 2009.

91. For further discussion, see Ramaswamy 2010b. On Husain's reticence—in words and works—about the Partition, see especially Kabir 2010.

92. Whiles 2010, 167, 241.

93. Dadi 2010, 31.

94. Sinha 2006, 39. Another admirer similarly draws attention to the anguished bodies painted on these canvases, "their faces contorted, their bodies writhing, and their eyes looking upwards uncomprehendingly. Blinded and with a pain beyond belief, each one of them appears inconsolable" (Malik 2002, 75–76). These works remain underanalyzed in the scholarship. Gujral himself wrote movingly of the impact of Partition on his own family in his autobiography, published in 1997.

95. I have not seen any sustained discussions of these works. Salima Hashmi, doyen of Pakistani arts scholarship, only notes that they "are unusual in their vision" (S. Hashmi 1997, 14). Akbar Naqvi is a little more forthcoming when he comments on their "necrophobic" content, and asks, "Was it the killing and suffering of millions of people that turned his art into stone? . . . We must leave the Ustad at his own inexplicable best. These paintings establish, however, that this art was invaded by a new power and occupied by thoughts and feelings intriguingly strange, to say the least. It has intruded into forbidden territory" (Naqvi 1998, 131–32).

96. Citron 2009, 69. The artist was witness to a brutal murder of a fellow Muslim on the streets of Bombay during the Partition riots that rocked that city in 1947.

97. Hoskote 2005, 18–19.

98. Citron 2009, 81–86.

99. Adapted from Mirzoeff 2011, 242.

100. Nasar 2007; and Dadi and Nasar 2011. In August 2013–February 2014, the show was also exhibited at the Nasher Museum of Art, Duke University.

101. Nasar 2007, 42.

102. The official passport and seal of Bangladesh bears the logo-form of the map, connecting citizenship in that country with Radcliffe's cartographic handiwork. Although the official flag of Bangladesh no longer includes its geo-body, "when Bangladeshi independence was declared in 1971, the new nation's flag showed a bottle green background with a red circle in the middle. In the circle was the yellow outline of the new country," also based more or less on the 1947 boundary (van Schendel 2005, 349–50). The use of the Radcliffe Line to give coherent form to the new nation is especially ironic given that "the Boundary Commission's territorial surgery of Bengal resulted not in the simple bisection that is usually imagined but in the creation of no less than 201 territorial units" (ibid., 43).

103. Kumar 2001, 47. As Chester notes, on Radcliffe's maps, the line was drawn "a quarter of an inch thick," which translates to a mile-wide line on the ground, far from the reality of the actual border (Chester 2009, 87). For more on the routine "performance of nationalism" at the

very point where the Radcliffe Line separates the thriving metropolises of Amritsar and Lahore, see J. Menon 2012.

104. Reproduced in Dadi and Nasar 2011, 140–41.

105. Nasar 2007, 42.

106. Nada Raza, quoted in Dadi and Nasar 2011, 140.

107. Chester 2009, 80. For a published account on how the riverine system of Punjab was affected by the Partition, see Michel 1967. I am very grateful to Susan Bean, David Gilmartin, Pika Ghosh, and Monica Juneja for discussing this complex work with me.

108. I thank Lee Schlesinger for this insight.

109. Dalmia and Hashmi 2007, 189.

110. Nasar 2007, 42–43.

111. Dadi 2000, 103.

112. Dadi 2000, 103. Amitava Kumar reminds us that at Wagah, it is a *white* line that marks the border separating the two countries, white arrows further pointing to the line, "as if you could miss it" (Kumar 2001, 47).

113. Mufti 2011, 94. There is now a fairly extensive scholarship on this fascinating artist, who signs off with her given name in professional contexts. See especially Samantrai 2004; Milford-Lutzker 2001, 2004, 2005; Patel 2007; and Milford-Lutzker 2013.

114. Z. Hashmi 2011. In an interview in 2007, in response to a question explicitly directed toward her view of Partition and why it took so long for her to visit that theme, she commented, "So the first line I did, *Dividing Line*, had been festering all these years" (Patel 2007, 77).

115. Milford-Lutzker 2005.

116. Milford-Lutzker 2001, 16.

117. Samantrai 2004, 185.

118. The statement was made in the context of a new round of peace talks between India and Pakistan, for the artist went on to say, "So if India and Pakistan are holding peace talks now, I'm glad. I just wish they had done it earlier." http://www.countercurrents.org/ipk-sarwar120304 .htm (accessed on March 10, 2011).

119. Samantrai 2004, 185. Note the fact as well that Zarina does not include the definite article "the" for the title of her piece, and also the fact that "we don't know where [the line] starts— nor where it ends" (Milford-Lutzker 2005). For subsequent works in which Zarina resurrects her "dividing line," see *Folding House*, a set of twenty-five collages that she completed in 2013, and *The line I cannot erase*, a pin drawing from 2014. I thank Renu Modi of the Gallery Espace in New Delhi for showing me these luminous works.

120. Milford-Lutzker 2001, 13. For some reflections on Zarina's use of Urdu calligraphy, see Mufti 2011.

121. Z. Hashmi 2011; emphasis mine. In a conversation, Zarina observed that she worked with an atlas that she purchased on one of her trips to Pakistan (interview with author, New York City, March 12, 2011). It is worth noting that *Atlas of My World IV* draws Kashmir in a manner that is closer to Pakistani than Indian visions of this disputed territory.

122. For an early assessment of this brilliant artist, see Desai 2001. See also Sangari 2013.

123. I am borrowing this formulation from Kabir 2009b. For a reproduction of these works, see Sangari 2013.

124. Kabir 2009b, 187.

125. Ali 1998, 22–23. For a discussion of Ali's poetry, itself shot through with spatial and cartographic imagery, and its relationship to Sheikh's work, see Kabir 2009b, passim; and Sangari 2013, 272–86.

126. I thank Rich Freeman for this observation. For an earlier work in which a similar figure recurs, see *My Hometown* (2008).

127. Kabir offers an alternative interpretation of this image when she identifies the "map-baring" citizen as "an Indian, imprinted with the image of the nation," appealing to "a Kashmiri with whom he is coupled" (Kabir 2009b, 199). I also wonder whether in this gesture, Sheikh's Muslim figure speaks back to another famous Kashmiri poet, Ghulam Ahmad Mahjoor, who wrote a verse in October 1947 that began, "Though I would like to sacrifice my life and body for India, yet my heart is in Pakistan." Chitralekha Zutshi writes that the governing National Conference Party "put him behind bars for this poem, [and] the poem itself cannot be located in the Indian part of Kashmir." After he recanted this statement, he was released from prison and even granted the status of "National Poet of Kashmir" (Zutshi 2004, 303).

128. Ramaswamy 2010a, 231–32.

129. Because of that, it also resonates with a line from another of Agha Shahid Ali's poems, "The Country without a Post Office," in which the poet writes of "that map of longings with no limit" (Kabir 2009b, 186). In a set of reflections on her work that the artist offered in Bangalore in November 2010, she observed that in her recent work on Kashmir, she has attempted "to take on ways of undoing the fixity of boundaries, using a scale and modes of extension that require other kinds of experiential relationship." I thank Nilima Sheikh for permission to quote her words.

130. Emphases mine.

131. Kabir 2009b, 187.

132. It is worth noting that two influential books that inaugurated the feminist discourse on Partition bring to visibility paintings by female artists responding to the events of 1947, although it is telling that neither of these works has itself been analyzed in these important volumes. Anjali Ela Menon's 1982 work *Mataji* (Mother) is reproduced as the cover image for Menon and Bhasin 1998 with the caption "She sits knitting in the sun dreaming of Lahore in the days before Partition." Nalini Malani's *Excavated Images to Stain an Old Quilt, Brought by My Grandmother from Karachi in 1947, version 2* (1997) provides the cover design for Butalia 2000.

133. Harley 1988, 301.

REFERENCES

Ahmed, Akbar S. 1997. *Jinnah, Pakistan, and Islamic Identity*. London: Routledge.

Ali, Agha Shahid. 1998. *The Country without a Post Office*. New York: Norton.

Anderson, Benedict. 1991. *Imagined Communities: Reflections on the Origin and Spread of Nationalism*. 2nd ed. London: Verso.

Ata-ur-Rehman, ed. 1998. *A Pictorial History of the Pakistan Movement: Taḥrīk-i Pākistān Kī Taṣvīrī Dāstān*. Lahore: Dost Associates.

Ayres, Alyssa. 2009. *Speaking like a State: Language and Nationalism in Pakistan*. Cambridge: Cambridge University Press.

Aziz, K. K., ed. 1978. *Complete Works of Rahmat Ali.* Islamabad: National Commission on Historical and Cultural Research.

Bayly, C. A. 1996. *Empire and Information: Intelligence Gathering and Social Communication in India, 1780–1870.* Cambridge: Cambridge University Press.

Bose, Sugata, and Ayesha Jalal. 1997. *Modern South Asia: History, Culture, Political Economy.* Delhi: Oxford University Press.

Brasted, Howard, and Carl Bridge. 1994. The Transfer of Power in South Asia: An Historiographical Review. *South Asia* 17 (1): 93–114.

Brückner, Martin. 2006. *The Geographic Revolution in Early America: Maps, Literacy and National Identity.* Chapel Hill: University of North Carolina Press.

Butalia, Urvashi. 2000. *The Other Side of Silence: Voices from the Partition of India.* Durham: Duke University Press.

Chatterji, Joya. 1999. The Fashioning of a Frontier: The Radcliffe Line and Bengal's Border Landscape, 1947–52. *Modern Asian Studies* 33 (1): 185–242.

———. 2007. *The Spoils of Partition: Bengal and India, 1947–1967.* Cambridge: Cambridge University Press.

Chester, Lucy. 2009. *Borders and Conflicts in South Asia: The Radcliffe Boundary Commission and the Partition of Punjab.* Manchester: Manchester University Press.

Citron, Beth. 2009. Contemporary Art in Bombay, 1965–1995. Philadelphia: University of Pennsylvania.

Copland, Ian. 1997. *The Princes of India in the Endgame of Empire, 1917–1947.* Cambridge: Cambridge University Press.

Cosgrove, Denis. 2005. Maps, Mapping, Modernity: Art and Cartography in the Twentieth Century. *Imago Mundi* 57 (1): 35–54.

Dadi, Iftikhar. 2000. Blood Lines. In *Independent Practices: Representation, Location, and History in Contemporary Visual Art*, edited by Bryan Biggs, Angela Dimitrakaki, and Juginder Lamba, 102–8. Liverpool: Bluecoat Aerts Centre.

———. 2010. *Modernism and the Art of Muslim South Asia.* Chapel Hill: University of North Carolina Press.

Dadi, Iftikhar, and Hammad Nasar, eds. 2011. *Lines of Control: Partition as a Productive Space.* London: Green Cardamom/Herbert F. Johnson Museum of Art.

Dalmia, Yashodhara, and Salima Hashmi. 2007. *Memory, Metaphor, Mutations: Contemporary Art of India and Pakistan.* New Delhi: Oxford University Press.

Desai, Vishakha. 2001. Engaging "Tradition" in the Twentieth Century Arts of India and Pakistan. In *Conversations with Traditions: Nilima Sheikh and Shahzia Sikander*, 6–17. New York: Asia Society.

Devji, Faisal. 2013. *Muslim Zion: Pakistan as a Political Idea.* London: Hurst.

Dhulipala, Venkat. 2015. *Creating a New Medina: State Power, Islam and the Quest for Pakistan in Late Colonial North India.* Delhi: Cambridge University Press.

Embree, Ainslee. 1989. Frontiers into Boundaries: The Evolution of the Modern State. In *Imagining India: Essays*, 67–84. New Delhi: Oxford University Press.

Flood, Richard, and Frances Morris, eds. 2001. *Zero to Infinity: Arte Povera, 1962–1972.* Minneapolis: Walker Art Center.

Ghosh, Amitav. 1988. *The Shadow Lines.* Delhi: Ravi Dayal Publishers.

Gilmartin, David. 1988. *Empire and Islam: Punjab and the Making of Pakistan*. Berkeley: University of California Press.

———. 1998. Partition, Pakistan, and South Asian History: In Search of a Narrative. *Journal of Asian Studies* 57 (4): 1068–95.

———. 2015. The Historiography of India's Partition: Between Civilization and Modernity. *Journal of Asian Studies* 74 (1): 1–19.

Golwalkar, M. S. (1939) 1947. *We or Our Nationhood Defined*. 4th ed. Nagpur: Bharat Prakashan.

Government of India. 1951. *Democracy on March*. New Delhi: Publications Division.

Guha, Ramachandra. 2007. *India after Gandhi: The History of the World's Largest Democracy*. New York: Harper Collin.

Harley, J. Brian. 1998. Maps, Knowledge, and Power. In *The Iconography of Landscape: Essays on the Symbolic Representation, Design, and Use of Past Environments*, edited by Denis Cosgrove and Stephen Daniels, 277–312. Cambridge: Cambridge University Press.

———.2001. *The New Nature of Maps: Essays in the History of Cartography*. Baltimore: Johns Hopkins University Press.

Harmon, Katharine. 2009. *The Map as Art: Contemporary Artists Explore Cartography*. New York: Princeton Architectural Press.

Harvey, P. D. A. 1993. *Maps in Tudor England*. Chicago: University of Chicago Press.

Hashmi, Salima. 1997. *50 Years of Visual Arts in Pakistan*. Lahore: Sang-e-Meel Publications.

Hashmi, Zarina. 2011. Conversation with My Self. In *Zarina Hashmi: Recent Work, Jan 15–Feb 15 2011*. New Delhi: Gallery Espace.

Hoskote, Ranjit. 2005. Images of Transcendence: Towards a New Reading of Tyeb Mehta's Art. In *Tyeb Mehta: Ideas, Images, Exchanges*, 3–48. New Delhi: Vadhera Art Gallery.

Jacob, Christian. 2006. *The Sovereign Map: Theoretical Approaches in Cartography through History*. Translated by Tom Conley. Chicago: University of Chicago Press.

Jalal, Ayesha. 2000. *Self and Sovereignty: Individual and Community in South Asian Islam since 1850*. London: Routlege.

Kabir, Ananya Jahanara. 2009a. Cartographic Irresolution and the Line of Control. *Social Text* 27 (4): 45–66.

———. 2009b. *Territory of Desire: Representing the Valley of Kashmir*. Minneapolis: University of Minnesota Press.

———. 2010. Secret Histories of Indian Modernism: M. F. Husain as Indian Muslim Artist. In *Barefoot across the Nation: Maqbool Fida Husain and the Idea of India*, edited by Sumathi Ramaswamy, 100–116. London: Routledge.

Kaur, Raminder. 2003. *Performative Politics and the Cultures of Hinduism: Public Uses of Religion in Western India*. Delhi: Permanent Black.

Khan, Yasmin. 2007. *The Great Partition: The Making of India and Pakistan*. New Haven: Yale University Press.

Khilnani, Sunil. 1997. India's Mapmaker. *Observer*, June 22, 7.

Khurshid, Abdus Salam. 1977. *History of the Idea of Pakistan*. Karachi: National Book Foundation.

Krishna, Sankaran. 1996. Cartographic Anxiety: Mapping the Body Politic in India. In *Challenging Boundaries: Global Flows, Territorial Identities*, edited by Michael J. Shapiro and Hayward R. Alker, 193–214. Minneapolis: University of Minnesota Press.

Kumar, Amitava. 2001. Splitting the Difference. *Transition* 89:44–55.

Lal, Vinay. 2012. Hindutva's Sacred Cows. In *Fear and Loathing*, edited by Ziauddin Sardar and Robin Yassin-Kassab, 65–84. London: Hurst and Co.

Malik, Keshav. 2002. Satish Gujral and His Paintings on Partition. In *Pangs of Partition*, edited by S. Settar and Indira Baptisa Gupta, 75–82. New Delhi: Manohar.

Menon, Jisha. 2012. *The Performance of Nationalism: India, Pakistan and the Memory of Partition*. Cambridge Studies in Modern Theatre. Cambridge: Cambridge University Press.

Menon, Ritu, and Kamla Bhasin. 1998. *Borders & Boundaries: Women in India's Partition*. New Brunswick: Rutgers University Press.

Menon, V. P. 1956. *The Story of the Integration of the Indian States*. New York: Macmillan.

Michel, Aloys Arthur. 1967. *The Indus Rivers: A Study of the Effects of Partition*. New Haven: Yale University Press.

Milford-Lutzker, Mary-Ann. 2001. Reflections on Mapping a Life. In *Zarina: Mapping a Life, 1991–2001*, 9–17. Oakland, CA: Mills College Art Museum.

———. 2004. Cities, Countries, and Borders: Recent Work by Zarina Hashmi. In *Cities, Countries and Boders: Prints by Zarina*. Bombay: Gallery Chemould.

———. 2005. Mapping the Dislocations. In *Counting 1977. 2005*. New York: Bose Pacia.

———. 2013. The Poetry of Zarina's Art. In *Themes, Histories, Interpretations: Indian Painting, Essays in Honour of B. N. Goswamy*, edited by Mahesh Sharma and Padma Kaimal, 411–21. Ahmedabad: Mapin Publishing.

Mirzoeff, Nicholas. 2011. *The Right to Look: A Counterhistory of Visuality*. Durham: Duke University Press.

Mufti, Aamir. 2011. Zarina Hashmi and the Arts of Dispossession. In *Lines of Control: Partition as a Productive Space*, edited by Iftikhar Dadi and Hammad Nasar, 87–99. London: Green Cardamom/Herbert F. Johnson Museum of Art.

Naqvi, Akbar. 1998. *Image and Identity: Fifty Years of Painting and Sculpture in Pakistan*. Karachi: Oxford University Press.

Nasar, Hammad. 2007. Lines of Control: State of the Art. *Art Asia Pacific* 54 (July/August 2007): 42–43.

Oondatje, Michael. 1993. *The English Patient*. New York: Vintage.

Patel, Samir. 2007. Zarina Hashmi: The Edges of Her World. *ArtAsiaPacific* 54 (July/August): 72–77.

Pickles, John. 2004. *A History of Spaces: Cartographic Reason, Mapping, and the Geo-coded World*. London: Routledge.

Pinney, Christopher. 2004. *Photos of the Gods: The Printed Image and Political Struggle in India*. New Delhi: Oxford University Press.

Pirzada, Syed S., ed. 1995. *Evolution of Pakistan*. Karachi: Royal Book Company.

Rahmat Ali, Choudhury. 1946. *Dinia: The Seventh Continent of the World*. Pakasia Literature Series, no. 11. 1st ed. Cambridge: Dinia Continental Movement.

———. (1935) 1947. *Pakistan: The Fatherland of the Pak Nation*. 3rd ed. Cambridge: Pakistan National Liberation Movement.

Ramaswamy, Sumathi. 2010a. *The Goddess and the Nation: Mapping Mother India*. Durham: Duke University Press.

———. 2010b. Mapping India after Husain. In *Barefoot across the Nation: Maqbool Fida Husain and the Idea of India*, edited by Sumathi Ramaswamy, 75–99. London: Routledge.

———. 2011. Midnight's Line. In *Lines of Control: Partition as a Productive Space*, edited by Iftikhar Dadi and Hammad Nasar, 25–35. London: Green Cardamom/Herbert F. Johnson Museum of Art.

Ramusack, Barbara N. 2004. *The Indian Princes and Their States*. Cambridge: Cambridge University Press.

Samantrai, Ranu. 2004. Cosmopolitan Cartographies: Art in a Divided World. *Meridians: Feminism, Race, Transnationalism* 4 (2): 168–98.

Sangari, Kumkum, ed. 2013. *Trace Retrace: Paintings, Nilima Sheikh*. New Delhi: Tulika Books.

Sherwani, Latif Ahmed, ed. 1969. *Pakistan Resolution to Pakistan, 1940–1947: A Selection of Documents Presenting the Case for Pakistan*. Dacca: National Publishing House.

Sinha, Gayatri, ed. 2006. *Satish Gujral: An Artography*. New Delhi: Roli Books.

Tan, Tai Yong, and Gyanesh Kudaisya. 2000. *The Aftermath of Partition in South Asia.*. London: Routledge.

Taussig, Michael. 1993. *Mimesis and Alterity: A Particular History of the Senses*. New York: Routledge.

Thongchai Winichakul. 1996. Maps and the Formation of the Geo-body of Siam. In *Asian Forms of the Nation*, edited by Hans Antlov and Stein Tonnesson, 67–91. London: Curzon Press.

van Schendel, Willem. 2005. *The Bengal Borderland: Beyond Nation and State in South Asia*. London: Anthem Press.

Whiles, Virginia. 2010. *Art and Polemic in Pakistan: Cultural Politics and Tradition in Contemporary Miniature Painting*. London: Tauris.

Wilcox, Wayne. 1963. *Pakistan: The Consolidation of a Nation*. New York: Columbia University Press.

Zutshi, Chitralekha. 2004. *Languages of Belonging: Islam, Regional Identity, and the Making of Kashmir*. New York: Oxford University Press.

CHAPTER EIGHT

SIGNS OF THE TIMES

COMMERCIAL ROAD MAPPING AND NATIONAL IDENTITY IN
SOUTH AFRICA

Thomas J. Bassett

Oppression was palpable. No detail was neutral.

WILLIAM FINNEGAN, *Crossing the Line: A Year in the Land of Apartheid*

1. INTRODUCTION

South African road maps look like most other road maps in the world. They
follow certain conventions or practices and have the recognizable form of a
folded sheet that can fit easily into a car's glove compartment (fig. 8.1). When
it's unfolded, the telltale signs really stand out: the thick and thin lines of road
networks; the names of towns and cities in different font sizes; the chart full
of numbers in the corner of the sheet showing the distance between major
locations; and, of course, the legend with its emphasis on road types, political
boundaries, and other symbols that assist travelers to reach their destination.

As in other countries, a variety of road maps exist to appeal to the widest

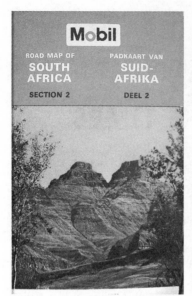
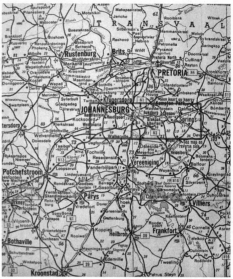

FIGURE 8.1. The *Road Map of South Africa/Padkaart van Suid-Afrika* produced in 1967 by MapStudio for the Mobil Oil Company. The map stands out for its English and Afrikaner place names. African place names are scant, which contributes to the apartheid-era ideology of South Africa as a "White man's country."

possible audience. There is the general-reference road map at the national scale, the regional or provincial map showing more detail, and the city map showing streets and sometimes buildings. Then there is the tourist map that guides the motorist to a specific destination like Kruger National Park or to a circuit like the Garden Route of the Western Cape. The national road atlas brings together this diversity of road maps into a single volume. This panoply of road map products looks very familiar to veteran map users. We recognize a road map when we see it. New editions show new subdivisions, new highways, and new towns so we can plan our trip with the most up-to-date information. We feel comforted by this geographic knowledge and proud of our map-reading skills. Our comfort is in part based on the assumption that the maps are accurate representations of reality, that the distance between points A and B is accurately recorded, and that we won't run out of gas en route.

South Africa's road maps *are* like road maps around the world. They replicate a system of signs that is widely shared among mapmakers, which tends to naturalize maps as stable, immutable objects that enjoy an "ontological security" (Akerman 2002, 182–83; Pickles 2004, 60–61; Kitchin and Dodge

2007). Map users assume a one-to-one correspondence between the sign and its signifier—a line stands for a road and a dot represents a locality along that road. But this is a false impression in that maps, as Rob Kitchin argues, "are never fully formed but emerge in process and are mutable (they are remade, as opposed to mis-made, misused, or misread)" (Kitchin 2008, 214). He suggests that the process of mapmaking involves both the mapmaker and the map user in contexts that are forever changing. Maps do work and have effects, but they do so only because of the existence of a map culture or "mapping practices" that give form and meaning to maps (Kitchin and Dodge 2007, 343).

The case of South African road mapping lends itself to this analytical framework for a variety of reasons. First, the history of extreme racial discrimination in that country resulted in an underdeveloped map culture. Whites have been the principal users of road maps, which has historically led mapmakers to design their maps in ways that appeal especially to whites. This is evident in the mapping practice of selecting and omitting certain features on the map. The 1967 road map of South Africa shown in figure 8.1 speaks clearly to a white audience. The map's text (title, legend, technical information) is in English and Afrikaans, the languages spoken by the ruling white population. Furthermore, English and Afrikaans place names dominate the map. The map does not speak to black Africans. Their history, culture, and language—in short, their presence is omitted from the map.

Second, in the South African context of racial apartheid, mapmakers continually debated the technical process of road mapping, especially the criteria for portraying the distribution of population centers on their maps. These discussions led them to upgrade small towns populated by whites and to downgrade large towns populated by blacks. Third, the transition from apartheid to postapartheid South Africa changed the political context of mapping practices. In this "postcolonial" moment, new maps are unfolding that differ from apartheid-era maps. The question is, how different are they? The answer to this question will become more evident in later sections of this essay. Fourth, the postapartheid transition is occurring simultaneously with a technical revolution in mapmaking associated with digital cartography. In this context, map literacy is expanding with the emergence of online mapping and personal navigation systems. The implications of this significant technological change for "conventional" road mapping are potentially far reaching but remain unknown.

In this essay I argue that South African mapmakers have been guided by similar sets of mapping practices both during and after apartheid, which give road maps a similar although changing look. At the risk of simplifying practices that span these historical periods, I ask and seek to answer the following two questions: Can we identify a set of mapping practices that characterized the production of apartheid-era road maps? Can we identify a set of mapping practices that distinguish the emergence of postapartheid-era maps? Before answering these questions, it is important to review the changing sociopolitical and technical contexts in which road mapping has been historically situated. Thus, the following section provides background information on the ideology and practice of apartheid. In the remaining sections I focus on specific mapping practices that characterize road mapping during these periods.

2. INTERNAL COLONIALISM AND THE GEOGRAPHY OF APARTHEID

Apartheid-era road maps can be read as ideological expressions and material artifacts of internal colonialism. Internal colonialism refers to a political-economic system in which a social group claims sovereignty over a territory and people *within the boundaries of a state* with the goal of controlling resources, labor, and markets (Hind 1972; Love 1989; Wolpe 1974). This system is characterized by the political-economic oppression of certain social groups by a repressive state apparatus and ruling class. Internal colonialism emerged in Africa in countries like South Africa and Zimbabwe where European settlers gained control of the state following a period of external colonialism when colonial authorities and settlers dispossessed indigenous peoples of their land and resources. Under external and internal colonialism, capitalist relations of production structured the exploitation of populations and resources.

In the case of South Africa, there was a close relationship between class exploitation and racial domination (Wolpe 1974). White settlers (Dutch and British) depended on the state to ensure the flow of black migrant workers to their farms, mines, and industries. Internal colonial policies such as land alienation, the creation of labor reserves called Bantu Homelands or Bantustans, and migrant labor regulations worked to facilitate capital accumulation by white property owners. Agrarian and industrial capitalists paid workers low wages because part of the costs of social reproduction of the work force (food, social security, education) were assumed by the families of migrant workers residing in the Bantustans (Wolpe 1974, 244–50). But by the 1940s, the lack of state investment, population growth, and declining agricultural productivity in the reserves had eroded the capacity of migrants' families to support themselves. The response of migrant workers to this crisis in simple reproduction was to organize strikes for higher wages (Lemon 1987; Wolpe 1972, 444). Such protests were severely repressed by the internal colonial state. This conflict between capital and labor and between white and black labor led many whites to support the apartheid platform of the Afrikaner Nationalist Party, which came to power in 1948 (Wolpe 1972, 446). The suite of apartheid laws introduced by successive Nationalist governments sought to maintain the flow of cheap labor from the Bantustans by regulating black residence and mobility in a more systematic and coercive manner (Wolpe 1972, 447). This mix of racial and capitalist relations of production structured the political economy of internal colonialism in South Africa, popularly known as "apartheid." Apartheid-era maps gave form and meaning to internal colonialism. They helped to create the territory of apartheid by reproducing its sociospatial relations and sustaining the ideology of a "White South Africa."

APARTHEID-ERA LAWS

"Apartheid" means "apartness" in Afrikaans—the language spoken by descendants of Dutch settlers who emigrated to southern Africa in the second half of the seventeenth century (Parsons 1983, 271). Apartheid was a policy of extreme racial discrimination based on a series of laws legalizing white supremacy in social, political, and economic affairs. This section highlights the key laws that buttressed this pernicious social system.

The *Prohibition of Mixed Marriages (1949) and the Immorality Act* (1950) forbade interracial sexual relations and marriage between blacks and whites.

The *Population Registration Act of 1950* assigned people to racial groups based on a classification scheme that placed individuals into four racial groups: black, colored, Indian, and white (including Japanese and Chinese).

The *Group Areas Act of 1950* designated separate residential areas for races based on the Population Registration Act. The enforcement of this law led to forced removals of people if they happened to be living in an undesignated group area. Some 3.5 million people were forcibly removed between 1960 and 1980 from undesignated areas.

The *Homelands Policy*, also known as "Grand Apartheid," emerged over the 1950s and 1960s. The foundation of this policy was the *Land Apportionment Act of 1936*, which set aside 87% of national land for whites and just 13% for blacks. This meant that a minority of the population (15%) controlled most of South Africa's territory while the majority (75%) of the population held just 13% of the land, most of which was of marginal quality. The Homelands Policy created ten "tribal homelands," to which blacks were assigned and given citizenship. Unless employed outside the homelands, blacks were required to reside in these homelands.

The *Pass Law of 1952* required all African men and women over sixteen years to be fingerprinted and to carry "reference books" containing personal identification and employment information. The law was designed to regulate the movement of people out of the homelands. The Sharpeville Massacre of 1960, in which sixty-nine unarmed blacks were killed by police, took place during a demonstration against the pass laws (Parsons 1983, 299–301).

The *Bantu Education Act of 1953* led to the strict segregation of elementary and high school education. There was some integration at the university level in English-speaking universities such as the University of Cape Town, the University of Witwatersrand, and Rhodes College. The Soweto Uprising of 1976, in which 576 blacks died, was initially a protest over a law requiring Afrikaans to be taught as the primary language in schools.

The *Separate Amenities Act of 1953* allowed cities to discriminate in the provision of public services such as parks, swimming pools, beaches, libraries, restrooms, and other facilities on the basis of race. The law was repealed in 1990.

The *National States Act of 1971* established a three-stage process in which the homelands would ultimately become independent states. The first stage involved the appointment of homeland leaders by the Department of Development Aid (formerly the Bantu Administration), who in turn formed their own governments by selecting ministers, and so on. During the second stage, the appointed governments would exercise greater economic and political control over the internal affairs of their homelands, culminating in the granting of full internal self-government within South Africa. The third and final stage granted homelands political independence from South Africa, making them foreign countries. The South African government proclaimed four homelands to be "independent republics" between 1976 and 1981: Transkei (1976), Bophuthatswana (1977), Venda (1979), and Ciskei (1981) (fig. 8.2). No government in the world outside of South Africa recognized the existence of these so-called independent republics.

The *Internal Security Act of 1982* consolidated all security laws passed since the 1950s. It allowed the government to ban individuals for political reasons; to restrict individual freedom to travel, speak, and work; to place individuals under house arrest; and to detain people for interrogation without the right to an attorney.

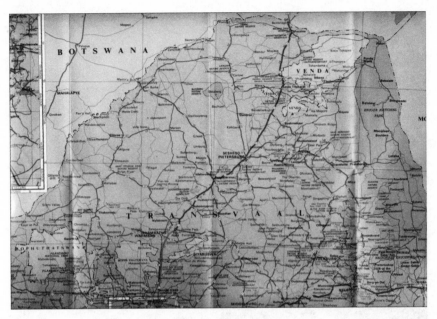

FIGURE 8.2. A section of the Automobile Association (AA) of South Africa's 1990 *South Africa/Suid Afrika* road map, showing the so-called independent republics of Bophuthatswana and Venda.

APARTHEID-ERA GEOGRAPHIES

The geography of apartheid demonstrated in a dramatic way the strong linkages between the organization of society and the organization of space (Smith 1985). The racial and class domination of blacks by whites in South Africa built upon spatial structures that derived from and reinforced the apartheid policies of the Nationalist Party, which ruled South Africa from 1948 to the end of apartheid in 1994.

The sociospatial structure of apartheid cut across three scales: the national, the local (cities and towns), and the individual. At the national level, the territory was unevenly divided between blacks and whites. The areas set aside for the ten black homelands constituted 13% of the national territory. The white minority controlled 87% of the territory. Road maps from the 1950s through the 1980s "invoke[d] the territory" (Wood and Fels 2008, 191) of apartheid at the national scale by delimiting these sociospatial relations through the use of color, international boundary lines, and font selection. Mapmakers showed these homelands as independent states because the government had declared them as such (personal communication, Automobile Association of South Africa [AA of SA], July 6, 2009).

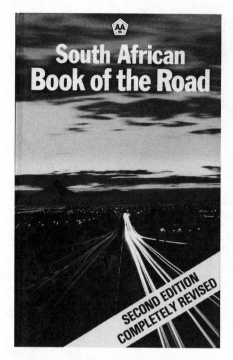
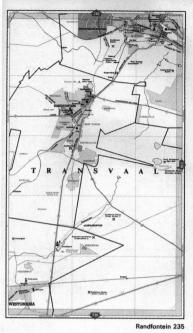

Randfontein 235

FIGURE 8.3. The black townships of Toekomsrus and Mohlakeng outside the white group area of Randfontein in the AA's 1989 *South African Book of the Road*. Apartheid-era planners designed townships to police their populations especially during periods of unrest. Buffer zones such as unbuilt areas, roads, and sports fields physically separated black residential areas from white-designated localities. Typically just one wide road led into townships, which allowed government security forces to monitor the comings and goings of township residents and visitors.

At the local scale, residential areas were segregated on the basis of the Group Areas Act and the Population Registration Act. Blacks employed outside the homelands were required to obtain passes and live in townships on the outskirts of white cities. Other "racial" groups defined by apartheid laws lived in similarly segregated areas. Open the *Automobile Association of South Africa's Book of the Road* and the townships appear on virtually every page (AA 1989). An estimated 50% of the black population lived in townships at the height of apartheid in the 1970s and 1980s, revealing both the dependence of white South Africans on black African labor and the contradictions of apartheid policies to separate racial groups (Lemon 1987, 47–49). Figure 8.3 shows the apartheid solution to this contradiction: the former black townships of Toekomsrus and Mohlakeng are spatially separated from the white group area of Randfontein by unbuilt areas, roads, and sports fields.

The Separate Amenities Act of 1953 allowed municipalities to segregate

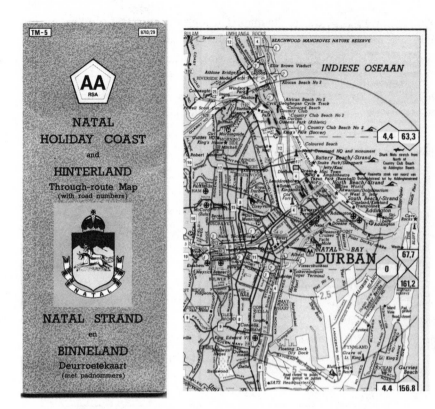

FIGURE 8.4. The AA's 1977 *Natal Holiday Coast and Hinterland Through-route Map* (1:100,000) showing the segregated beaches of Durban. A note on the middle right notes that "shark nets stretch from North of Country Club Beach to Addington Beach." Black-designated beaches lacked shark nets.

access to public facilities on the basis of race. Blacks and whites did not go to the same schools, frequent the same beaches, or ride in the same train cars. The AA's *Natal Holiday Coast and Hinterland Through Route Map* dating from 1977 maps the geography of petty apartheid by labeling beaches reserved for blacks, coloreds, and whites (fig. 8.4).

In summary, the social and political geography of South Africa reflected and reinforced apartheid policy. The political challenge of the apartheid state was to keep blacks in their subordinate place. The strict segregation of arbitrarily defined racial groups was a classic divide-and-rule strategy. Road mapping was inextricably linked to the sociospatial problems generated by apartheid. On the one hand, mapmakers mimicked the official sociospatial order by carefully drawing the spaces of apartheid on their maps, as the above examples illustrate. In a more subtle manner, mapmakers created the whites-only territory through manipulating map design in a way that made the outnumbered whites appear dominant on the map. Through the processes of selection and omission, mapmakers conjured up the white "fanatic-segregationist vision" of dispossessing the majority of the population of their freedoms and citizenship (Finnegan 1986, 266). That is, road maps did more than assist the traveler to get from one town to the next. They also worked to create a predominately "white man's country" by ensuring that white history, culture, and identity dominated the map. To achieve this, mapmakers used cartographic sleight of hand to make black Africa disappear from the map.

SOUTH AFRICAN COMMERCIAL ROAD MAP COMPANIES

There are currently seven major commercial road map companies operating in South Africa (table 8.1). The oldest is Brabys Maps, founded in 1903; the youngest is MapIt, a digital mapping firm created in 2002 by Ray Wilkinson, the former managing director of MapStudio. Brabys is best known for its business directories but had nearly 100 maps for sale on its website in 2010. MapIt is located in Centurion, a suburb of Pretoria that was until recently named Verwoerdburg, named after Hendrik Verwoerd, South Africa's

TABLE 8.1 Commercial Map Makers of South Africa (2010)

Company	Headquarters	Year Established	Maps/ Atlases	Website
AA of South Africa	Kyalami (Johannesburg)	1930	69	www.aa.co.za
Brabys	Pinetown (Durban)	1904	99	www.brabys.com/
GeoGraphic Maps (Map Graphix)	Benoni (Johannesburg)	2003	25	
Globetrotter	Cape Town	1990	96	www.newholland publishers.com/globe trotter.asp
MapIt	Centurion	2002	Digital maps	www.map-it.co.za/
MapStudio	Cape Town	1958	143	www.mapstudio.co.za/
Sunbird Publishers	Cape Town	1998	17	www.sunbirdpublishers .co.za

prime minister from 1958 to 1966 and one of the primary architects of apartheid (Lemon 1987, 47). MapIt supplies digital map data to businesses that use geospatial data for such applications as web mapping and navigation devices. It is owned by TeleAtlas (49%) and New Holland Publishers (51%). Globetrotter is also a subsidiary of New Holland Publishers. The company had 96 different maps and guide books for sale in 2010. MapStudio was founded in 1958 in Johannesburg by E. G. "Bill" Buckley and Lionel Miller. Today it is a subsidiary of Struik Publishers based in Cape Town and sells 143 different map products, two-thirds of which are road maps, road atlases, and street guides. MapStudio published the promotional oil company maps for Total, Shell, BP, Mobil Oil/Esso, and Caltex. Each company had its own color specifications, symbols, and fonts. But the maps consistently portrayed a white South Africa that was recognizable to its privileged motorists. MapStudio, MapIt, and Globetrotter are all part of Avusa Limited (http://www.avusa.co.za), a South African media and entertainment conglomerate that owns the parent companies (New Holland Publishers and Struik Publishers) of these mapmaking firms.

The Automobile Association of South Africa was created in 1930 out of a federation of South African automobile clubs. It acquired its two main rivals, the Royal Automobile Club of South Africa in the 1960s and the Rondalia Touring Club in the 1980s. The AA historically lobbied the government for improved roads, the provision and pricing of oil and gasoline, and nationwide motoring legislation and road signage. It pioneered roadside emergency services to its members and promoted tourism by offering accommodation reservations and especially road maps. Membership was restricted to whites and peaked at 650,000 motorists in the 1980s. The growing demand for travel-related services coincided with the nation-wide state of emergency, when road travel could be dangerous. AA travel specialists provided its members with detailed guides and route maps in which travel safety was a top priority. In 2010 the AA published sixty-nine different road map products with an average monthly consumption of twenty-five thousand publications (AA of South Africa 2010).

GeoGraphic Maps and Sunbird Publishers are relative newcomers to the road mapping scene in South Africa. Their small but growing output of maps and atlases is focused entirely on South and southern Africa.

3. THE COPRODUCTION OF MAPS

The theoretical footing of this essay is a diverse set of writings by critical cartographers who share a common interest in understanding how maps work in the world. The works of J. Brian Harley, Matthew Edney, Denis Wood and John Fels, John Pickles, and Rob Kitchin and Martin Dodge are particularly influential in drawing attention to the postrepresentational character of maps. A key idea is that maps are constituitive, not merely descriptive. They produce territory, shape identities and conceptions of nature, and have other effects (Harley 1989; Pickles 2004; Wood 1992, 2010; Wood and Fels 2008). But they only work in this way when "mapping practice such as recognizing, interpreting, translating, and communicating, are applied to the pattern of ink" (Kitchin, Perkins, and Dodge, 2007, 21). In the absence of a map culture, maps cannot do their work; they are simply "a lot of confusing lines and dots," as one taxi driver summed it up as we drove through Cape Town with the aid of an in-vehicle navigation system. Thus, maps can only work if there is a population with the knowledge and skills to read them.

Following Kitchin and Dodge, it is productive to view maps as "contingent, relational, and context-dependent" (Kitchin and Dodge 2007, 342). Maps are contingent on a shared culture of mapmaking and map-reading practices. The makers of road maps generally follow an international set of norms that guide their decisions to show roads and towns in relation to each other within a larger territory. Knowledgeable map readers will connect the lines and dots symbolizing roads and towns based on their experience in solving spatial problems. These engagements linking mapmakers and users around a map pivot around sociospatial problems that maps are called upon to solve. Road maps are well known for their utility in solving navigational problems, for assisting motorists to get from point A to point B. They work, as John Pickles argues,

> by naturalizing themselves by reproducing a particular sign system and at the same time treating that sign system as natural and given. But, map knowledge is never naïvely given. It has to be learned and the mapping codes and skills have to be culturally reproduced so that the map is able to present us with a reality that we recognize and know. This known reality is differentiated from the reality we see, hear and feel, and this is the magic and the power of the map. (Pickles 2004, 60–61)

Road maps are the product of such conjuring, in which mapmakers and users "invoke the map" to solve a host of sociospatial problems (Wood and Fels 2008, 191). This paper argues that road maps of South Africa do more than assist the traveler to get from one town to the next. They also *work*, through technical and ideological practices, to create the spaces of apartheid and postapartheid society. In this way, maps (re)produce social systems, territory, identities, and authority by normalizing power relations through their propositional character and everyday use. They perform these functions in the case of South Africa in at least three ways: (1) by delimiting racialized spaces; (2) by classifying settlements on the basis of services and infrastructure rather than by population; and (3) by demarcating a postapartheid political geography, including the renaming of cities and streets with reference to multicultural and historical considerations.

The focus on mapping practices during the apartheid and postapartheid eras illustrates how maps *emerge in process* through technical and ideological practices "*to solve diverse and context dependent problems*" (Kitchin and Dodge 2007, 340, 342; emphasis in original). This view of maps as "processual, as opposed to representational," as emerging "through contingent, relational, context-embedded practices" (Kitchin and Dodge 2007, 342), is illustrated in the following pages through the words of South African mapmakers as they describe their decision-making processes and techniques in making road maps. The discussion shows that a coproduction of maps has taken place primarily among white mapmakers and users within a very specific context.

4. APARTHEID ROAD MAPPING

Road maps tell us something about the geography of a country. At the very least, they should inform the reader about the distribution of cities and roads. The phenomenon of "invisible towns" on South African maps challenges such assumptions (Stickler 1990). Like all maps, road maps are rhetorical in that they make claims to territory, sovereignty, and ownership through the deployment of basic cartographic elements such as boundary lines, color, place names, and other symbols (Wood 1992). In the case of South Africa, road maps participated in the territorialization of apartheid by delimiting homelands and the so-called independent states. Every time an apartheid-era road map was unfolded, the spaces of apartheid were re-created in the eyes and minds of map readers. The geography of "white" and "black" South Africa was (re)etched in the minds of map readers when they navigated through the highly fragmented KwaZulu Homeland along National Road 3 (N3) or N2 en route to Durban (fig. 8.5). The map and the territory it depicts were brought into being during these moments. This engagement with the AA map may have taken place at the height of apartheid in the 1980s or fifteen years later, but maps are

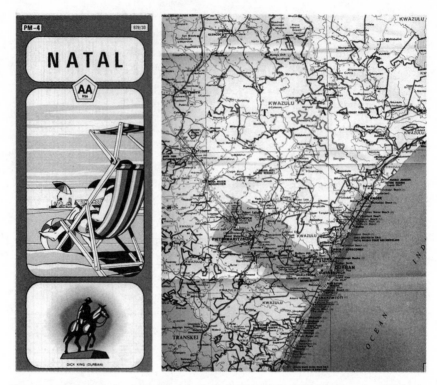

FIGURE 8.5. The AA's 1987 provincial road map *Natal*, showing the fragmented boundaries of the KwaZulu homeland. KwaZulu was one of ten "tribal homelands" in South Africa in which black Africans were required to reside unless legally employed in a white group area. KwaZulu was the official homeland for the Zulu ethnic group.

"*always* remade every time they are engaged with" (Kitchin and Dodge 2007, 335; emphasis in original). Each time I open the Automobile Association of South Africa's 1987 map of Natal, the spaces of apartheid reappear.

MAPPING FOR A WHITE AUDIENCE

Mapmakers always write with an audience in mind. Whether tourists or business travelers, the audience of South African road maps is typically white, relatively wealthy, mobile, reads English or Afrikaans, and grew up in a map culture (personal communication, AA of SA, July 6, 2009; personal communication, MapStudio, July 13, 2009; personal communication, former research director, MapStudio, December 18, 2009). It could hardly be otherwise. Apartheid severely limited travel by blacks. To travel from one's assigned homeland to an official white area, blacks had to possess a pass book that proved that they were employed in a white area. The employer's name, address, and signature had to appear in the designated section of the pass book. The pass laws were ruthlessly enforced. South African police arrested millions of blacks for pass law violations (Smith 1985, 21–22). If authorized to be in a white group area, blacks were required to live in townships on the outskirts of "white cities." Most blacks commuted to their jobs by bus or by train from their townships and then back again. They didn't need a road map (personal communication, eThekwini Municipal Library, Durban, July 9, 2009).

Even if blacks owned automobiles, they could not join the AA. The AA of South Africa was an exclusively white auto club. Petty apartheid laws forbade blacks from joining the organization. A former general manager of AA explained: "Because of government policy, we could not have black members. Even if you could have them, what could you offer them? A tour of a black township? We couldn't offer them hotel bookings because they weren't allowed to stay in tourist resorts" (personal communication, AA of SA, July 6, 2009). The covers of apartheid-era road maps illustrate this white map-using audience. The 1990 AA map of South Africa/Suid Afrika (fig. 8.6) shows a white nuclear family of four consulting a map after a picnic lunch alongside the road. Its emphasis on leisure travel and mobility speaks to the class and racial privileges of whites. Apartheid-era restrictions on employment and mobility denied blacks these options. The only images of blacks that appear on road maps are of individuals in native dress who appear as part of the white tourist landscape. Since booking hotel reservations was one of AA's remunerative activities, the company sought to make travel enticing. The production manager in charge of publications at AA explained that the reason for placing "lovely images" on road map covers is "to arouse a desire in people to travel. Part of what we do is sell dreams" (personal communication, AA of SA, July 6, 2009).

A MapStudio mapmaker listed the classic set of tourist road maps. They include maps of the Garden Route, the Winelands, the Cape Peninsula, Kruger National Park, the Natal Coast, and the Drakensberg Holiday Resorts. When asked about the audience for this set of maps, he explained that it was not oriented toward "whites" as much as toward "middle class" tourists. But in light of the history of institutionalized racism in South Africa, "white" and "middle class" were largely synonymous in that country (CSVR 2007, 31). The absence of tourist maps focused on black African historic sites lends further support to the view that South African road maps speak to a predominately white audience.

FIGURE 8.6. The 1990 *South Africa/Suid Afrika* road map published by the AA of South Africa. In addition to AA's corporate logo, the cover presents an image of the idealized consumer of AA's products and services—a white nuclear family on vacation.

FEAR OF THE ROADS · Cartographers were particularly attentive to the safety concerns of travelers in their mapmaking. At AA, mapmakers prepared custom route maps for their members based on a number of criteria such as destination, time constraints, whether they were pulling a trailer, and so on. "Our job was to recommend the best route. They would come to us and say that they wanted to go from point A to point B. We would give them the route maps, tell them the condition of the road, give a route description and all the facilities. And we would say this is the recommended route. That was our job" (personal communication, AA of SA, July 6, 2009). The recommended route

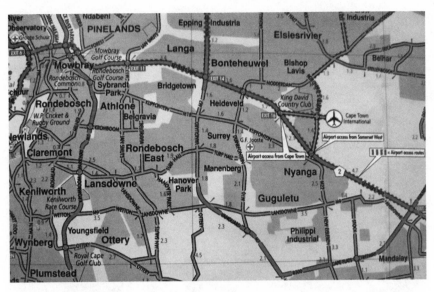

FIGURE 8.7. "Airport access routes" shown in the *Globetrotter Travel Map, Cape Winelands*, 1st edition (1:130,000), published in 2000, emphasizing the safest auto routes connecting central Cape Town to the international airport. Mapmakers have highlighted these routes by outlining them in a crenulated pattern.

was not necessarily the most direct route but it was always the safest route. Travelers were warned against straying from the main roads if they were in the vicinity of townships. If they entered a township by mistake, they might not come out alive (personal communication, AA of South Africa, July 6, 2009). This was especially true in the 1970s and 1980s, when civil unrest forced motorists to seek alternate routes. William Finnegan, the young American writer who taught in a black high school on the Cape Flats in the early 1980s, described the heightened insecurity on the roadways during a boycott against a bus company in response to its exorbitant fare increases. "By this stage, stones flung by black youths were beginning to loom large in the fears of white Capetonians. White motorists driving to the northern suburbs via Elsies River were being stoned so frequently that traffic was rerouted" (Finnegan 1986, 222).

Motorist's safety remains a major concern of postapartheid mapmakers. The most direct route from the southern suburbs of Cape Town to the international airport passes through the former black township of Manenberg. On the Globetrotter Cape Winelands road map produced by MapStudio in 2000, cartographers highlighted the recommended and safest routes to the airport with a crenulated design (fig. 8.7). Thus, tourists visiting the wine district of Constantia were visually directed to follow the "airport access route" north

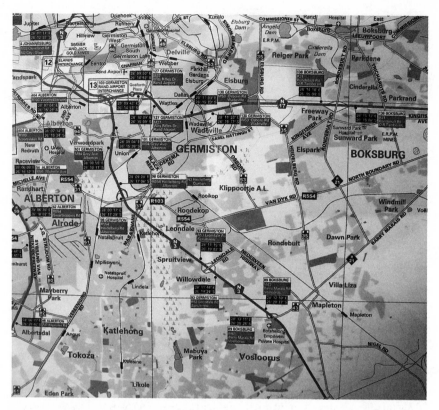

FIGURE 8.8. The first edition of MapStudio's *Map of Gauteng Roads* (1:100,000), published in 2008, displaying the GPS coordinates (waypoints) of main highway exits. This innovative mapping practice addresses a long-standing concern of South African mapmakers to show the safest route between locations.

toward Mobray and then east along National Route 2 to Cape Town International Airport.

In 2008, MapStudio developed a new technique to address the safety concerns of motorists. It placed global positioning system (GPS) coordinates on its maps. On its *Gauteng Roads* map, the waypoints appear in black rectangles next to blue miniature exit signs at main highway intersections (fig. 8.8). The idea is that travelers can enter these coordinates into their in-car navigation systems to plot a course from one part of the city to another. High crime rates, particularly vehicle hijacking, appear to be the greatest fear addressed by this new practice. More than half of South Africa's carjacking and truck hijacking crimes take place in Gauteng Province (CSVR 2007, 80–84). A MapStudio mapmaker exclaimed: "If you are in Jo'burg, it can be quite frightening if you don't know where you are in the middle of the night." He believed that "fear of the roads" was a major factor in the brisk sales of the *Gauteng Roads* map (personal communication, MapStudio, July 13, 2009). National crime statistics indicate that middle-class white households are the most common victims of such crimes. The national media, which are concentrated in Gauteng, give disproportionate attention to these incidents in comparison to other types of aggravated assault that disproportionately affect low-income black households (CSVR 2007, 84).

This combining of traditional road maps with digital mapping devices is changing the look and use of road maps in at least two ways. First, the road map is no longer the primary source of information for motorists. Hand-held navigation devices and in-car navigation systems are increasingly popular. By placing waypoints on paper maps, commercial mapmakers seek to complement rather than compete with these new geospatial technologies. Second, the display of geographic coordinates on the map lends an unprecedented level of geographic accuracy to road maps, at least to highway intersections. This simultaneous loss and gain in geographic knowledge is a commercial risk for mapmakers, since accuracy is a major selling point. Finally, the art of mapmaking has definitely taken a step backward. With its road signs and waypoints littering the sides of the road, the *Gauteng Roads* map is aesthetically unattractive.

DOWNGRADING LOCALITIES · Mapmakers deployed a second technique to steer travelers away from potential danger. They would "downgrade" certain localities and "upgrade" others. A large African township was downgraded by making its font size smaller. White localities, even if they were much smaller than a township, were upgraded with a larger font size. The classic case is Soweto, the South West Townships near Johannesburg. Although it is one of South Africa's largest cities, its font size makes it appear equivalent to much smaller areas such as Standton and Germiston, not to Johannesburg or Pretoria. A MapStudio cartographer explained this practice with reference to the interests of the map reader: "Soweto was deliberately downgraded because you wouldn't want to indicate it as a tourist site (personal communication, MapStudio, July 13, 2009).

Mapmakers commonly took the safety concerns of their audience to an extreme. They simply omitted some localities from the map. The rationale for this extreme form of downgrading was simple in the context of apartheid: "This is somewhere you need not to go" explained a senior cartographer at MapStudio (personal communication, MapStudio, July 13, 2009). Even if you wanted to visit a township, it was difficult to get there if you couldn't find it on the map. This was William Finnegan's experience when he attempted to travel to Soweto in 1980.

> We bought a city street map and went for an inaugural drive. I was in my own little Whitmanic heaven. New car, new continent. *Allons! The road is before us* . . . Soweto, the great black township, was the only place in Johannesburg we knew, so we made that a first destination. Yet we could not find Soweto on the city map. The area where one would expect to find it, southwest of the city center—Soweto is an acronym for "South West Townships"—was just a large blank on the map. We found that astonishing. Soweto was, after all, the largest city in southern Africa. Did they have segregated street maps, as well as segregated newspapers? (Finnegan 1986, 13)

The mapmakers' blank spaces had their desired effect. Map users were steered away from the townships. As geographer Cuan Bowman of Statistics South Africa summed it up, "If it is not on the map, I am not going there" (personal communication, Cape Town, July 13, 2009).

This technical practice of upgrading, downgrading, and omitting localities on road maps came to dominate South African road mapping. This practice was debated among mapmakers because they knew it did not conform to internationally recognized cartographic standards.

CLASSIFYING CITIES

Map readers are accustomed to mapping conventions in which the width and color of the line symbolizing roads is proportional to the number of lanes on that road (e.g., "single carriage," "dual carriage"). The same thing goes for population centers. The standard mapping practice is to vary the font size of a place name in proportion to that city's population size. Large font and bold letters indicate major population centers. But the population geography of apartheid South Africa presented a distinctive set of sociospatial problems for mapmakers. Should they follow international cartographic norms and depict townships in a way that represented their relative demographic importance? Or should they be downgraded or simply made to disappear from the map? With few exceptions, mapmakers chose the latter option because it addressed the concerns (safety, identity, hegemony) of their clients. The maps that resulted from this decision to make white areas prominent and black areas invisible meant that road maps did not "represent" South Africa as if they were a mirror of reality. They reflected what whites wanted to see, which was themselves in the mirror. Road mapping reinforced white identity, history, and power by exaggerating their place on the map. Maps served to reinforce the apartheid ideology, articulated by Prime Minister Hendrik Verwoerd, who insisted that "South Africa is a White man's country and he must remain master here" (cited in Parsons 1983, 293). In the process, road mapping diminished the importance of black identity, history, and power by marginalizing their place on the map.

Mapmakers justified this upgrading and downgrading of cities with reference to an alternative system of classifying settlements. In contrast to international norms in which population numbers determine whether and how a settlement appears on the map, apartheid-era road maps are based on different criteria—the existence of services and infrastructure within a community. Thus, a small Afrikaaner town of two hundred that possessed a library and fire and ambulance services appeared on the map while a black community of two hundred thousand that did not have these services was rendered invisible. This grading of settlements was influenced by the South African government's ranking of local authorities based on a set of weighted factors such as local government revenues, the number of water and electricity meters, roads maintained, public housing, and even the number of library books checked out in a year (Stickler 1990). Mapmakers at AA and MapStudio stated that the government published a list of ranked cities in the *Government Gazette* based on this grading system and that they used this ranking to select and omit cites in road maps (personal communication, former research director, MapStudio, December 18, 2009). In some cases, mapmakers felt conflicted over using this alternative classification system as opposed to following the international standard based on population.

> The original government classification constantly came up within the research department [at MapStudio]; that is, how to handle it. We wanted to follow international guidelines but in order to show someone that you had a huge town along the road, and then for you not to find anything there, not even a shop, would have been misleading. In a way, I don't know if it was misleading to put them on or not to put them on. We did our best to show as much as possible. We would show the township in smaller type. We would use a small font than what the population warranted but what the infrastructure warranted. For example, Germiston would have a large font type. The reason for this was that 40 years ago it had the biggest railroad yard in South Africa. . . . Outside of Germiston was a location or township where blacks

lived who worked in Germiston. Katlehong and Natalspruit had more people living there than in Germiston. I doubt on the maps of 40 years ago that these were ever mentioned. (Personal communication, former research director, MapStudio, December 18, 2009)

Despite their qualms in following the government's classification system, mapmakers realized that to sell maps, they had to be "recognizable" as well as useful to the white map-buying public. "We could never switch it completely to population because even if you had accurate data, *a map of South Africa showing places based on population wouldn't be recognizable.* In the sense that you have well-known historic towns in rural areas that were typically shown on maps (e.g. Afrikaner farming communities) that would disappear. But historically they are incredibly important to the people buying maps" (personal communication, former research director, MapStudio, December 18, 2009; emphasis added). The AA of South Africa similarly defended its decision to adopt the infrastructure-and-services system of city classification rather than basing it on population. A mapmaker who had been with the company for more than thirty-five years questioned the logic of showing black townships on maps simply because a lot of people lived there. "Why would you want to show those things [black townships] on a map if there was nothing there for people to go there for, and that if you went there you would be at risk? We do what the members want. There are no members there. We make maps for our members. There are no hotels there, no tourist facilities. . . . It is dangerous to go in there so what is the point of showing them on the map? They don't appear on any of the signs on the highway" (personal communication, AA of SA, July 6, 2009).

Mapmakers also had the business traveler in mind. In apartheid South Africa socioeconomic development was concentrated in white areas and underdevelopment in black areas. To appeal to business people, mapmakers emphasized the principal commercial towns and cities. They stated that there was no point in drawing attention to areas that had few business services or little market interest. The discrepancy between developed and underdeveloped areas of South Africa was likened to the First and Third Worlds.

> In Third World countries, infrastructure is not an issue. But South Africa is unique, it has a weird setup of it being part First World and part Third World in the same country. Infrastructure is hugely important for the First World section of South Africa. You have to show Cape Town and Johannesburg because they are centers of commerce. And in a map that relates

to the values and principles that hold, you cannot show all places based on population. (Personal communication, former research director, MapStudio, December 18, 2009)

The "weird setup" of a country containing Third World and First World characteristics refers to the geography of internal colonialism. Like colonialism, internal colonialism is not simply a political-economic system that produces uneven development; it is also an ideology and set of beliefs, or "the values and principles that hold." The road maps of pre-1994 South Africa are the product of this internal colonial system and mentality

Map companies were not forced to show cities according to the government's ranking system. They adopted the scheme based on their own rationale to show or not to show a city. "There was never a rule book that said 'This is what you will do.' [Rather,] decisions were made within groupings of the company at the time" (personal communication, MapStudio, July 13, 2009). That is, road-mapping practices evolved within companies in the context of apartheid and became habitual. Mapmakers knew their audience and the kind of maps it desired. In this sense, road mapping was a "co-constitutive production between inscription, individual and world" (Kitchin and Dodge 2007, 335). In the end, mapmakers did not see any contradiction between their mapping practices and the phenomenon of invisible towns. They were committed to an image of South Africa that made it reasonable to conclude that "some towns disappeared because they didn't fit the perfect picture of a town—tarred roads, fuel, and a safe environment" (personal communication, MapStudio, July 13, 2009).

In summary, apartheid road mapping was characterized by technical and ideological practices that engaged the map reader and the mapmaker in a process of constant coproduction of the territory. The broad racial and political divisions of apartheid society were writ large on the map, just as they were in the minds of mapmakers and map users. From the tortuous boundaries of fragmented homelands to the segregated beaches of Durban, mapmakers and readers inscribed their identity as a nation divided along the lines of petty and Grand Apartheid in which white history, culture, and places mattered most. In the context of apartheid, the unconventional system of ranking cities based on infrastructure and services became conventional. Mapmakers knew that a road map of South Africa that showed places based on population "wouldn't be recognizable" to their audience. It would seem logical then that postapartheid maps would look different.

The end of apartheid in 1994 signaled a tectonic scale shift in power relations and a political need to form a democratic society with a new and more inclusive national identity. To what extent is this new South Africa inscribed in the country's road maps? In other words, can we identify a distinctive postapartheid road map whose dots, lines, and colors have been rearranged to represent a new political geography that speaks to a larger multicultural audience?

5. MAPPING THE NEW SOUTH AFRICA

When Nelson Mandela became president of South Africa in 1994, Bill Buckley, the cofounder of MapStudio, knew what he had to do. He compiled and produced the first map of the "New South Africa" (scale: 1:2,400,000) and sent copies to the president (personal communication, July 24, 2007). Reflecting the monumental change in the political arena, the map showed the transformation in the country's political geography. The fragmented homelands and independent states were gone, their territories blended into the nine new provinces that emerged with the Constitution of the Republic of South Africa. The map is also striking for naming three of the new provinces in African languages. Gauteng is the Northern Sotho and Tswana name for Johannesburg. Mpumalanga is derived from Zulu; it means "the sun comes out" or simply "sunrise." KwaZulu refers to the place or home of the Zulu people (Raper 2004). These changes in the political geography and toponymy of South Africa are major characteristics of postapartheid road mapping. What also stands out is the extent of continuity in mapping practices that were characteristic of the apartheid era. There is little change, for example, in the classification of cities based on the criteria of infrastructure and services. The new South Africa looks a lot like the old South Africa in terms of the prominence given to English and Afrikaner localities. In short, the map remains "recognizable" to white map-buying clients. It continues to "show places where you could stop for a cup of coffee or tea" (personal communication, former research director, MapStudio, December 18, 2009). But even these mapping practices have begun to change. In the fifteen years since the advent of a democratic South Africa, mapmakers are taking tentative but demonstrable steps in drawing a map that is becoming increasingly multicultural. We see this in the emergence of satellite towns, African spellings of place names, and the initiatives to rename the streets of cities like Durban and Cape Town after the heroes of the antiapartheid struggle. Postapartheid road mapping is producing in an incremental manner a different, more inclusive territory in which identity, memory, and culture intersect in a more inclusive way.

SATELLITE TOWNS

On the back cover of MapStudio's *South Africa Road Atlas—2009–2010* (Map-Studio 2009), a new atlas feature is advertised: ALL SATELLITE TOWNS SHOWN (NEW). "Satellite towns" refers to the former black townships that were down-graded and often rendered invisible in both apartheid and postapartheid maps. These satellite towns are shown for the first time and labeled by their African place names. They are all found adjacent to apartheid-era white towns (fig. 8.9). A comparison with the 2008–9 edition of the *South Africa Road Atlas* illustrates this notable step toward recognizing the existence of black localities on the map. This initiative was taken by a former director of research at MapStudio who decided, on her own initiative and with the support of the production manager, to simply "acknowledge the existence" of black communities. The production manager explained: "She took it on by herself. She drove it. She probably said that other companies are not doing it so we should. She also put the original KwaZulu place names on the map. Lower case letters and all. They look like typos" (personal communication, MapStudio, July 13, 2009). The decision to note the existence of black settlements emerged from a desire to rectify the past practice of omitting them on maps. This was a sociospatial problem for which the research director sought a solution. Her group repeat-edly discussed the issue, asking themselves, "Do you go out of your way to downgrade or do you start showing townships and routes going to them? We did a lot in the research department giving them names, and showing them on the maps. . . . Our goal was to simply show they exist. We had to acknowledge their existence" (personal communication, former research director, Map-Studio, December 18, 2009). One can only speculate that the political realities of the new South Africa inspired this innovation. Interestingly, the symbols used to represent satellite towns, a circle with a dot in the center and the name of the settlement, do not vary in size. It is thus impossible to determine the demographic importance of one satellite town from another. For example, the font and symbol for the satellite towns of Nduli (Western Cape) and Bohlokong (Guateng) do not vary in size. But when we look at satellite images of these two localities on Google Earth, Bohlokong appears to be much larger. The former research director at MapStudio acknowledged this limitation when she stated:

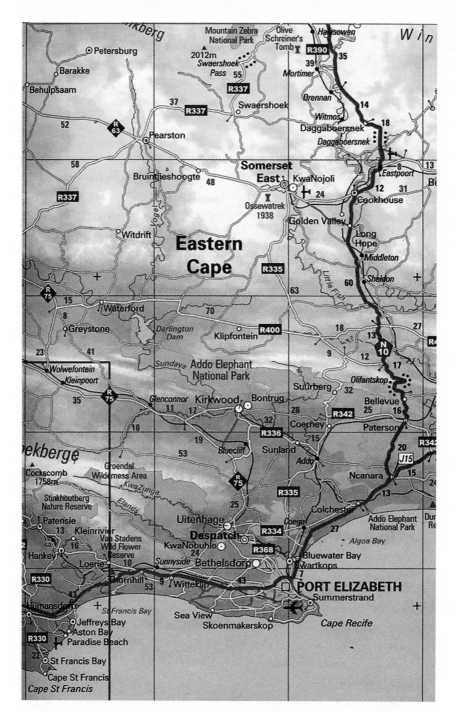

FIGURE 8.9. Many former black African townships appear for the first time on road maps in Map-Studio's 2009–10 *South Africa Road Atlas*. Three examples of these so-called "satellite towns" show up in the Eastern Cape. KwaNobuhle appears by the former white town of Despatch. Bontrug is shown by Kirkwood, and KwaNojoli by Somerset East. The appearance of these towns on road maps indicates the growing importance of black African history, identity, and power in the New South Africa.

"We were trying to introduce changes over a period of time. It was the best we could do in the research department. With each new edition we would show more of this and more of that" (personal communication, former Research Director, MapStudio, December 18, 2009). The road map of South Africa is changing as a result of these new mapping practices. Indeed, it is possible that if mapmakers continue to search for better ways to represent the actual distribution of South Africa's population, the map of South Africa will some day become unrecognizable to some of its readers.

Despite the appearance of "satellite towns" and other black African localities on current road maps of South Africa, the old practices of upgrading and downgrading settlements continue to take place. The case of Katlehong and Germiston on the AA's 2008 *Touring Map of Gauteng* (1:300,000) is illustrative (fig. 8.10). Katlehong, a former black township, has an estimated population of 500,000 people. Germiston is half that size but appears to be a much bigger city on the map. The same is true of Alberton, a city of some 215,000 people. Map readers who are accustomed to interpreting the font size of a city as indicative of its population size will think that Alberton is larger than Katlehong. AA mapmakers suggest just the opposite. While they now place black settlements on the map, they continue to represent cities on the basis of the apartheid-era criteria of infrastructure and services. This is apparent in both the font sizes of city names and in the symbols used to locate the city on the map. The simple circle representing Katlehong is interpreted in the legend as "No facilities." The black-in-white circle symbol for Germiston indicates "Hotel and garage."

When asked why these practices persist, mapmakers offered a number of reasons. First, they consider tourists and business travelers to be the principal buyers of road maps. Since these map users rely upon business and travel facilities, it is important to highlight those localities where these services can be found. Second, mapmakers view the former black townships located near the historically white group areas as "suburbs" of these main cities. It is sufficient, they argue, to show the main city only. Third and related to the second point, mapmakers stated that the scale of the map precluded showing "every little dot" (personal communication, former research director, MapStudio, July 6, 2009). That is, "the amount of detail you can show is determined by scale" (personal communication, AA of SA, July 6, 2009). But the omission of black settlements in MapStudio's *South Africa Road Atlas—2008* (MapStudio 2008) and their appearance in the 2009 edition, which is at the same scale of 1:250,000, suggests that the will of the cartographer is more important than scale per se.

In summary, a postapartheid/postcolonial map is emerging in which (some) black localities can finally be found. This recognition of black identity and his-

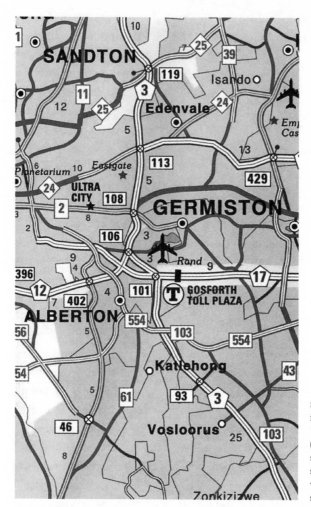

FIGURE 8.10. The Germiston area in the 2008 AA *Touring Map of Gauteng* (1:300,000). The road map shows Katlehong to be much smaller than Germiston when in fact it is twice its size.

tory is muted, however, by the persistent practice of upgrading and downgrading settlements. That is, the demographic importance of localities continues to take a back seat to the overriding criteria of infrastructure and services. In short, in the minds and computer screens of mapmakers, the map-reading audience continues to be white middle-class motorists on vacation or engaged in business travel.

THE AFRICANIZATION OF PLACE NAMES

A second major change in the (re)making of road maps in the context of the New South Africa centers on the spelling of African place names. MapStudio's maps of KwaZulu-Natal stand out for their use of Zulu orthography in the

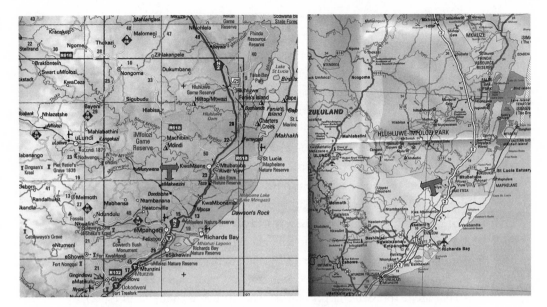

FIGURE 8.11. A comparison of the AA's 2009 *Touring Map of KwaZulu-Natal* (*right*) and MapStudio's map of the province in its 2009 *South Africa Road Atlas* (*left*) showing differences in toponymy in the KwaZulu Natal area of Mtubatuba. Note the more widespread use of KwaZulu place names and orthography in the MapStudio road atlas.

spelling of place names. The use of lowercase letters at the beginning of Zulu place names on MapStudio maps contrasts with AA's use of European spelling conventions. A comparison of the AA's 2009 *Touring Map of KwaZulu-Natal* with MapStudio's maps of the province in its 2009 *South Africa Road Atlas* is illustrative (fig. 8.11). The MapStudio map also displays more KwaZulu place names than the AA map of the same area.

Like the appearance of satellite towns, the Africanization of existing place names gives greater recognition to African history, culture, and identity. These remappings are closely linked to the political and ideological shifts that have taken place over the past fifteen years in South Africa. The process of including black Africans on the map is the outcome of decades of political struggle, a struggle that continues in the battles currently taking place in municipalities across the country where there is increasing pressure to change the names of streets.

TOPONYMIC STRUGGLES: TAKING BACK SOUTH AFRICA STREET BY STREET

The controversy over street renaming in Durban illustrates how maps can be viewed as inscriptions whose meaning and relevance to society are subject to contestation, modification, and reinscription. The South African Geographical Names Council oversees place name changes at the national and provincial levels. At the municipal level, the authority to modify street names resides with municipal councils. In 1999, the Durban City Council recommended changing the names of two main streets named after colonial-era figures (West St. and Smith St.) to two heroes of the struggle for democratization in South Africa, Nelson Mandela and Chris Hani. But when the city council made its recommendation, it was met by "a storm of criticism from business and opposition leaders who felt it made poor business sense and that it could alienate other communities" (Mchunu 2003). Business leaders claimed that the street name changes would confuse clients and increase operating expenses, notably the cost of new stationary and business cards. Leaders of two main political parties, the white-dominated Democratic Alliance (DA) and the Zulu-based Inkatha Freedom Party (IFP) decried the name changes because of their biased commemoration of African National Congress (ANC) party heroes. One DA provincial leader expressed a larger cultural and historical concern felt especially by the former white ruling population "with any idea that would attempt to obliterate the past" (Mchunu 2003). As a result of this furor, the place name changes for Durban were temporally shelved.

Not to be dissuaded from his vision of replacing several icons of the apartheid era with those of the democratic struggle, Durban's mayor revived his street-renaming plan in 2003. After a long consultative process, a special road-naming committee recommended eight street name changes. The new names included those of democratic-struggle icons Chief Albert Luthuli, Nelson Mandela, and Steve Biko as well as "lesser known historical heroes" (Makhanya 2005a). But continued opposition stalled the implementation process. The city council's proposal to rename the Mangosuthu Highway Moses Mabhida Highway, after an ANC liberation-struggle hero, drew a response from IFP president Chief Mangosuthu Buthelezi, who wrote: "I feel obliged to caution the ruling party against their rush to rewrite history of this province and country by giving prominence to only ANC-affiliated freedom fighters over everyone else involved in the struggle for liberation; especially those from the minorities" (Mail and Guardian Online, April 13, 2007). The DA leadership stated that it only supported new names for newly constructed streets and buildings. The DA caucus leader Lyn Ploos Van Amstel told the press, "New names must rectify imbalances and reflect the cultural diversity as historically and demographically applicable. Unless they are hurtful or offensive, existing names should be retained" (Makhanya 2005b).

The removal of offensive place names from maps began in 1994 when the

Directorate of Maps and Surveys systematically removed "the K word" from its maps. "The K word" refers to "kaffir," a derogatory racial term for a black person commonly used by whites during the apartheid era. "Native Yard" (or NY) is another offensive term that is increasingly being removed from maps (Maposa 2007). The NY was widely used by apartheid-era urban planners to designate streets in black townships. Streets were named in consecutive order with the NY prefix. In contrast to the acrimony over street changes in Durban, renaming the NY streets has not been controversial. Black residents of Guguletu renamed more than sixty streets in 2007. The renaming is popularly viewed as an act of recovering local history and culture. Their inscription on maps plays a part in commemorating and instilling a sense of identity, citizenship, and belonging in the New South Africa.

COUNTERMAPPING

In museums and art galleries, a countermapping is challenging apartheid-era mappings by reinscribing black history, culture, and memory with reference to specific places. At the Durban Art Gallery, an exhibit titled *No Longer at This Address: Navigating Post-apartheid Identities* ran from September 28 to November 30, 2007. It was inspired by the controversy over the renaming of Durban's streets. The exhibit combined photographs of the old and new street signs with portraits of the individuals whom they commemorated. The goal of this form of countermapping was to show how the "forefathers of Durban stamp[ed] their brand of history upon the city streets" and how the new street signs represent an attempt to write "'new' histories represented by the new names" (Durban Art Gallery 2007).

One of the most powerful examples of countermapping is found within the District Six Museum in Cape Town. District Six was a mixed racial neighborhood established in 1867 near the center of Cape Town. In 1966 it was declared a White Group Area under the Group Areas Act of 1950. By 1982 the population of some sixty thousand people had been forcibly removed to the distant Cape Flats, and the once vibrant community was bulldozed to the ground. Through testimonials, photographs, and an enormous hand-drawn map on the museum's main floor, the memory of that community and forced removals is retained (fig. 8.12). The original street signs of District Six rise above the museum floor. They were donated to the museum by the bulldozer operator who collected them from the rubble. Former residents of District Six are encouraged to interact with the floor map by writing their names at their

FIGURE 8.12. Street plan of District Six on the floor of Cape Town's District Six Museum. Former residents interact with the map and sustain the memory of their interracial community by writing their names on the streets where they used to live (see detail). Some of the district's street signs salvaged by the bulldozer operator can be seen at the top of the photograph. Photo by T. Bassett.

old street addresses. Walking on the map reconnects one to a community that no longer exists but refuses to be forgotten. The site of District Six remains vacant, except for traces of street pavement and the Holy Cross Church, which was not razed. Some of the district's residents hope to return and rebuild on the vacant land. If they do so, they will instigate a new mapping of Cape Town that will keep mapmakers busy.

Mapmakers at MapStudio talked about the difficulties of keeping track of and verifying the flurry of street name changes. MapStudio is best known for its detailed and up-to-date street guides (personal communication, AA of SA, July 6, 2009). The production manager of MapStudio recounted how the company was excoriated in a *Cape Times* editorial for renaming roads on its maps before the city council had officially approved the changes. Since then, the research department only makes street name changes after they appear in the *Government Gazette*. The number and frequency of place name changes is so great that the AA of South Africa gives the following notice in its 2008 Touring Map series (1:700,000): "Due to the rapidly changing South African scene, the information [contained in the map] could change and you are advised to check the details locally" (AA 2008).

One of the selling points of commercial road-mapping companies is to show that their maps include the most recent street name changes. *Durban CBD Streets and Avenues Map* published by GeoGraphic Maps indicates the city's old and new street names. The inset map on the map cover shows the apartheid-era street names in parentheses next to the new street names (fig. 8.13). The company had a lot of work to do. Starting with eight names in 2007, the Durban City Council had renamed a total of ninety-four streets by 2009. This map suggests that the new names have been accepted, and that a new multicultural mapping is emerging that differs from previous maps. On the other hand, the map seems highly unstable; what will the next edition show? The beguiling certainty of the map masks the tensions and conflicts that accompany all mappings. Resistance to the new street names persists, as is evident in the widespread vandalism to the new street signs (fig. 8.14). The backlash appears to be coming from two directions, although in some cases they may converge. The first is from white South Africans who feel threatened by the street name changes because they diminish their sense of identity and belonging. A senior mapmaker at MapStudio seemed to express a general view of many white South Africans when he stated, "The heroes of the struggle don't really mean anything to us" (personal communication, former research director, Map-Studio, July 6, 2009). This sentiment may be driving some people to protest the new place branding through everyday acts of resistance like vandalism.

FIGURE 8.13. The cover of GeoGraphic
Maps's *Durban CBD Streets and Avenues Map*.
The inset map and street index highlight the
new street names in postapartheid Durban. The
map's slogan, "Your GPS Back-Up," indicates
that the folded map is now playing a secondary albeit complementary role to GPS-enabled
mapping devices.

FIGURE 8.14. Vandalized street signs in Durban, July 2009. Opponents of the Durban City Council's
street name changes sprayed black paint on the new street signs. Photo by T. Bassett.

The second source of discontent is linked to rivalries between political parties. The IFP and DA view the ANC's remapping of Durban as a way of consolidating its political power via symbolic capital in the streetscape. This hypothesis gains ground when one compares the ANC's prorenaming position in Durban and its antirenaming position in Cape Town, where the DA has a strong following (Kinahan 2010) and supports street renaming there. In summary, the politics of street naming and, by extension, the politics of mapping cannot be easily reduced to a black-and-white group conflict or process of national reconciliation.

6. CONCLUSION

To conclude, four points can be made about road-mapping practices in the New South Africa. First, there is much continuity with past practices. The criteria of services and infrastructure still guide mapmakers in selecting and omitting localities on their maps. This persistent feature is linked to the legacy of internal colonialism and the historically uneven development of South Africa's map culture, in which whites have been the dominant clients of commercial road-mapping companies.

Second, there is a multicultural dimension to today's road maps that has been absent in the past. The advent of democracy and black majority rule has empowered black politicians and ordinary citizens to reclaim their history, culture, and identity in multiple arenas. On road maps it appears in the spread of African place names at the national (provincial), local (municipal), and street block levels (street renaming), in the adoption of African orthography in place names, and in the increasing visibility of African settlements on the map.

Third, the past fifty years of road mapping in South Africa illustrate that "maps are contingent, relational, and context dependent; they are always mappings; spatial practices to solve relational problems" (Kitchin and Dodge 2007, 335). The stories told by mapmakers of why they chose to show this and not that town, the anxieties they expressed over not following international conventions when classifying cities, the concern about the safety of their white audience in highlighting specific routes and rendering certain settlements invisible, the individual initiative to modify past practices by showing satellite towns and spelling place names following black African conventions—all of these anxieties and desires, conventions and innovations demonstrate that mapping is a continuous process that is never complete (Kitchin and Dodge 2007). It is always "employed to *solve diverse and context-dependent problems*" (Kitchin

FIGURE 8.15. A computer screen shot from the AA of South Africa's website on July 29, 2011. A black African couple is seen driving happily, while another black family calls for AA assistance after breaking down. This featuring of black families on its websites reflects AA's efforts to increase membership size.

and Dodge 2007, 342; emphasis in original). How those problems are framed and who participates in their definition takes us to the heart of the politics of mapmaking. For many South Africans who feel alienated from the culture of maps because of unequal educational and travel opportunities, the road map may never assume a place in their daily lives. The spread of geospatial technologies and digital mapping in cell phones and in-vehicle navigation systems is already making the folded road map a "back-up" for GPS-enabled mapping (fig. 8.13).

We might also expect to see more hybrid road maps like MapStudio's *Gauteng Roads*, which combines digital and paper maps, and its *South Africa Road Atlas*, which displays black "satellite towns" but still exaggerates the importance of former white localities based on infrastructure and services. These developments in road-mapping technologies and practices suggest that South African commercial mapmakers are making incremental changes to their maps in an effort to reach a wider audience. Such initiatives are evident on the AA's website, in which black middle-class families are prominently featured (fig. 8.15). The AA's appeal to black motorists to join the association contrasts sharply with its apartheid-era catering to a whites-only audience. It is in the light of such overtures that one can begin to imagine that the road map of South Africa will someday become "recognizable" to an even larger segment of the population.

REFERENCES

AA (Automobile Association of South Africa). 1987. *Natal Holiday Coast and Hinterland Through-Route Map (with Road Numbers)*. Scale: 1:100,000.

———. 1989. *South African Book of the Road*. Cape Town: AA The Motorist Publications.

———. 1990. *South Africa/Suid Afrika*. Scale: 1:100,000.

———. 2008. AA Touring Map. *Mpumalanga*. Scale: 1:570,000.

———. 2010. *Master List of AA Maps and Touring Publications*. AA of South Africa internal document.

Akerman, J. 2002. American Promotional Road Mapping in the Twentieth Century. *Cartography and Geographic Information Science* 29 (3): 175–91.

CSVR (Center for the Study of Violence and Reconciliation). 2007. *The Violent Nature of Crime in South Africa*. Braamfontein: CSVR. http://www.info.gov.za/issues/crime/violent_crime .pdf.

Durban Art Gallery. 2007. *No Longer at This Address*. Exhibit catalog. Durban: Durban Art Gallery.

Finnegan, W. 1986. *Crossing the Line: A Year in the Land of Apartheid*. New York: Harper & Row.

GeoGraphic Maps. 2014. *Durban CBD Streets and Avenues Map*. Scale: No scale. Benoni: Map Graphix.

Harley, J. B. 1989. Deconstructing the map. *Cartographica* 26 (2): 1–20.

Hind, R. J. 1984. The Internal Colonialism Concept. *Comparative Studies in Society and History* 26 (3): 543–68.

Kinahan, O. 2010. Letter to the Editor. *Cape Times*, October 14.

Kitchin, R. 2008. The Practices of Mapping. *Cartographica* 43 (3): 211–15.

Kitchin, R., and M. Dodge. 2007. Rethinking Maps. *Progress in Human Geography* 31 (3): 331–44.

Kitchin, R., C. Perkins, and M. Dodge. 2009. Thinking about Maps. In *Rethinking Maps: New Frontiers in Cartographic Theory*, edited by M. Dodge, R. Kitchin, and C. Perkins, 1–25. New York: Routledge.

Lemon, A. 1987. *Apartheid in Transition*. Boulder, CO. Westview Press.

Love, J. 1989. Modeling Internal Colonialism: History and Prospect. *World Development* 17 (6): 905–22.

Makhanya, P. 2005a. Unsung Heroes Find a Place in Durban Streets. *Mercury*, March 8.

———. 2005b. Durban to Rethink Street Name Changes. *Mercury*, March 16.

Maposa S. 2007. Erase Offensive "Native Yards"—PAC. *Cape Argus*, June 6.

MapStudio. 2008. *South Africa Road Atlas—2008*. Cape Town: MapStudio.

———. 2009. *South Africa Road Atlas—2009–2010*. Cape Town: MapStudio.

Mchunu, V. 2003. Plan to Rename Durban's Streets Revived. *Mercury*, June 10.

Parsons, N. 1983. *A New History of Southern Africa*. New York: Holmes & Meier.

Pickles, J. 2004. *A History of Space: Cartographic Reason, Mapping and the Geo-coded World*. London: Routledge.

Raper, P. E. 2004. *New Dictionary of South African Places Names*. Johannesburg: Jonathan Ball Publishers.

Smith, D. 1985. *Update: Apartheid in South Africa*. Cambridge: Cambridge University Press.

Stickler, P. J. 1990. Invisible Towns: A Case Study in the Cartography of South Africa. *GeoJournal* 23 (3): 329–33.

Wolpe, H. 1972. Capitalism and Cheap Labor Power in South Africa: From Segregation to Apartheid. *Economy and Society* 1 (4): 425–56.

———. 1974. The Theory of Internal Colonialism: The South African Case. In *Beyond the Sociology of Development: Economy and Society in Latin America and Africa*, edited by I. Oxaal, T. Barnett, and D. Booth, 229–52. London: Routledge & Kegan Paul.

Wood, D. 1992. *The Power of Maps*. New York: Guilford Press.

———. 2010. *Rethinking the Power of Maps*. New York: Guilford Press.

Wood, D., and J. Fels. 2008. The Nature of Maps: Cartographic Constructions of the Natural World. *Cartographica* 43 (3): 189–202.

CONTRIBUTORS

JAMES R. AKERMAN, Curator of Maps and Director of the Hermon Dunlap Smith Center for the History of Cartography at the Newberry Library, Chicago

RAYMOND B. CRAIB, Associate Professor, Department of History, Cornell University

MAGALI CARRERA, Chancellor Professor of Art History, University of Massachusetts Dartmouth

LINA DEL CASTILLO, Assistant Professor of History, University of Texas at Austin

JORDANA DYM, Professor of History/Latin American Studies, Director, John B. Moore Documentary Studies Collaborative (MDOCS), Skidmore College

JAMIE MCGOWAN, Assistant Director, Global Collaborations, Committee on Institutional Cooperation, University of Illinois at Urbana-Champaign

KAREN CULCASI, Associate Professor of Geography, Department of Geology and Geography, West Virginia University

SUMATHI RAMASWAMY, Professor of History and International Comparative Studies, Duke University

THOMAS J. BASSETT, Professor of Geography and Geographic Information Science, Director, LAS Global Studies, University of Illinois at Urbana-Champaign

INDEX

Abbasid Empire in Egypt, 256
Abdications of Bayonne, 114, 124
Abu, Alhaji Iddrisu, 237–38, 243
Abu Simbel (Egypt) relocated, 274
Abyssinia. *See* Ethiopia
Academia de Historia (Spain), 80
Academia de Mathemáticas (Barcelona), 78
Academia Real de Historia (Spain), 78–80
Academy of Science (Paris), 102, 103f
Accra (Ghana), 213; cadastral survey (1924), 233; government printing office, 228; surveying school, 236; surveys of, 228
Acuña, Rudolfo, 26
Adelman, Jeremy, 16
administrative convenience as basis for internal boundaries, 17
administrative mapping: in formation of nation-state, 255; in Mexico, 73
adventure, sense of, as an attraction of surveying, 232
aesthetics in road maps of South Africa, 357
Afghan Province, part of proposed Muslim home-land, 296

Afhani, Jamal al-Din, 259
Africa, 2, 3, 5–6; British and French colonies in, 113; decolonization in, compared with Asia, 13; demise of colonial rule, 12; European colonies in, 11; Fanon on independence in, 24; French colonies in, 40; native empires within, 222–23; "neocolonialism" in, 42; number of independent countries in, 13; "scramble for," 44; surveying departments established, 225; surveying school established, 231; tax-free zones in, 54
Africa, Northwest, 24
African-Americans on road map cover art, 7
African National Congress (ANC, South Africa), 368, 373
African peoples: as "accomplices" and intermediaries in mapping, 210–13; as part of Guatemalan racial heritage, 196; "peculiar aptitude" for mapping, 226
African personnel: crucial for topography on British colonial maps, 226; in Gold Coast, 225–33; increasing anonymity of, 234; limited opportunities for promotion, 211; outranked by Europeans, in Gold Coast, 233; productiv-

Bontrug (South Africa), 364f
book fair in Guatemala, 160, 161f
Bophuthatswana (South Africa), 344, 345f
Bougainville Rebellion (1988), 26
border defined as a "line that calls attention to itself by violence," 286
borders, 8; artificiality of, 164; as central to mapping, 285; of Colombia, 121, 129, 132; declared "false," 89; disputes spurred mapping in Mexico, 80; in Egyptian atlas, 263; in Gold Coast and Ghana, 220, 237; in India, had to be "permeable and flexible," 301; in India/Pakistan, 6–7; international, 17; of Kashmir denied or erased, 311; on local maps in Mexico, 87; on map of "Arab Homeland" (1965), 267; and mapping, 16–18; mapping of, aptitude of Africans for, 228; in Middle East, 265f; post-colonial, 31; in South America, 120
borders, internal, 17; in Guatemala, 164–65, 172, 178; local knowledge of, in Ghana, 238–39; in Mexican land maps, 87–96, 88f, 89f, 104; racial, on map of Sudan, 27f; on road maps of South Africa, 345
borders, political, 17
borders, regional, 217
Boston (MA), 116
Boundary Commissions (India), 291–94
Bourbon dynasty, 75, 77–78, 82, 104
Bowman, Cuan, 358
BP (British Petroleum), 349
Brabys Maps (South Africa), 348
Brazil, 4; free trade in, 121; as part of Gran Colombia, 110; persistence of slavery in, 24
Breme, Cape, 124f
Bremer, L. Paul, 54
Bretton Woods Agreement (1944), 53
bridges as boundary markers in Mexico, 89
Brión, Luis, 138, 145
British. See also Great Britain
British Association for the Advancement of Science, 226
British Baluchi States, 299
British Broadcasting Corporation (BBC), 54
British Columbia, native peoples in, 33
British Gold Coast. See Ghana; Gold Coast
British Honduras, 5, 18
British India, 6–7; on colonial maps of India, 290; map of, as starting point for maps of India and Pakistan, 299; Muslim state envisioned in, 296; partition announced, 291. See also India
Brower, Ben, 56
Brussels (Belgium), 47
Bryan, Joe, 32–33

Bryan, William Smith, *Our Islands and Their People* (1899), 44
Buckley, E. G. "Bill," 349, 362
Buenos Aires, 117, 124
Buksh, Ustad Allah, *Anthropomorphic Landscape*, 313
Burbank, Jane, 16, 39–40
bureaucracies, 31, 184, 210–11, 245
Burkina Faso, border with Ghana, 238
Burnett, D. Graham, 207
Buthelezi, Mongosuthu, 368

Cabo Gracias a Dios (Honduras and Nicaragua), 130, 132
Cabrera, Francisco, 168
Cabrera, Licenciado, 90–91
cadastral mapping: aptitude of Africans for, 228; in colonial societies, 31; in formation of nation-state, 255; in Ghana, 238; in Gold Coast, 230–31; and growth of nation-state, 210
Cadastral Survey of Accra, 233
Cádiz (Spain), 125, 140
Cairo (Egypt), 255, 257, 274
"Caironese" as an Egyptian identity, 256
Calcutta/Kolkata (India), 19
Caldas, Francisco José de, 130, 135
California, 28f; map of (1791), 80; University of, 198
"Called My Son Out of Egypt," 276, 277f
Caltex (oil company, South Africa), 349
Cambridge (UK), 261; University, 296
caminos reales (Mexico), 78–79
Campbell-Copeland, Thomas, *American Colonial Handbook* (1899), 44
Campomanes, Pedro Rodríguez de: *Discurso sobre la educación popular* (1775), 79–80; *Reflexiones sobre el comercio español a Indias* (1762), 79
Canada, 30, 33; First Nations, 34f; trajectory from colony to independence, 38
Canonicus of Chili, 138
Cape Breme (Barima Point, Venezuela), 124f
Cape Coast (Ghana), 215
Cape Coast Wesleyan School (Ghana), 215
Cape Flats (South Africa), 369
Cape of Honduras, 130, 132
Cape Peninsula (South Africa), 353
Cape Times (Cape Town, South Africa), 371
Cape Town (South Africa), 348–50, 360; International Airport, 355–56; renaming of streets in, 362, 369–73; University of, 344
capital flight, 42
capitalism: and Cartesian perspective, 51; and colonialism, 42–43
Capuchin friars, 143, 145–48

Coalition Provisional Authority (Iraq), 54
coasts and harbors, maps of Mexican, 80
coats of arms: on *Atlas of Egypt*, 260–61; on map of Mexico, 75, 81, 82
cochineal insect, 99–100, 102
coexistence and inclusion as part of art aesthetic, 285
coffee on maps of Guatemala, 185, 171
cofradía (guild, Mexico), 90
Cojti, Narcisco, 193
Cold War, 26, 39, 47–48, 52
Colegio de Mexicoédicos y Cirujanos de Guatemala, 196
collaboration and resistance as responses to colonialism, 210–11
Collection of Topographical Maps of Egypt (1929), 260–62
Colom, Alvaro, 160, 197
Colombia, 4; border with Central America, 132; border with Dutch Guyana, 133f; circulation of maps of, 127; vs. Columbia, 118; "erasures" and "forgetting" in maps of, 129; geo-body of, 115; as independent country, 110–11; manuscript map of, 129, 131f, 132, 133f; mint, 129; relations with France after restoration, 138–39; as signifier for America, 116; slaveholders in, 146, 149. *See also* Colombia Prima
Colombia Prima, 4, 114, 127, 136, 150–51; boundaries of, 121; equivalent to South America, 110; meaning of term, 126. *See also* Colombia
"Colombia Prima or South America" (1807), 110, 111f, 112, 115, 118, 119f, 120, 123, 130, 151
Colombo vs. Columbus, 117
colonial: boundaries, 5; eras, treatment in national atlases, 22; expansion, centrality of mapping to, 255; governmentality, literature survey, 208–10; mores, persistence of, 26; orders not supplanted by independence, 113; peoples, supposed passivity of, 23; peoples as "accomplices" and intermediaries in mapping, 210–13
colonialism: "continuum of unequal relationships," 51–56; definitions, 36–38; from external to internal, 22–27; internal, 3, 23–27, 361; and mapping, literature summarized, 206–13; perpetuated by nation-state, 28
Colonialism on Trial (1991), 34f
Colonial Office List, 13–15f, 38, 40–41f
colonies vs. "outposts," 39
"colonization of everyday life," 51
colonized vs. colonizer, 42
Colorado, 28f
"colored," South Africa racial classification, 343
colors: on Gold Coast maps, 216; green used for Pakistan, 303; red associated with Great Britain, 12–13, 38, 40, 41f, 220; red indicating Radcliffe Line and blood, 318; red showing intent to colonize, on Gold Coast maps, 220; red used for Indian national territory, 304; on road maps of South Africa, 345; yellow used for princely states in India, 304
Columbia (SC), 116
Columbia, District of, 116
Columbia vs. Colombia, 118
Columbus, Christopher, 116; portrait on map of Mexico, 76
Columbus vs. Colombo, 117
Comanchería (Comanche Empire), 36, 37f
Comaroff, John and Jean, 35
Comisión Guatemalteca de limites con Mexico (1889–), 177f, 179
Comisión para el Esclarecimiento Histórico (CEH), 194
Comité de Unidad Campesinas (CUC, Guatemala), 194, 195f
Comité Voluntario de Autodefensa Civil (Guatemala), 191–92
commercial: projects in Guatemala, 175; road mapping in South Africa, 339–76
Commission on Geography (Pan American Institute of Geography and History), 164
commonwealths, 38–39
communication maps of Guatemala, 185
communism, map showing threat to Latin America, 48–49f
communities. *See* cities and towns
community mapping projects, 33
Como River (Africa), 220
Compact of Free Association, 39
"company-state," 54; map of, 55f
compass roses: in Gold Coast, 216; in Mexico, 87, 90, 93
compass traverses taught in Gold Coast, 231
Comunero Rebellion (Colombia, 1781), 117
CONAIE (Confederacíon de Nacionalidades Indigenas del Ecuador), 164
Concessions Bill (Gold Coast), 225
concessions in Gold Coast, 224–25
concrete relief map of Guatemala, 179, 180f, 181, 196
Confederación de la Tierra Firme, 125–26
Confederacíon de Nacionalidades Indigenas del Ecuador (CONAIE), 164
confederation as an alternative to empire, 16
Congress of Cariaco (1817), 138
Congress Party (India), 300
Consejo de Indias, 79
conservation as subject of alternative mapping, 273
Constantia (South Africa), 355

Contador General de la Real Azogues (New Spain), 97

Continent of Dinia and Its Dependencies (1946), 298f

"continuum of unequal relationships," 51–56

conventional signs in Gold Coast and Ghana, 234, 239, 240f

Convention People's Party (CPP, Ghana), 241

Convergencia por los Derechos Humanos (Guatemala), 160, 161f

"Conversation with Myself," 320–21

Cook, James, 20–22, 21f

Cook Islands, status of, 39

Cooper, Frederick, 16, 28; *Empires in World History*, 39–40

Coptic: Cairo, 276, 277f; Christians as Egyptian minority, 6, 255–56, 276; language, 276

Coptic Archeological Society (Société d'Archéologie Copte), *Atlas of Christian Sites in Egypt* (1962), 276

copyright law, 54

cordeles (Mexico), 96

Cornell University, 325

Coro (Venezuela), 117–18, 151

Corollary to the Monroe Doctrine (1905), 44, 45f, 53

Coromina, Ignacio Rafael, *Mapa y Tabla Geografica de Leguas comunes* (1775), 76, 77f, 80–82

corporations and decolonization, 52

Correo del Orinoco (Angostura, Colombia), 150

Cortés, Hernán, portrait on map of Mexico, 76

Cortés y Madariaga, José Joaquin, 125–26

Cosa, Juan de la, 121

Costa Cuca (Guatemala), 171

Costa Ferreira, João da, 111f, 120

Costa Rica, 179; on map of Colombia (1825), 132; omitted in Colombian atlas (1827), 134; as part of Gran Colombia, 110

costs of topographical surveys in Gold Coast, 233

countercartographies. *See* countermapping

countermapping, 54–56; in Egypt, 6, 260, 272–73; in Ghana, 221; and imperial cartography, 3; in South Africa, 369–73

"Country between Say & Bontuku," 218, 219f

coups d'état, 47

cover art on road maps, 7, 353, 354f

CPP (Convention People's Party, Ghana), 241

Craib, Raymond B., 1–3, 11–71, 74, 168

Crampton, Jeremy, 56

Creole elites, 130; in Colombia, 137; in Latin America, 24; mapping practices of, 3–4, 8, 100–101; in Mexico, 96–97; in South America, 127

Croni River, 143

Croquis para accompañar unos estudios (1882), 177f

Crossing the Line (1986), 339, 355, 358

Cruz, Pedro de la, 89

Cruz Cano y Olmedilla, Juan de la: *America meridional* (1791), 80; *Mapa geográphico de América Meridional* (1775), 118, 121, 122f, 123f

Cuba, 47; map in *Atlas de América* (1791), 80; persistence of racism and slavery in, 24

CUC (Comité de Unidad Campesinas, Guatemala), 194, 195f

Culcasi, Karen, 5–6, 56, 252–83

cultural: equality and decolonization, 23; minorities in decolonization, 24; repression as goal of nation-state, 30

Cultural Distribution of India (1938), 299, 300f

culture, colonial view of, 33

Curfewed Night (2009), 324

currency devaluation, 53

customs, colonial view of, 33

Cuzco (Peru), 80

Cynthius (Apollo), as symbol of Mexico City, 81

Daboya community (Ghana), 219

Dadi, Iftikhar, 312, 325; *Bloodlines* (artwork, 1997–), 316, 318, 319f

Dagomba (African people), 218, 220

Dalmia, Yasodha, 316

Damascus (Syria), 256

dances as "title deeds," 32–33

Danquab, J. B., 242

D'Anville, Jean Baptiste Bourguignon de, 117

Darmet, engraver, Paris, 142f

Das, Sushil, 306f

data sources for study of mapping in Ghana, 212–13

David, printer, Paris, 136f, 142f

Dawn (Pakistani newspaper), 294

"Dawn of Freedom," 295f

Daza, Maximiliano Gómez, 90, 92f

Debord, Guy, *Naked City* (1957), 51, 52f

decolonization: and colonization, tension between, 74; "continuum of unequal relationships," 51–56; definition, 7, 36–38, 42; and independence, 16; Lefebvre on, 51; and mapping, 2, 8, 11–71; and maps of Guatemala, 162, 172; Prasenjit Duara on, 13; and social and political exclusion, 23

Delarochette, Louis Stanislas Darcy, 121, 123, 136; "Colombia Prima or South America" (1807), 111f, 119f, 124f, 126; monograph on South American independence, 120; relationship with Miranda, 114; works for Faden, 119f

Del Riego Rebellion (1820), 127

Demerary (Dutch Guyana), 130

Democratic Alliance (DA, South Africa), 368, 373

democratization and maps of Guatemala, 162, 184

Gandhi, Mohandas, 288, 304; portraits on maps, 305f, 306f

García Cubas, Antonio, 74, 76, 82, 104, 178; *Atlas pintoresco e histórico de los Estados Unidos Mexicanos* (1885), 24, 25f, 74–76, 75f; *Carta histórica y arqueológica* (1885), 25f, 74; *Reyno de la Nueva Espana a principios del siglo XIX* (1885), 75f

García Peláez, Francisco de Paula, 168

Garden Route (South Africa), 340, 353

Garifuna peoples of Guatemala, 196

Gauteng (South Africa), 356, 362; map of (2008), 356f, 357, 365, 366f

Gauteng Roads, 374

Gavarrete, Francisco, 179, 186; *Geografía de Guatemala* (1860–74), 173; map of Guatemala, 178

Gavarrete, Juan Justo, 173–75; *Carta de la República de Guatemala* (1880), 173, 174f; map of Guatemala (1878), 181, 183

GDP (gross domestic product), 56, 57f

Gehüchte, Agustin van de, 171

gender and mapping, 325

Gentleman's Magazine, 116

geo-bodies, 7–8, 302; of Colombia, 5, 115; defined, 5; of Guatemala, 6, 162–65, 172–73, 175–76, 178–98; of India, 290, 304; of Pakistan, 297, 299

"geocoded" realms, 285

geodesy in Guatemala, 186

geodetic: framework in Gold Coast, 233; surveys in colonial societies, 31

Geografía de Guatemala (1860–74), 173

Geografía de Guatemala: Cuaderno de mapas (2007), 189f

Geografía elemental de Guatemala (1936), 185–88

Geografía Visualizada (1976–), 188–90

Geographic Information Systems (GIS), 54

GeoGraphic Maps (South Africa), 348–49; *Durban CBD Streets and Avenues Map*, 371, 372f

geography: effects on Mexican people, 100; of Egypt, 256; exam in, in Gold Coast, 231; and foreign policy, 54; of New Spain, by creole writer, 98

geological maps of Guatemala, 186

geometrical drawing in Gold Coast, 231

German surveyors in Guatemala, 171

Germany: claims in Gold Coast, 214, 218; colonial border with Gold Coast, 215; competition with Great Britain and France in Africa, 219; Ferguson's map of Gold Coast published in, 219; map of Guatemala printed in, 181

Germiston (South Africa), 357, 359–60, 365, 366f

Ghana, 3, 5–6; boundary with Burkina Faso, 238; branch of RICS in, 236; Cartography Section, 239; chieftaincy boundaries in, 238; differences between colonial and post-colonial maps, 243, 244f; independence, 205; internal boundaries in, 238–39; mapping of, 205–47; surveying agencies (*see* Survey Department; Survey of Ghana); surveyors in, 31. *See also* Gold Coast

Ghana Institution of Surveyors, 237

Ghosh, Amitav, 285

Ghosh, Durba, 56

Gilmartin, David, 325

GIS (Geographic Information Systems), 54

Gitksan people, 33

global capital and postcolonial states, 17

Globetrotter (South Africa), 348–49

Globetrotter Travel Map, Cape Winelands (2000), 355f

go-betweens, indigenous Australians as, 211

Gold Coast, 211; administrative mapping, 233–34; border with German territory, 215; colonial administration and development (1901–30), 224–34; colonial conquest and expansion (1874–1900), 213–24; colonial consolidation (1931–57), 234–45; colonial consolidation and decolonization (1931–57), 234–45; Constabulary, 223; decolonization (1931–57), 234–45; development mapping, 233–34; end of colonial status, 235; forces opposing colonialism, 234–35; founding as British colony, 206; government printing office opened, 228; internal borders, 215; map of colonial divisions (1907), 214f; mapping and colonial expansion, 218–21; mapping of, 205–47; Mines Survey Department, 224–27; Northern Territories, 224; Office of Colonial Secretary, 225; Public Works Department, 227; Public Works Office, 215; scales of maps in, 233–34; surveying agency (*see* Survey Department); surveying school in, 229–33; treaties with ethnic groups, 218. *See also* Ghana

Gold Coast Colony, 231; on map of Ghana (1907), 214f; topographic framework begun, 224; topographic maps of, 228

gold deposits in Gold Coast, 221, 224–25

Golfo Dulce (Costa Rica), 132

Gómez Daza, Maximiliano, 90, 92f

Gonja (African people), 218, 220

Goubaud, Emile, 173, 186

Goubaud Carrera, Antonio, 186; linguistic map of Guatemala (1946), 187–88

governmentality, literature on summarized, 208–10

Government Gazette (South Africa), 359, 371

Government of India Bill, 296

GPS-enabled mapping: in South Africa, 356f, 372f; traditional road map a "backup" for, 374

Gracias a Dios, Cabo, 132

Grafton Street (London), 125

Guggisberg, Frederick Gordon, 226–29, 231; *Handbook of the Southern Nigeria Survey* (1911), 227

Guguletu (South Africa), 369

Guiana, Dutch, 121, 123, 130, 132, 133f. *See also* Guyana

Guiana, French, 121, 124f. *See also* Guyana

Gujarati artists, 313

Gujral, Satish, "Partition Series" (1950–57), 313

Gulf of Guinea, 224

Gulf of Mexico, 80

Gurunsi (African people), 218

Gutiérrez Ardila, Daniel, 128–29

Guyana, 4, 110, 121, 123, 124f, 130, 132, 133f, 138, 145, 147–49, 207; border of (1827), 134f; mapping in, 207; as part of Gran Colombia, 110; Piar attacks, 138. *See also* Guiana, Dutch; Guiana, French

Guyana la Vieja (Venezuela), 145, 147

Hacienda de Primo y San Miguel el Grande (1723), 91f

Hacq, engraver, Paris, 142f

Haiti, 146, 163

Haitian Revolution, influence in Colombia, 137

Hala'ib Triangle, tenuous status of, 263–64

Hamalainen, Pekka, 36; *Comanche empire*, 37f

Hamburg, 171, 182f

Hamilton, Alexander, friendship with Miranda, 116

hamzah (Arabic letter) in Oceanic toponymy, 19

Handbook of the Southern Nigeria Survey (1911), 227

hand-held navigation devices in South Africa, 357

Han empire, 30

Hanging Church (Cairo, Egypt), 276, 277f

Hani, Chris, 368

Hanlon, David, 56

Hanson, F. O., 233

Hanuman (devotee of Rama), 323

Hapsburg: dynasty, 75, 77; empire, 30

Hardt, Michael, 28

Harley, J. B., 207, 285, 292, 325, 350

Hashmi, Salima, 316

Hashmi, Zarina, 325; *Atlas of My World IV* (2001), 320–21, 321f; "A Conversation with Myself," 320–21; *Dividing Line* (woodcut, 2001), 318, 320–21, 320f

Hau'ofa, Epeli, 19

Havana (Cuba), 80–81, 116

Hawai'i, University of, 52

Haya de la Torre, Víctor Rául, 47

Heaney, G. F., 294

Henry, Patrick, 286

Herzog, Tamar, 101

higher education in New Spain, 98

highway signs in South Africa, 360, 372f

hijacking: of cars, in South Africa, 356; of cartographic language, 31–32, 51, 113

"Hindoostan" on colonial maps of India, 290

"Hindu India," 290

Hindu-Indic thought, 316

Hindus: artists' responses to partition of India, 312–25; iconography, 323; in Kashmir, 307, 310; nationalism of, 288; and partition of India, 293; relations with Muslims, 321–24; religion, 288

"Hindustan" on *Atlas of My World IV* (2001), 320

Historia de Guatemala: Desde un punto de vista crítico (2007), 194, 195f

Historia de la revolucion de la Republica de Colombia (1827), 112f, 115

historical: geographies, 23; maps and pro-Nubian discourse in Egypt, 275; sites of black Africans, 353

Historical Atlas of Canada, 38

history: colonial and national, 22–23; of Egypt, 256; erased from maps of Colombia, 121, 123, 137–39; exam in, in Gold Coast, 231; on Gavarrete's map of Guatemala (1880), 175; on map of Mexico, 76; narratives of, 22; of New Spain, by crillo writer, 98; a prerequisite for the nation-state, 76; in Valle's description of Guatemala, 167

History of Cartography, 58

hojitas (blank maps) of Guatemala, 187, 190

Holdich, Thomas, *How Are We to Get Good Maps of Africa?* (1901), 226

holograph lenticular print, 315

Holy Alliance (1815), 127

Holy Cross Church (Cape Town, South Africa), 371

Holy Family in Egypt, 276

Holy Land, views of, by Roberts, 275–76

homelands (South Africa). *See* Bantustans (South Africa)

Homelands Policy (South Africa), 343–44

Hondo River (Guatemala, Mexico and Belize), 130

Honduras, 132; border with Guatemala, 169, 179, 181–82, 196; languages in, 193; and Miskitu people, 166; as part of Gran Colombia, 110; on Urrutia's maps of Guatemala, 181

hoofprints as symbols for roads, 85

Hopkins, Anthony, 38

Horrabin, James Francis, *An Atlas of Empire* (1937), 11–12f, 36

Hoskote, Ranjit, 314

house arrest in South Africa, 344

houses on Aztec maps, 85

as part of proposed Muslim homeland, 296; religious makeup, 307, 310

Kashmiri poets, 323

Katlehong (South Africa), 360, 365, 366f

Kekchi language, 194

Kennedy, Dane, 56

Kenneth Nebenzahl, Jr. Lectures, 1, 56, 198

Keynesian paradigm, 53

Khan, Yasmin, 293, 299

Khartoum, 26

Khyber Pass, 299

King, Rufus, 119, 123

Kingdom of Egypt (ca. 3200 BCE), 256

"Kingdom of Guatemala," 183

King's College (New York), 116

Kiribati, status of, 39

Kirkwood (South Africa), 364f

Kitchin, Rob, 341, 350

Knight, Alan, 56

Kolkata/Calcutta (India), 19

Kosovo, 316

Kruger National Park (South Africa), 340, 353

Krygier, John, 56

Ku Klux Klan, 28f

Kumaon (India), 210

Kumar, Amitava, 315

Kumasi (Ghana), 236

Kumasi College of Technology (Gold Coast), 235–36, 241–42

Kuranchi, S. W., 238

Kwame Nkrumah University of Science and Technology (Ghana), 235–36, 241–42

KwaNojoli (South Africa), 364f

KwaZulu Homeland (South Africa), 351, 352f

KwaZulu-Natal (province, South Africa), 362; map with Zulu orthography, 366–73; on road maps (2009), 367f

Kyalami (South Africa), 348

labor: capture, as goal of nation-state, 30; movements in Guatemala, 163; reserves in South Africa, 342

laborers, Africans as, in Gold Coast, 226

Lacandon people, 169, 172, 174, 178

La Concepcion (Mexico), 89

Ladino people in Guatemala, 161

Lahore (Pakistan), 293, 313, 315

Lahore Resolution, 288, 296

Lake Nasser (Egypt), 268–69, 274, 275f

Lake Nicaragua, 132

Lakota logo-map, 31, 32f, 35–36

Lakshmanan, V., 309f

land: alienation in South Africa, 342; Apportionment Act of 1936 (South Africa), 343; areas on Mexican maps, 90; claims in Mexico, 104; grants in New Spain, 80, 90; maps in Mexico, 82–96; owners, in Mexico, 104; privatization, 209–10; reform, 31, 32; tenure, 4, 31, 33, 209–10; titles in Mexico, 88; use in Guatemala, 186; use in Mexico, 83–84

Landívar, Rafael, 99–101

languages: in Guatemala, 147, 162, 167, 174f, 175, 183, 187–88, 193–94; of minorities, and decolonization, 24; persistence of colonial policies, 26; as site of anticolonial struggle, 19; and toponymy, 19

"Languages of Guatemala and Belize" (1988), 193

Lanz, José, 135

lapel button, 32f

Lapierre, Dominique, 293

Lara, Jacinto, 147–48

Lasso, Marixa, 145

Lat Dior, 24

Latif, Syed Abdul: *Cultural Distribution of India* (1938), 299, 300f; *Federation of Cultural Zones for India* (1938), 299, 300f

latifunda, 46

Latin America, 2, 5, 16; abolition of slavery in, 24; borders of, 17; British hegemony in, 44; cartoon map of, 45f; decolonization in, compared with North America, 13; insurgent politics in, 28; loss of caste distinctions, 24; map showing communist threat to, 48–49f; "neocolonialism" in, 42; privatization in, 53; role of creole elites, 24; south-at-the-top map of, 49, 50f; struggle for economic independence, 43–44; US hegemony in, 44

Latour, Bruno, 301–2

La Vela de Coro. *See* Coro (Venezuela)

Lawrance, Benjamin, 211

laws: and First Nations, 34f; and land tenure, 32–33; relating to apartheid in South Africa, 343–48

"Law vs. Ayook," 34f

lawyers on Indian Boundary Commissions, 292

League Against Imperialism, 47

Lebanon, 256

Lefebvre, Henri, 51

Lemuria, 31

lenticular print, 315

lettering. *See* typography

"Letter to Spanish Americans," 120

leveling in Gold Coast, 231, 233

Liberator Party (Colombia), 4, 113–14, 127–29, 132, 134, 136, 139, 142, 145–47, 151

Library of Congress (US), 198

Libreria Americana, 136f

Libya, border with Egypt, 263, 267

National States Act of 1971 (South Africa), 344
nations: "born in the map," 255; defined, 254; vs. states, 163; "yearn for territorial form," 301f
nation-states, 8; as antithesis of colony, 36; artificiality of, 23; and "company state," 54; and decolonization, 36–38; defined, 254; "efficacy" of, 74; and Egyptian national narrative, 278; history a prerequisite, 76; number of, in Africa, 13; as perpetuation of colonialism, 28; rapaciousness of, 25; seek to homogenize the nation, 254
Native Americans. *See* American Indians
native heritage on map of Mexico, 25f
native peoples. *See* indigenous peoples
native vs. foreign, 33
"Native Yard" removed from maps of South Africa, 369
naturaleza in Mexico, 87, 96, 101
natural resource maps in formation of nation-state, 255
natural resources: in Gold Coast, 224; in New Spain, 80
Nava, Joseph, 77f, 81
navigation chart, indigenous, 20f
Nebenzahl Lectures, 1, 56, 198
Negri, Antonio, 28
Negroid peoples in Sudan, 26
Nehru, Jawaharlal, 290, 294, 304; on announcement of partition, 292; portrait on map, 307, 310f
neocolonialism, 16, 42
neoliberalism, 53
Netherlands arrival in Guyana, 121
"neutrality" of surveying, 237–39
Neutral Zone (Gold Coast), 218–20
Newberry Library, 325
New Delhi (India), 292, 299, 315
"New Elementary Tamil" (1958), 309f
New England, 80
New Granada. *See* Nueva Granada
New Holland Publishers, 349
New India (map, ca. 1950), 302, 303f, 304, 305f
New India No. 2 (map, ca. 1950s–60s), 306f
New Kingdom (Egypt, ca. 1600–1000 BCE), 256
Newman, Barnett, 314
New Mexico, 28f
New Orleans (LA), 80
"New South Africa" (map, 1994), 362
New Spain, 3–4; defined by metaphor, measurement, and map, 82; history of, by crillo writer, 98; intellectual life in, 98
newspaper maps during partition of India, 294
newspapers in Guatemala, 163
New York (NY), 116, 171, 318; Public Library, 198

New Zealand, territories of, 39
Nicaragua: map by Sonnenstern, 171; and Miskitu people, 166; as part of Gran Colombia, 110
Nicaragua, Lake, 132
Nigeria, 31
Nigerian surveyors, 207
Nile: Delta, 262, 267, 271; River, 263, 274, 275f; Valley, 256; watershed, 267
Nkrumah, Kwame, 241–42
nobility, insurgent leaders aspiration to, 114
Noguera, Calixto, 144
No Longer at This Address (exhibit, 2007), 369
non-European peoples. *See* indigenous peoples
"nonstate" actors, 164
nopal plant as boundary marker, 89
North Africa: Arab states in, 258; map of (1922), 263; Miranda tours, 116
North America, 116; decolonization in, compared with Latin America, 13; dominant empire in, 36; map in *Atlas de América* (1791), 80; map of (1755), 77f, 81
North American Free Trade Agreement (NAFTA), 53
North Carolina, 116; State University, 325; University of, in Chapel Hill, 56
Northern Areas (Pakistan), 311
Northern Mariana Islands, status of, 39
Northern Rhodesia, surveyors in, 31
Northern Sotho language, 362
Northern Territories (Ghana), 214f, 224
north vs. south, 42
North-West Frontier Provinces: on Latif's map of India (1938), 299; as part of proposed Muslim homeland, 296
Now or Never: Are We to Live or Perish Forever? (1933), 296
ntornel, as boundary marker in Ghana, 238
Nubia, 273; in *Arab Atlas* (1986), 269; distinct from Egypt in 1800, 275; relief map of, 274, 275f; views of, by Roberts, 275–76
Nubian General Club (Egypt), 274
Nubian Museum (Egypt), 274, 275f
Nubians: displaced by Aswan Dam, 274; in Egypt and Sudan, 273; as Egyptian minority, 6, 256; and mapping of Egypt, 255–56
"Nubia Submerged" (exhibit), 274
"Nubia Today" (website), 275
Nueva Galicia. *See* Guadalajara (intendancy, Mexico)
Nueva Granada, 151; as part of Colombia, 127; union with Venezuela, 140; Viceroyalty, 110, 127
Nuevo mapa geographico de la America septentrional (1792), 102, 103f

Sociedad Geológica de Guatemala, 191f
Société d'Archéologie Copte (Coptic Archaeo-
    logical Society), *Atlas of Christian Sites in Egypt*
    (1962), 276
society, organization of in South Africa, 345
Society Islands, map of, by Tupaia, 21f, 22
Soconusco (region, Mexico), on maps of Guate-
    mala, 172, 174f, 180
soldier on logo map of Guatemala, 192f
Somalia as part of "Arab Homeland" (1986),
    267–68
Somerset East (South Africa), 364f
Sonenstern, Maximiliano von, 171–72, 178
songs as "title deeds," 32–33
Sonsonate (department, El Salvador), 166, 168
sound traverses taught in Gold Coast, 231
South Africa, 7, 30; arrests for pass law violations,
    353; Bantustans, 342, 343; commercial road
    mapping, 339–76; constitution (1994), 362;
    countermapping in, 369–73; culture of map-
    ping in, 341, 350; Department of Development
    Aid, 344; Directorate of Maps and Surveys,
    369; discrimination in public services, 344; eco-
    nomic factors underlying apartheid, 342; *Gov-*
    ernment Gazette, 359, 371; internal colonialism,
    342–48, 361; land alienation, 342; laws under-
    lying apartheid, 343–48; national atlases of, 25–
    26; organization of space and society in, 345;
    population percentages, 343; post-apartheid
    rivalries, 373; post-apartheid road maps, 362–
    73; racial groups in, 343; road map companies
    in, 348–49; undeveloped map culture in, 341;
    white audience for road maps, 352–62
South African Book of the Road (1989), 346f
South African Geographic Names Council, 368
South Africa Road Atlas (2009), 363, 364f, 365, 367f,
    374
South Africa/Suid Afrika (1990), 345f
South America: equivalent to "Colombia Prima,"
    110; indigenous rebellions in, 117; map in *Atlas*
    de América (1791), 80; map of, 49, 50f; south-at-
    the-top map of, 49, 50f
South Carolina, 116
Southeast Asia, 30
Southeastern Europe and decolonization, 53
Southern Nigeria, surveying school in, 227, 231
south vs. north, 42
Southwest Asia: Arab states in, 258; decoloniza-
    tion in, compared with Oceania, 13; persistence of
    colonial boundaries, 17
South West Townships (South Africa). *See* Soweto
    (South Africa)
sovereignty, 31
Soviet Union, 26, 47, 53, 273

Soweto (South Africa), 357–58
Soweto Uprising (South Africa, 1976), 344
Soyaniquilpan de Juárez (Mexico), 86–87
space: administrative vs. localist, in Mexico, 73;
    conceptualized through travel, 99; depiction
    on indigenous and European maps, 83; organi-
    zation of, in South Africa, 345; and territori-
    ality, 82
space exploration, 54
Spain, 24; border with Portugal in South America,
    120; Bourbon reforms, 78; colonies of, 39;
    competition with other European nations, 79;
    dissolution of South America colonies, 112;
    educational institutions, 78; land tenure sys-
    tem, 103; loss of empire, 77; maps of, in *Diccio-*
    nario histórico-crítico . . . de España, 78; Miranda
    tours, 116; poor roads in, 78; population, 78;
    "recolonization" of American territories, 79;
    royal abdication at Bayonne, 114; tenurial sys-
    tem of land use, 84; treaty with Great Britain
    (1786), 166
Spanish-Cuban-American War (1898), 39, 44
Spanish Royal Guards, 76
Spanish Succession, War of the (1701–14), 77
Sparke, Matthew, 17
spatial fixity vs. mobility, 35
Splendour that is India (map, ca. 1947), 308f
Srinagar (India), 310, 324
staff reductions in Survey Department (Gold
    Coast), 235
stamps, maps on, 184, 185f, 311f
standards: in South Africa, 358–59; in Survey
    Department (Gold Coast), 230, 233, 245
Standton (South Africa), 357
state: defined, 254; vs. "nation," 163
state-run economy and transition to free-market
    economy, 53
statistical studies spurred mapping in Mexico, 80
Statistics South Africa, 358
steel tapes used in Gold Coast, 234
stick chart, 20f
Stone, Jeffrey, 207
stool boundaries in Ghana, 237–38
stories as "title deeds," 32–33
Strange Maps: An Atlas of Cartographic Curiosities
    (2009), 55f
street: maps, personal names inscribed on, 369,
    370f, 371; names in South Africa, 362, 367–73;
    signs, vandalism of, in South Africa, 371, 372f
Struik Publishers, 349
student unions, as opponents of colonialism in
    Gold Coast, 235
subordination vs. autonomy, 38
Sucre, Antonio José de, 148

Toekomsrus (South Africa), 346f
Tokelau, status of, 39
Tonga, status of, 39
topographical: framework in Gold Coast, 224; mapping in Gold Coast, 228, 231; maps, 34f, 255; surveying, textbook on, 227; surveys, cost of, in Gold Coast, 233; surveys in colonial societies, 31
topography: on African maps dependent on native input, 226; training in, in Spain, 78
toponyms, 8, 18–22; in Egypt, 265; glyphs representing, 83; on Gold Coast maps, 217; as indigenous aids to European mapping, 207; on map of Guatemala, 193; on maps of UAR, 272; as mnemonic devices, 19; on road maps of South Africa, 351, 366–73; of streets, in South Africa, 362, 367–73; used by Survey Department (Gold Coast), 234
Torre, Víctor Rául Haya de la, 47
Torres-García, Joaquin, 49, 50f
Total (oil company, South Africa), 349
Totonicapán (department, Guatemala), 170
Touré, Samori, 222–23
Touring Map of Gauteng (2008), 365, 366f
Touring Map of KwaZulu-Natal (2009), 367f
"Touring Map series" (South Africa, 2008), 371
town-dwellers and pastoralists, 30
towns. *See* cities and towns
townships (South Africa), 346f
town surveyor in Gold Coast, 230
trade: colonial, with Spain, 80; contraband, 116; and decolonization, 38; unions, as force opposing colonialism in Gold Coast, 235
Transkei (South Africa), 344
Translations (1981), 19
transnational corporations and decolonization, 52
transportation: in Guatemala, 171–72, 185; and postcolonial states, 17
Transylvania, 53
travel: accounts of Guatemala, 173; vs. grid as way of mapping space, 99, 104; reports as indigenous aids to European mapping, 207; space conceptualized through, 99; writing, accounts of Egypt in, 256–57
Travers, Robert, 56
treaties with ethnic groups in Gold Coast, 218
Treaty of Munster (1648), 130
trees as boundary marker in Ghana, 238
triangulation: in Gold Coast, 224, 233; in Guatemala, 186
Trouillot, Michel-Rolph, 163
Trust Territory of the Pacific Islands (US), 39
Tswana language, 362

Tulane University Latin American Library, 198
Tupac Amaru rebellion (1780–82), 117
Tupaia, 21, 21f, 22
Turkey: borders of, in 1922, 266; historical maps, 22
Turnbull, David, 21f, 211
typography: on Gold Coast maps, 216; on road maps, 358–59; on road maps of South Africa, 345, 357, 365; and significance of cities and towns, 144, 146–47, 150; on Urrutia's maps of Guatemala, 181

UAR. *See* United Arab Republic (UAR)
Ubico, Jorge, 185
Umayyad Empire rules Egypt, 256
underdevelopment vs. development, 43–44
UNESCO, 274
United Arab Republic (UAR), 258; on map of "Arab Homeland" (1965), 267; maps of, 270–72; name changed to Arab Republic of Egypt, 269
United Commercial Bank (India), 294
United Empire Society, 38
United Fruit Company, 47
United Nations, 47; Development Programme, 190; number of member states, 13; Special Committee on Decolonization, 39
United Provinces (India), 294
United States, 30, 47, 130; Agency for International Development, 186; black militants in, 26; Central Intelligence Agency, 47; Chicano nationalists in, 26; as colonial power, 11; colonies of, 39; commonwealths and territories, 39; Congress establishes District of Columbia, 116; economic power over Latin America, 43; engagement in Caribbean and Central America, 16; First Cavalry Division, 54; funding for Aswan Dam, 273; Indian lands, 32f; Indians (*see* American Indians); Jim Crow laws in, 26; map of territorial expansion, 42f; marginalized populations in, 26–27; military interventions in Mexico, 46; possessions in Oceania, 39; rapaciousness of, 25; recognizes Colombian Republic, 151; Revolution (1775–83), 116; road map cover art, 7; slavery perpetuated after independence, 24; support for Guatemalan IGN, 186; Supreme Court ruling, 39; territories of, 39; treaty with France (1800), 118, 123; urged to liberate "Colombia," 117; urged to recognize Colombian independence, 141
Universidad Francisco Marroquín (Guatemala), 193
University College, London, 242
University of California Bancroft Library, 198